About Island Press

Since 1984, the nonprofit Island Press has been stimulating, shaping, and communicating the ideas that are essential for solving environmental problems worldwide. With more than 800 titles in print and some 40 new releases each year, we are the nation's leading publisher on environmental issues. We identify innovative thinkers and emerging trends in the environmental field. We work with world-renowned experts and authors to develop cross-disciplinary solutions to environmental challenges.

Island Press designs and implements coordinated book publication campaigns in order to communicate our critical messages in print, in person, and online using the latest technologies, programs, and the media. Our goal: to reach targeted audiences—scientists, policymakers, environmental advocates, the media, and concerned citizens—who can and will take action to protect the plants and animals that enrich our world, the ecosystems we need to survive, the water we drink, and the air we breathe.

Island Press gratefully acknowledges the support of its work by the Agua Fund, Inc., The Margaret A. Cargill Foundation, Betsy and Jesse Fink Foundation, The William and Flora Hewlett Foundation, The Kresge Foundation, The Forrest and Frances Lattner Foundation, The Andrew W. Mellon Foundation, The Curtis and Edith Munson Foundation, The Overbrook Foundation, The David and Lucile Packard Foundation, The Summit Foundation, Trust for Architectural Easements, The Winslow Foundation, and other generous donors.

To Conserve Unimpaired

To Conserve Unimpaired

THE EVOLUTION OF THE NATIONAL PARK IDEA

ROBERT B. KEITER

Island Press, 2000 M Street NW, Suite 650, Washington, DC 20036

Library of Congress Cataloging-in-Publication Data

Keiter, Robert B., 1946-
To conserve unimpaired : the evolution of the national park idea / Robert B. Keiter.
pages cm
ISBN 978-1-59726-659-8 (cloth : alk. paper) -- ISBN 1-59726-659-0 (cloth : alk. paper) -- ISBN 978-1-59726-660-4 (pbk. : alk. paper) -- ISBN 1-59726-660-4 (pbk. : alk. paper)
1. National parks--United States--Philosophy. 2. Nature conservation--United States--Philosophy. 3. National parks--Government policy--United States. I. Title.
SB481.6.K45 2013
363.6'8--dc23

2012041661

Printed on recycled, acid-free paper

Manufactured in the United States of America
10 9 8 7 6 5 4 3 2 1

Note: Portions of chapter 10 previously appeared in the fiftieth anniversary issue of the Natural Resources Journal. The full citation is Robert B. Keiter, "The National Park System: Visions for Tomorrow," *Natural Resources Journal* 50(2010): 71-110.

Keywords: Island Press, national parks, National Park Service, national park system, National Park Service (System) Organic Act, the national park idea, nature conservation, wildlife management, Leopold Report, tourism and national parks, gateway communities, national park concessioners, national parks and Native Americans, science and national parks, recreation and national parks, wildlife and national parks, history of national parks, national parks and wilderness, new national park designations and proposals, external threats and national parks, education and national parks, natural regulation policy, climate change and national parks, ecosystem management, ecological restoration

To Linda,
for everything over the years

CONTENTS

PREFACE

Growing up in the midst of a national park controversy was, in retrospect, an extraordinary if sometimes painful experience, even though I had only vague notions about what a national park was or how one was created. The Chesapeake and Ohio Canal and the nearby Potomac River, to me and my childhood friends, was just an uninhabited natural setting where we could fish, wander, and daydream, a place we visited regularly after school or just on a whim. Situated less than a hundred yards from my back door, the canal and river and the thick hardwood forests bordering them conjured wilderness images, representing our own retreat where we could test our mettle against raw nature. We knew little about the canal's history or that a battle to preserve it was brewing. New road proposals deemed necessary to connect the area's sprawling suburbs with the nation's capital city were on the drawing board and moving forward.

By the time the battle was over and the new parkway was built, however, I understood, even as a youngster, that nature conservation could not be taken for granted. I observed with sadness the bulldozers slash through the steep woods that separated my backyard from the canal and my wilderness stronghold, and I watched with puzzlement as several neighbors abandoned their condemned homes that were then razed to make way for the roadway. It took a bit longer to realize that my quiet sojourns along the canal's towpath would never be the same once cars started whizzing by just a few dozen yards away.

I also learned that the parkway proposal had provoked controversy and opposition, enough to scale back the original plans and to launch a high-profile campaign to save the canal. Some of my despair dissipated when I heard that President Dwight D. Eisenhower had proclaimed the new Chesapeake and Ohio Canal National Monument, although his declaration did not include my stretch of the canal. Congress then completed the job in

1971 when it established the Chesapeake and Ohio Canal National Historical Park, protecting the lands along the entire length of the canal and thus ensuring that they were safe from further incursions. Today, the park serves more than four million visitors annually, who revel in its natural beauty and recreational opportunities.

Since that time, national parks have remained an important element in my life, both personally and professionally. Childhood trips to the Great Smoky Mountains and Shenandoah National Parks are still etched in my consciousness, including the peaceful time a friend and I spent as fourteen-year-olds camping in solitude atop Hawk's Nest. My first encounter with the rugged and stunning beauty of Yellowstone, Grand Teton, and Glacier National Parks, coming on a summer trip across the West during my college years, set the stage for my later life-changing sojourn westward in search of professional opportunity, wild nature, and the chance to live some of my childhood wilderness dreams. When a research opportunity addressing the external threats confronting Glacier National Park during the 1980s presented itself, I jumped at the chance to apply my professional legal training to the problem. My Glacier research, however, also reminded me of the C&O Canal controversy of my youth and reinforced that nature conservation, even in our largest and most remote national parks, cannot be taken for granted.

To Conserve Unimpaired: The Evolution of the National Park Idea addresses the American national park idea, which has long captured the world's imagination. It does so by exploring the anomalous fact that national parks and controversy go hand in hand. While pondering the diverse and often shrill conflicts that have engulfed the parks over the years, I was struck that these conflicts reflected fundamental changes in our view of what a national park is or should be. Perhaps this revelation is not surprising given the profound societal changes that have transpired during the past one hundred and fifty years. At the same time, though, it is noteworthy that the basic law governing the national parks—the so-called Organic Act—has not changed since 1916, nor has the revered status that the parks enjoy among the American public. The Organic Act's fundamental mandate—to conserve these special places in an unimpaired condition while also enjoying them—still sets the standard for how the parks are to be conceived and managed. What has changed, though, is the degree to which the larger world has come to the parks, in the form of unremitting visitor pressures, persistent commercial demands, and recurrent external development proposals, all of which have inexorably reshaped the national park idea.

Although regularly proclaimed "America's best idea," the national park idea is actually not a single idea, but rather an amalgam of ideas that have evolved over time. The principal ideas or concepts that define a national park do so in several quite distinctive terms, conceiving of the national park as a wilderness area, a tourist destination, a recreational playground, a commercial commodity, an ancestral homeland, a natural laboratory, a wildlife reserve, and, more recently, a vital ecological cornerstone. Several of these ideas have long generated much attention and discussion, whereas others have remained obscure in most national park circles. No single one of these images fully captures the essence of a national park in today's world, but collectively these ideas help us better understand what the parks add to our culture and the challenges they face on the eve of the national park system centennial.

In reviewing how these ideas have influenced our understanding and management of the parks, one fact stands out: the national parks have never been secure and isolated nature reserves. From their inception, national parks have been interconnected with the surrounding world. These connections are manifold, as the ensuing chapters explain, including deep economic and cultural connections to adjacent communities, strong historical ties to a business community that understands the parks in market terms, and ever more fragile ecological linkages that are vital to park wildlife and its long-term survival. In a world where private enterprise is revered and where park visitors present an unrelenting array of recreational and other demands, the challenge of conserving our national parks in an unimpaired condition for future generations is becoming more difficult each passing year. As science has come to play a more prominent role in park management and as our understanding of ecosystem processes improves, however, we are expanding our definition of nature conservation to embrace the broader landscape and thus our view of the national park idea itself.

One might ask why the national park idea evokes controversy when the value of these remarkable places is so widely acknowledged. My answer is straightforward. As highly valued and visible public places, the national parks are inherently political entities, mostly carved from a public domain that itself has generated conflict aplenty since the nation's founding. Many—if not all—national parks have been forged in controversy, reflecting part of the larger national dialogue about nature conservation and its role in our civic life. This conversation has involved such matters as national recreation policy, endangered species, social justice, and the respective roles of private enterprise, science, and economics. As time has passed, the dialogue has expanded to now include such matters as climate change, biodiversity conservation, large predators, wildfire policy, wilderness preservation, tribal neighbors, and the management of ecosystems. To continue to meet the challenge of conserving national park resources in an unimpaired condition for future generations, all these topics must be foremost in our ongoing conservation policy debates.

From the outset, the national parks have occupied a central position in nearly any conservation policy debate. Although national parks constitute only a part of the nation's extraordinary conservation systems, national park policies have long set an important if sometimes imperfect standard for what it means to preserve our natural heritage, both in professional and in personal terms. When most Americans think of nature conservation or wish to experience the natural world, they usually think of the national parks. Although set aside as protected reserves where wild nature reigns, the parks are nonetheless accessible to most people (regardless of their income level or outdoor skill set) and provide opportunities to observe nature in its splendor and fragility. Given their prominence and popularity, the national parks have long been and will likely remain at the forefront of this discussion. As a result, national park resource management policies will undoubtedly continue to define the basic principles that will guide our nature conservation efforts for the next century.

The goals of this book are several. One goal is to encourage us to see the national parks as they are, not just as magnificent settings but also as imperfectly protected venues subject to an array of political, economic, and other pressures. Another goal is to recount the origins and evolution of the various ideas that shape our view of a national park and thus our expectations of what they are and should aspire to be. Yet another goal is to demonstrate the tensions and highlight the controversies that regularly confront park managers, examining how these matters are being addressed on a day-to-day basis. With this more complete understanding of the national park idea, we can begin to think more clearly about what it means to conserve these places in an unimpaired condition and how this enduring standard should shape our future view of the national park idea. My overarching goal, then, is to present an unvarnished view of our most iconic landscapes, revealing the manifold challenges that confront these special places and explaining why our concept of the parks is the key to preserving them unimpaired for future generations.

To address these concerns, this book takes a topical approach, examining each major idea that has shaped our view of the national park. In separate chapters, the book traces the historical evolution of each idea, examines principal controversies involving application of the idea in practice, and reflects on what the idea may mean for national parks in the future. In addition, after examining these seminal ideas, the book explores what our evolving view of a national park means for the system as a whole and how these views may affect the opportunity to expand existing parks or establish new ones. In short, the pages that follow reassess the fundamental purpose of the national parks to better understand what the future may hold for them.

By adopting this organizational structure, there is necessarily some repetition in the narrative so as to present a complete picture of how and why particular ideas have evolved over time. For instance, the groundbreaking 1963 Leopold report not only reshaped our view of how park wildlife should be managed but also helped strengthen the role of science in national park policy and management decisions. The Park Service's Mission 66 building binge not only gave tourism top billing within the agency, but it directly affected park wildlife habitat and concession relations. Also, the controversies that are recounted sometimes have implications that extend across the national park idea. Watershed restoration efforts at Grand Canyon and Olympic National Parks, for example, speak to both the role of science in park policy and the need to think of parks as part of larger landscapes. In addition, key sources like the 1918 Lane letter and the Park Service's management policies occupy a central role in any discussion of national park history or agency resource management policies. Repeat references have been kept to a minimum without oversimplifying important ideas or illustrative controversies. The book's overall focus, given inevitable space limitations, is necessarily on the large natural parks—or "crown jewels"—that represent the heart of the national park system, with only passing reference to the quite different historical, cultural, and other sites that are also part of the system.

To address the national park idea, the book necessarily engages in critical analysis of the National Park Service, its approach to the Organic Act, and its natural resource management policies. In part, I follow in the footsteps of others—Dick Sellars and Joe Sax, to name just a couple—whose books have insightfully analyzed the agency and its history, accomplishments, and failures. In part, I am also building on my own prior work that has sought to identify and understand the laws and policies that drive park managers and the decisions they make as well as those they might make. To the extent that my analysis is critical, it is intended as friendly criticism, coming from one who has great respect for the men and women of the Park Service who oversee these extraordinary repositories of our natural and cultural heritage. They labor long hours for modest compensation in an often unfriendly and at times partisan environment, both at the national and local levels. That they persist in the face of hostility and ill-informed criticism is a tribute to them and to the devotion that our national parks inspire. That they learn and adjust from their experiences, as we shall see, is a testament to their wisdom and commitment to ensuring the integrity of these special places. We are, as a nation, deeply indebted to them, and my observations are not meant to diminish the vitally important work that they do on our behalf and for the generations that will follow.

ACKNOWLEDGMENTS

No book project of this scope is the sole work of the author. In my case, many others have contributed, both directly and indirectly, to the final product as well as to my understanding of America's best idea. I would be remiss not to recognize them and their roles in this endeavor. Of course, any errors or omissions are my responsibility alone.

Two individuals went far beyond the bounds of collegiality and friendship by reading the entire manuscript and providing extremely helpful comments, suggestions, and corrections. Ron Tipton, who has truly devoted his career to the national parks at the National Parks Conservation Association (NPCA) and other organizations, brought his extensive historical insights and policy perspectives to bear on the manuscript and set me straight in several instances. John Ruple, a research associate at the University of Utah S.J. Quinney College of Law's Wallace Stegner Center and a keen student of the public lands, not only reviewed the entire manuscript but also fact-checked my assertions and references. To both, I am extraordinarily grateful.

Over the years, I have benefited from the unstinting research assistance provided by Quinney Fellows at the University of Utah S.J. Quinney College of Law. Becky Holt, April Cobb, Melanie Stein Grayson, and Landon Newell unfailingly located materials that I requested and usually found much more; their efforts have improved my historical narrative and policy analysis. Before them, College of Law graduate Rob Dubuc provided me with early research assistance as well as encouragement that the national parks merited further examination. The staff at the Quinney Law Library also regularly lent me a helping hand to find and secure historical sources and other materials that have improved the book. In particular, I am indebted to John Bevan, Laura Ngai, Linda Stephenson, Rita Reusch, and Suzanne Darias for their stalwart assistance. My faculty colleague Alex Skibine graciously reviewed the chapter on Native Americans and the National Parks.

My knowledge about the national parks, park history, and park policy has been enriched through my engagement with the dedicated board and staff of the National Parks Conservation Association. My fellow association trustees who give of their time and treasure to protect our national park system have not only instilled in me a deep appreciation for how important the national parks are to the nation, but also a clear sense of what is necessary to carry the national park idea forward. Both Deny Galvin and Gretchen Long have given me their time and attention to discuss my ideas and to read parts of the manuscript. The NPCA staff, for whom I have the utmost respect for the tireless work they do to safeguard the national parks, have diligently answered my questions and requests without complaint, fact-checked parts of the manuscript, and introduced me to new and sometimes perplexing issues that confront our parks. My most sincere gratitude goes to David Nimkin, Don Barger, Tony Jewett, Michael Jamison, John Adornato III, Lynn McClure, Jim Stratton, Alex Brash, Joy Oakes, Ron Sundergill, Tim Stevens, Suzanne Dixon, Kristen Brengel, Mark Wenzler, and Sean Smith. I also owe thanks to the NPCA senior leadership—Tom Kiernan, Theresa Pierno, Craig Obey, and Libby Fayad—as they, too, have responded to my queries and helped keep me pointed in the right direction.

Few of us in academic or other professional pursuits would be where we are without the support, guidance, and encouragement of mentors along the way. For me, several individuals have played crucial roles in my understanding of the national parks and public land policy generally. Early on, when I was just entering the field, Joe Sax took me under his intellectual wing and nurtured my knowledge and meager talents; the two research projects we did together on Glacier National Park remain a high point in my career. I am also much indebted to Charles Wilkinson and John Leshy, whose landmark works on public land law and policy continue to serve as an inspiration, and to Bill Lockhart, whose stalwart efforts defending our national parks merit broad recognition and demonstrate the need for meaningful park protection laws and policies. The late historian Robin Winks unselfishly shared with me and others his encyclopedic knowledge about the national park story as well as his unbounded enthusiasm for the parks and those who oversee them. My natural resource law teacher colleagues, who hail from law schools across the nation, sat through an early presentation of my ideas for the book at Chico Hot Springs, and I benefited greatly from their insightful questions and suggestions.

Many other individuals with deep ties to the national parks have elevated my knowledge and grasp of Park Service history, policies, and practices. I have been privileged to engage with former Yellowstone superintendent Mike Finley at several different levels, where I have observed his deep commitment to the Organic Act's nonimpairment mandate and his understanding of what it takes to meet that standard. Over the years, several other Park Service leaders have likewise shared their knowledge and experiences with me, which has improved the book and my policy analysis. For this help I extend my enduring gratitude to the late Roger Kennedy, Bob Barbee, Destry Jarvis, John Varley, Richard Sellars, Paul Schullery, Gil Lusk, Alan O'Neill, Cliff Martinka, Brace Hayden, John Reynolds, Mike Soukup, and Warren Brown. Outside the Park Service, I have received both guidance and assistance from Ed Lewis, Mike Clark, Michael Scott, Scott Christensen, the late Ken Diem, Mark Boyce, Susan Clark, Wayne Hubert, Gerry Wright, J. Michael Scott, Fred Wagner, and the late Ted Smith.

An author's efforts are ultimately only as good as his editor and publisher. I could not ask for a more engaged and thoughtful editor than Barbara Dean, whose editorial insights and judgments have strengthened the book immensely. That Barbara shared her skills so graciously only made the entire experience that much more enjoyable. Erin Johnson provided much-appreciated guidance in bringing the manuscript and references to completion. Sharis Simonian helped me navigate the copyediting process without misstep. In fact, the entire Island Press team has given me their unstinting support and assistance in producing the final book and introducing it to the marketplace.

Finally, to my dear and patient wife, Linda Keiter, I extend my enduring gratitude and love for seeing me through another book project. It would not have been possible without her support and encouragement. Perhaps this book will play some small role in ensuring that our extraordinary national parks remain secure and unimpaired so that our grandchildren and their children can enjoy them as much as we have.

CHAPTER 1

What Is a National Park?

For much of its existence, Yellowstone National Park spent the winter months quietly under a blanket of snow. Winter was a time of restoration, for the bears that hibernated on isolated mountain slopes, for the elk and bison relieved from the attention of visitors, and for the park rangers who used the time to recover from the hectic summer season. That is no longer the case, however. The park is now a beehive of activity during the winter months. Snowmobiles dash around the park roads, rangers are on frequent patrol to control wayward visitors, and the winter-stressed wildlife must endure regular encounters with snow machines and cross-country skiers. Along with this new winter season has come controversy that strikes at the heart of what our national parks are and supposed to be.

The Yellowstone snowmobile controversy is a bitter protracted battle over park management policy that has reached the highest levels of government. It pits an avid and growing motorized recreation constituency against an equally avid and entrenched environmental community, and it shows few signs of abating. During the 1960s, in a little-noticed decision, Yellowstone's superintendent, himself an avid snowmobile rider, decided to open the park to these new machines, observing that it would allow the public to see and experience their national park even during the harsh winter months. Besides, the park already was crisscrossed by roads heavily plied by noisy automobiles during the summer months. In a few decades, Yellowstone snowmobile numbers escalated dramatically, growing to more than eight-five thousand annually during the mid-1990s, introducing toxic air pollutants and the roar of two-cycle engines to the previously quiet park, and displacing wildlife from their familiar haunts. Much of this growth was fueled by West Yellowstone town officials who seized upon the iconic national park as a perfect vehicle to promote winter recreation in this wonderland setting and to thus jump-start the town's moribund winter economy.[1]

Eventually, faced with mounting environmental concerns, persistent visitor conflicts, and its distinctive preservation responsibilities, the National Park Service took action. In 2000, after an extensive environmental study, Yellowstone officials announced that snowmobiles would be banned from the park but that visitors could still enter the park in snow coaches, essentially minibuses on tank treads. The decision, reciting the Park Service's obligation to maintain park resources in an unimpaired condition, explained that snowmobiles created unacceptable environmental effects that threatened the park's integrity and that resource protection must take precedence over recreational activities. A firestorm of protest ensued along with litigation designed to keep the park open to snowmobiles. Soon after the administration of Bill Clinton gave way to the George W. Bush administration, the original ban was lifted, and the Park Service has since sought to justify why it has changed direction to continue allowing snowmobiles in the park daily.[2]

Exactly how this intractable controversy will be resolved remains to be seen, but it is merely one example among many in which the Park Service must reconcile resource protection with competing visitor access, recreational opportunity, and external economic pressures. Whether the issue is whitewater rafting permits on the Colorado River in Grand Canyon National Park, off-road vehicle travel in Death Valley National Park or Cape Hatteras National Seashore, or implementation of a new bus system in Zion National Park, the controversies highlight the divergent and often-conflicting views over the fundamental purpose of our national parks. Should snowmobiles be permitted in the parks during the winter months given the inevitable environmental effects and visitor conflicts? Does the Park Service have an obligation to protect park resources and the natural soundscape, or should it seek to accommodate as many visitors and activities as possible? What is the agency's obligation to gateway communities and concessioners? How should the agency interpret its governing legislation that speaks of conserving park resources in an unimpaired condition while also providing for public enjoyment of the parks? Is the agency constrained by past practices and decisions, or is it free to redefine park management policies to take account of new knowledge, experience, or demands?

Other park management decisions can have equally significant, even devastating, repercussions on park resources as well as the surrounding landscape. Take what occurred in northern New Mexico during June 2000, when the Park Service ignited a prescribed burn at Bandelier National Monument in an effort to reduce dangerous fuel loads. Although within the bounds of the agency's existing fire management policies, the blaze escaped containment and roared across the park boundaries, consuming thousands of acres of adjacent national forest and burning down four hundred homes in nearby Los Alamos. Not unlike what had happened twelve years earlier in the aftermath of the Yellowstone fires, the Park Service was once again forced to defend its ecologically based resource management policies, which sought to emulate rather than control nature and natural processes. One of the most troublesome criticisms directed toward the agency was whether igniting the park landscape was actually consistent with its mission to preserve the park for the benefit of the visiting public. Even if it was, how could park officials justify placing the park's neighbors in harm's way through a conscious decision to use natural processes as a resource management tool?[3]

The national parks have not always been managed with nature as the foremost concern. Throughout much of its early history, the Park Service actively sought to control nature, primarily to improve the visitor experience. Park officials routinely suppressed all wildfires to protect the scenery, eradicated wolves and other major predators to safeguard more desirable wildlife, fed the bears to create an evening spectacle for park visitors, and constructed hotels and roads near attractive venues without regard to the environmental effect. According to the Park Service's own historian, the agency was practicing "façade management," not ecological conservation. That finally changed during the mid-1960s when, in response to the influential Leopold report, the Park Service embraced the idea that the national parks should be managed to represent a "vignette of primitive America." Such management meant allowing fires, predation, and other natural processes to operate with minimal human interference so as to maintain a more historically representative ecological condition. Paradoxically, it also contemplated more human intervention to achieve restoration goals, including the use of controlled burns, the transplantation of missing predators, and the removal of dams that blocked free-flowing rivers and other natural processes.[4]

Almost from the outset this policy shift proved controversial. Critics questioned what the word *natural* means and whether it is possible to re-create long-past ecological settings, especially when we now understand that nature is regularly in an often unpredictable state of flux. The Park Service has nonetheless stood firm in its ecologically based approach to resource management, committed to minimizing and carefully directing human intervention on the landscape. The results are remarkable in several locations: the reintroduction of wolves in Yellowstone National Park, restoration of the Giant Forest in Sequoia National Park, the current removal of two dams on the Elwha River to restore historic salmon runs into Olympic National Park's interior, and periodic flooding events on the Colorado River in an effort to return the Grand Canyon river corridor to a more natural state. Agency officials, portraying the national parks as expansive outdoor laboratories, view these initiatives as part of a larger resource management experiment that enables us to better understand and thus better manage nature. Whether such a vision of the national parks can be sustained in today's ever more crowded and interconnected world is very much open to doubt. Also open to question is whether this experimental laboratory role for the parks can be squared with the agency's conservation responsibilities.[5]

Another icon—Glacier National Park—is beset by more distant but nonetheless equally challenging problems that could imperil its integrity as well. Here the problem is not how the Park Service is managing Glacier itself, but rather how the park's neighbors are managing the lands adjacent to Glacier, which can adversely affect park wildlife, water quality, and other resources. For more than three decades, Glacier officials have worried that oil and gas exploration on a neighboring national forest and tribal lands will bring industrial development to this still-pristine region in northwestern Montana, and they have bemoaned the heavy-handed timber-cutting practices on another nearby national forest. Park officials have also kept a wary eye northward on mining proposals in the remote Canadian North Fork region, where British Columbia politicians have seemed intent on expanding coal mining into this sensitive drainage. Although an earlier mine proposal was killed in the late 1980s, several more projects have since surfaced, rekindling concern about the fate of this yet undeveloped region. Add the unbridled growth occurring in the Flathead Valley immediately west of the park, and the prospects that new subdivision developments will eventually extend right up against the park boundary become very real. Park officials are trying to address this growing list of external threats through new regional partnerships to promote more coordinated planning and decision making, but whether this approach will succeed over the long term is far from certain.[6]

Glacier's plight is not unique among our national parks. Ever since the 1970s, when Redwood National Park found itself under assault from unrestrained upstream logging, the Park Service has identified external threats as a key resource management concern. Although Congress eventually intervened in the Redwood controversy, it has not seen fit to adopt legislation addressing the wider external threats problem, leaving Glacier and other parks to individually confront these issues. Yellowstone, like Glacier, also faces an array of energy and subdivision threats from adjoining public and private lands; the Everglades may lose their distinctive freshwater features unless upstream water diversions can be reversed through an unprecedented regional ecological restoration effort; and air quality at the Grand Canyon is often so bad that visitors can only view the storied chasm through a gauzy haze.

Today, national parks can no longer be viewed as isolated islands. Rather, they are part of larger ecosystems subject to ongoing human development pressures that often degrade the regional environment as well as the visitor experience. The Park Service, however, has no explicit legal authority over what occurs on the surrounding landscape, even when park resources may be at risk. To address this conundrum, agency officials and their allies have begun to cast the national parks as the vital cores—or cornerstones—of larger regional ecosystems that should be managed with restraint to ensure overall ecological integrity. But it is a hard argument to make in settings where private property rights and economic development interests have regularly held sway. This fact was brought home forcefully in northwestern Wyoming during the early 1990s, when proponents of the Greater Yellowstone Ecosystem concept saw a promising multiagency coordination effort dissolve in the face of local political pressures.[7]

One way to protect sensitive lands and resources is to designate a new park; another is to expand existing park boundaries. Over the years, Congress has seen fit to use both these means on several occasions. Not only has the national park system grown to include close to four hundred units, but it also covers more than 84 million acres in forty-nine states and several territories. Most of this acreage is embraced within the fifty-eight large natural parks, including those in Alaska, where the Park Service oversees more than fifty-five million acres. However, these new national park designations, as wonderful as everyone may feel about them today, have rarely come easily. Indeed, the political intrigue and legislative battles that underlie even our most treasured park lands are full of the same triumph, resistance, and compromise that otherwise permeate national politics. The case of Grand Teton National Park is instructive. The park was bitterly opposed by local residents and had to be gradually pieced together over several decades through a combination of presidential proclamations, clandestine land purchases engineered by philanthropist John D. Rockefeller Jr., and hard-nosed legislative horse-trading that rivaled the stratagems of Rockefeller's father in assembling the Standard Oil Company. Although most Wyoming citizens today laud the park and its role in the local economy and community life, the struggle to create and then expand it is testament to the heavy political lifting that has long been part of growing the national park system.[8]

The same concerns still surface regularly across the national landscape whenever park creation or expansion is mentioned. Even though recent research indicates that the presence of a national park strengthens the local economy, nearby residents still regularly resist new park creation or expansion proposals. As reflected in the seven-year battle over California's desert lands, opponents of the park expansions in Death Valley and Joshua Tree National Parks recited a litany of concerns, including existing mining claims, grazing privileges, and military training overflights. Not only do opponents routinely decry the loss of traditional mining, logging, ranching, and other development opportunities, but they also object to the strict protective management standards associated with a national park, notably restrictions on off-road driving, hunting, and other recreational activities. Sister agencies have historically resisted losing some of their most prized lands to the Park Service, and this resistance has even more resonance today when all the federal land management agencies have wilderness management responsibilities. Even the Park Service has proven to be a reluctant partner when it comes to new park designations, not infrequently objecting that the proposed area lacks "national significance" or that the proposal is driven by local politics and will divert badly needed resources away from existing parks.[9]

All these concerns are evident in the ongoing efforts to expand Canyonlands National Park in southern Utah. As originally drawn by Congress in 1964, the park's boundaries represented a political compromise to appease local ranching and mining interests who opposed the new park. Since then, however, the park has become central to the local economy, ranching and mining have faded in importance, and off-road vehicles have invaded the area. Expansion of the park boundaries across the entire geologic basin would address these problems and restore natural integrity to the park. Similar expansion proposals are on the table for Mount Rainier, Carlsbad Caverns, and Saguaro National Parks, and new park proposals could add Mount St. Helens, the Valles Caldera in New Mexico, Maine forestlands, and other unique landscapes to the system. The larger question, of course, is whether the national park system is essentially complete. Are we, as a nation, content with the lands that our forebearers had the foresight and self-restraint to set aside as national parks? Or are we prepared to increase the size and scope of the national park system to fully protect our existing parks and to see that future generations have the opportunity to experience the wonder of nature in a setting where people are mere visitors on the landscape?[10]

The national parks are where most Americans come face to face with wild nature. Their legendary names resonate with historical and cultural significance: Yellowstone, Yosemite, Denali, Zion, Grand Teton, and Grand Canyon. Just mention the term *national park* and a kaleidoscope of images flood the mind, whether of a solitary trapper first encountering the Yellowstone geyser basins, one-armed John Wesley Powell navigating unknown rapids in the Grand Canyon, a grizzly sow with her cubs ambling across the Alaskan tundra, or springtime waterfalls cascading down Yosemite Valley's granite cliffs. These iconic symbols of our unsurpassed natural heritage represent our first and still most visible commitment to protecting wild places. As James Bryce, an early-twentieth-century British ambassador to the United States, supposedly put it, "The national park is the best idea America ever had."[11]

That America's national parks have survived relatively unscathed for nearly a century is no accident. Fierce political battles have been fought over them, not only to establish new parks but to protect the integrity of the existing ones. These conflicts—snowmobiling in Yellowstone, restoration of the Everglades, dam construction in the Grand Canyon, energy exploration adjacent to Canyonlands, to name a few—have regularly generated front-page news coverage and called into question the very notion of nature conservation. Indeed, these controversies bring the basic preservationist impulse that underlies the national park idea into sharp contrast with the utilitarian instinct that has long fueled the nation's growth and development. Even as galloping urbanization spreads across the landscape, though, the American national park system stands as a powerful testament to the depth and strength of our commitment to nature conservation.

The enduring presence of the national parks masks, however, a more fundamental and important question: What exactly is a national park? This question has bedeviled us from the beginning, it persists yet today in more pressing form, and the answers have evolved over the years. In fact, the only constant in our national park heritage is the reality of change: change in how we conceive of national parks, change in how we manage them, change in what we seek from them, and change on the landscape surrounding them. These myriad changes reflect even-deeper-seated shifts in American thought and society, including our perception of wild nature, our growing scientific knowledge base, our increasingly disparate leisure-time activities, and the relentless pressures of population growth and economic prosperity. In the aggregate, these changes have affected not only our view of a national park but our approach to nature conservation as well.

At the beginning of the national park system, Congress gave expression to the preservationist ideal in the National Parks Organic Act of 1916, which remains the basic charter—or Magna Carta—governing the parks. The Organic Act charges the National Park Service "to conserve the scenery and the natural and historic objects and the wild life therein and to provide for the enjoyment of the same in such manner and by such means as will leave them unimpaired for the enjoyment of future generations." Since then, Congress has not only refused to alter the basic Organic Act mission, but it has reaffirmed that "the protection, management, and administration of these areas shall be conducted in light of the high public value and integrity of the National Park System and shall not be exercised in derogation of the values and purposes for which these various areas have been established, except as may have been or shall be directly and specifically provided by Congress." For the National Park Service as well as its erstwhile allies in the conservation community and elsewhere, this legislative language establishes a near-sacred responsibility to exercise the utmost diligence and skill to safeguard the special landscapes and resources under its care.[12]

Since its enactment, the Park Service has sought to clarify the Organic Act mandate and frame its conservation policies accordingly. The agency's seminal effort toward this end came in 1918 in the form of the so-called Lane letter, which was widely viewed as the new agency's definitive interpretation of its management responsibilities. Actually penned by Horace Albright, trusted assistant to Stephen Mather, founding director of the Park Service, the letter was released over Interior secretary Franklin Lane's signature to give it added force. The letter, still widely regarded as a key source for understanding the Park Service's fundamental mission, set forth three broad principles that continue to resonate yet today: "First, that the national parks must be maintained in absolutely unimpaired form for the use of future generations as well as those of our own time; second, that they are set apart for the use, observation, health, and pleasure of the people; and third, that the national interest must dictate all decisions affecting public or private enterprise in the parks." Since then, an extraordinary assortment of policy statements, secretarial decisions, and reports have further refined and revised national park conservation policies. Perhaps none is more important today than the Park Service's revised *Management Policies*, which further perfect the agency's view of its legal obligations and management responsibilities.[13]

The Organic Act language, however, presents the agency with a nearly impossible mission, obscuring an array of hard judgments that the Park Service confronts on an almost daily basis. It must safeguard these special places from environmental injury in an ever more complex world while also making them available for an ever more demanding general public. The critical question is simply, how should we go about ensuring that the national parks are conserved unimpaired? The answer to this question has changed dramatically over time, just as the issues that it raises have grown in number, scope, and complexity. When it comes to off-road vehicles, concessioner demands, wildfires, predators, wildlife population pressures, Native American treaty claims, and encroaching development, what exactly should the Park Service be doing to protect park resources and values? How the Park Service and our political institutions perceive and respond to these matters account in large measure for the national parks as we know them today and as subsequent generations will experience them in the years ahead.

The national park story is one of changing ideas and values, freighted with controversial judgments, missed opportunities, and sometimes visionary inspirations. The image of a national park speaks to how its keepers, visitors, and neighbors relate to these iconic landscapes. What was clear from the beginning and what remains clear today is that a national park cannot be all things to all people, nor can it be isolated from the society surrounding it. Guided by the Organic Act's enduring injunction to conserve the parks unimpaired for the benefit of posterity, each generation must define anew precisely what this simple but beguiling mandate means and how it applies on the ground. In doing so, we must be able to defend these judgments in the court of public opinion, the halls of Congress, and the federal courthouse. Anything less puts our national parks in peril and demeans the extraordinary contributions that those who came before have made to ensure our collective heritage.

We are, however, in danger of taking the national parks for granted and losing the essence of what these special places represent. Although everyone agrees that we need national parks and that they should be adequately protected, there is astonishingly little agreement on exactly what a national park is or might be. Not only does the average park visitor often lack any real appreciation for the difficult management judgments involved in maintaining these special places, but the Park Service itself still struggles with the true meaning of its conservation mission and how to implement it. Scholars, politicians, environmental groups, concessioners, and others are likewise divided over the national park idea. It is easy to extol perpetual conservation, but quite another thing to practice it in an increasingly crowded and contentious world.

National parks are at once not a single idea, but rather a complex assortment of ideas. In fact, multiple conceptions of the national park idea have held sway over the decades, often depending on whether the venue was Congress, the National Park Service, particular interest groups, or the general public. The very name "national park" conjures up a full spectrum of images: an unsullied wilderness, an attractive tourist destination, a vast playground, an economic engine, an ancestral homeland, a natural laboratory, a wildlife reserve, a vital ecosystem core. Each of these images has had some resonance over the past century; none fully captures the essence of a national park, yet each plainly evokes an important dimension of national park history, experience, and aspiration. It is these images—or ideas—that give expression to the policies we employ to create and manage the parks.

The chapters that follow explore the question of what a national park is and should aspire to be. The answer may be more important today than at any earlier time. The pressures on the nation's once-isolated national parks are intense, ranging from creeping subdivisions to energy exploration on the borders to vociferous recreational constituencies and commercial entities intent on exploiting park resources, let alone the alarming specter of climate change. No single incursion or decision will seal the fate of the national parks, but the cumulative effect of these incursions over time will change the face of the national park system forever. Hardly anyone expects Congress to decommission Yellowstone or its sister parks, but our national parks are quite different places today than they were just fifty years ago. They not only play much different roles today than they did at their inception, but they face much different external pressures than in prior times.

The challenges now are to define anew the national park idea and to inspire a new generation to meet our looming conservation challenges. If we cannot safeguard our national parks against the relentless forces of humanity and progress, then we will have failed to ensure that our grandchildren will inherit a world where wild nature continues to inspire, instruct, and amaze. And if the national parks face an uncertain and tenuous future, then what will the future hold for our other protected landscapes, which enjoy much less popular support and appreciation? In short, the national park idea goes to the very core of our commitment to nature conservation in a changing world.

CHAPTER 2

"Nature's Cathedrals"

A Wilderness Sanctuary

National parks and wilderness are practically synonymous, at least in the minds of most visitors. Never mind the roads that penetrate into the parks or the lodges scattered about them, one popular vision of the national parks is of untamed wilderness with miles of unbroken backcountry and legions of wild—and sometimes fearsome—animals. That would accurately describe Yellowstone, Glacier, and several other national parks, and it is an image the National Park Service has endorsed from its earliest days. Since 1964, however, following passage of the Wilderness Act, portions of only a few parks have been designated official wilderness areas, and the prospect of such a legal designation has ignited intense controversy in others. Given the ever-mounting visitor and recreation pressures coming from within the parks as well as development pressures coming from outside park boundaries, the challenges involved in maintaining an undisturbed natural setting have grown increasingly more difficult. As a result, whether the parks are true wilderness strongholds or whether the Park Service can consistently manage them as such is open to question.

WILDERNESS AND THE NATIONAL PARKS

Wilderness was certainly an apt description of the early national parks, most of which were carved from the sparsely settled western public domain. The Yellowstone country in 1872 and the Grand Canyon a few decades later generally fit this wilderness profile. The earliest and perhaps most evocative image of Yellowstone is attributed to trapper John Coulter, whose description of the area's vast thermal features was translated into "Coulter's Hell," an image that squared with the prevailing view of wild nature at the time. Ferdinand Hayden of the U.S. Geological Survey, who led an early survey expedition into the Yellowstone region, penned a similar description: "For fifty miles in every direction there is a chaos of mountain-peaks. . . . For grand rugged scenery I know of no portion of the West that surpasses this range."[1] Although John Wesley Powell had traversed the Grand Canyon on the Colorado River in 1869, the canyon landscape itself remained mostly unknown over the ensuing decades except to a handful of early entrepreneurs and a few intrepid visitors. Similar observations can be made for the other early parks, which resembled wilderness sanctuaries more than anything else at the time of their designation.

Conserving Wild Nature

To be sure, the early national parks were not without evidence of a human presence. Native Americans lived and hunted across many of these landscapes, and a few vanguard pioneers—trappers, miners, boomers, early settlers, and others—were busy exploring these lands, mostly with an eye toward claiming and exploiting them. Government survey parties chronicled the settlement and mining activities then under way in the Yellowstone country at Cooke City, Montana, and elsewhere well before the national park idea surfaced. Farther west, John Muir furiously lamented the domestic sheep—"hoofed locusts" he called them—that were overrunning his beloved Sierra Nevada mountains in California before either Yosemite or Sequoia attained national park status. In fact, it was the threat of wholesale settlement and development of these unique scenic settings that prompted the early movement to set them aside as "pleasuring grounds" to be enjoyed by everyone, not just the few who had arrived first and laid claim to the area.

Once convinced of the national park idea, Congress endorsed the notion that these newly protected lands should be maintained in a wilderness-like state. Although eschewing the term *wilderness* (after all, the frontier had just officially closed in 1890), the early legislation creating Yosemite, Glacier, and other national parks called for maintaining them in their "natural state." For early park proponents, this language plainly meant that park lands were henceforth closed to mining, logging, dams, or other development activities. This interpretation was not universal, however, as became evident during the early 1900s in the bitter struggle over construction of the O'Shaughnessy

Dam in the Hetch Hetchy Valley inside Yosemite National Park. Indeed, questions concerning the preservation of pristine nature would be debated at many points in the next hundred-plus years. Similar battles soon ensued over proposals to open the parks for strategic metals and timber to support the military effort during World War I. Faced with these recurrent early controversies, park advocates concluded that the nation's inaugural parks required additional legal protection to safeguard them in their "natural state," as the early legislation intended.[2]

The result was the aforementioned 1916 National Parks Organic Act, which created the National Park Service to oversee a new park system and to protect park scenery and wildlife in an unimpaired condition. Of course, the Organic Act's classic language—conservation and enjoyment—does not speak explicitly in wilderness preservation terms. In fact, the legislation sanctioned various intrusions, as reflected in provisions allowing some timber cutting, livestock grazing, and wildlife removals across the system, which would provide the basis for future conflicts. Moreover, the original 1872 Yellowstone legislation had characterized this first park as a "pleasuring ground," and officials from the Department of the Interior—along with customer-hungry railroad executives who supported the national park idea—were eager to begin attracting visitors to these spectacular yet undeveloped settings.[3]

The Park Service and Nature Conservation

Indeed, the fledgling National Park Service had no intention of treating these protected landscapes as wilderness sanctuaries. Under the leadership of former Borax marketing executive Stephen Mather, the agency's pathbreaking first director, the Park Service set about making the parks readily accessible to the public. Nowhere is this view more evident than the historic Lane letter. After instructing the Park Service "to faithfully preserve the parks for posterity in essentially their natural state," the letter then characterized the parks as a "national playground system," singled out "motoring" as a "favorite sport" in the parks, approved "luxurious hotels," and encouraged collaboration with automobile highway associations to promote park visitation.[4] The rationale underlying Mather's commitment to building roads, lodging, and other facilities in these wild settings was simple: the new national parks would remain vulnerable to exploitation unless they had a strong political constituency to ensure congressional protection, and that constituency would primarily be the American citizens who visited the parks and developed lasting ties with them. To create this relationship, Mather believed that it was the Park Service's job to make the then-remote parks accessible to the general public, to provide them with accommodations once they arrived, and to ensure their safety. In an era that predated the modern environmental consciousness and contemporary ecological knowledge, wilderness preservation simply was not a central concern. Given that the parks were a political creation, Mather's strategy was ingenious and has helped sustain the public support necessary to maintain and expand the park system over the years as other park directors followed Mather's lead.[5]

The agency's early commitment to roads, lodges, and automobile tourism plainly belied any notion that the national parks, as a whole, were being managed as wilderness. In a few short years, the early parks were literally transformed in appearance. In Yellowstone, by the 1920s, key lodges—Old Faithful, Mammoth, and Lake—had been built, and the basic loop road system was in place. The same held true for Yosemite, Glacier, Mount Rainier, and Grand Canyon National Parks, where lodges were completed, access roads constructed, and other visitor facilities put in place. But even with this robust commitment to development, the park superintendents in 1922 jointly called for more: "Roads and trails should be improved and extended, ample accommodations should be provided for visitors, and other improvements carried out, so that the parks may better fulfill their mission of healthful recreation and education to a larger number of people." Their rationale was revealing: "If there were no development, no road or trails, no hotels or camps, a national park would be merely a wilderness, not serving the purpose for what it was set aside, not benefitting the general public."[6]

The idea of the parks as wilderness settings nonetheless continued to surface during these early years, even at the highest levels. Clearly, not everyone subscribed to the Park Service's early penchant for new roads or visitor facilities. As part of the campaign to expand Yellowstone's boundaries southward, agency officials in 1919 tentatively suggested a new road through the undisturbed Thorofare Basin to link the towns of Cody and Jackson, Wyoming, and facilitate additional visitation. Jackson-area dude ranchers vigorously opposed the suggestion, however, arguing against "overflowing the country with tourists, and other encroachments of civilization that would rob it of its romance and charm." Their opposition convinced Director Mather to oppose the idea and to assure them that "a part of the Yellowstone country should be maintained as a wilderness for the ever increasing number of people who prefer to walk and ride on trails in a region abounding in wildlife." Further, Mather indicated that this ban on new roads would extend to other parks: "In the Yosemite National Park, as in all of the other parks, the policy which contemplates leaving large areas of high mountain country wholly undeveloped should be forever maintained."[7]

Beyond the question of roads and facilities in the parks, the Park Service was nevertheless committed early on to resource management policies radically inconsistent with the concept of wilderness. Rather than leaving park landscapes in a natural or undisturbed state, the Park Service actively sought to control nature so as to provide park visitors with a more aesthetically pleasing and less threatening experience. During the early 1900s, it meant fighting wildfires aggressively to avoid blackening park forests and marring the scenic backdrop. It also meant eliminating predatory animals, like the wolf and cougar, which preyed on such visitor-preferred animals as elk, deer, and bison. By the late 1920s, the wolf was eradicated from Yellowstone, leaving a noticeable void in this wilderness setting and an impoverished ecosystem. Although grizzly bears were also targeted, they survived federal eradication efforts only to be converted into a visitor spectacle. In Yellowstone during the early 1900s, the Park Service constructed viewing platforms at several garbage dumps so that visitors could see bears up close in a nightly feeding display. Yellowstone's bison were corralled and herded early on just like domestic livestock rather than allowed to roam free as they had for millennia. Yosemite went as far as to treat its visitors to the nightly spectacle of a firefall at Bridal Veil Falls, where each evening Park Service employees pushed a bonfire from the canyon rim over the lip of the falls in an effort to improve on the natural appearance of the canyon walls. In short, the active manipulation of nature to overcome the reality of wild nature was the original order of the day in the early national parks.[8]

The 1930s, with the onset of the Great Depression, provided another opportunity to enhance park visitation. President Franklin D. Roosevelt's vaunted Civilian Conservation Corps (CCC) became a ubiquitous presence throughout the national park system: CCC employees helped build roads like Yosemite's Wawona Road and Glacier's Going to the Sun Road; they added new automobile campsites; they constructed and refurbished visitor centers, park housing, fire towers, and other park buildings; they laid trails and built bridges and dams; and they built ski areas and other winter sports facilities. Their work also opened park backcountry areas, enabling the Park Service to begin fighting more-remote wildfires. Before the decade ended, the CCC had constructed more roads and other facilities in the national parks than had been built during the preceding fifty years.[9]

An Emerging Wilderness Movement

As the 1930s unfolded, the pace of development was setting off alarm bells with conservationists. Wrote one prescient activist: "Can a wilderness contain a highway? . . . No one who knows the National Parks is so naïve as to believe them to be wilderness areas. . . . Some primitive areas, however, still exist in almost all Parks. These should be guarded as the nation's greatest treasure; and no roads should be permitted to deface their beauty."[10] New road construction in the parks along with mounting incursions into undeveloped backcountry areas of the national forests finally prompted Aldo Leopold, Bob Marshall, and other prominent conservationists—including Robert Sterling Yard, former executive secretary of the National Parks Association—to create The Wilderness Society, a new conservation organization dedicated to protecting primitive landscapes across the public lands. During this same period, ironically, Congress decided to expressly inject the notion of wilderness preservation into the national park system. In 1934, when adding Everglades National Park to the fledgling system, Congress decreed in the enabling legislation that this large new water-bound park "shall be permanently reserved as a wilderness, and no [visitor] development . . . shall be undertaken which will interfere with the preservation intact of the unique flora and fauna and the essential primitive natural conditions now prevailing in this area."[11]

World War II triggered another assault on the parks, again under the patriotic banner of supporting the war effort. The best known of these attempted incursions was the Defense Department's request to open Olympic National Park to logging to harvest the park's straight-grained Sitka spruce trees that were critical components for aircraft production. The effort ultimately died when it became apparent that British Columbia timber was potentially available, and the Air Force soon shifted to aluminum for its airplane wings. To its credit, the Park Service was adamant in resisting this effort to industrialize Olympic's intact old-growth forests, arguing that logging should only be done as a last resort and only after alternative sources (including Canadian timber) were exhausted.[12] Otherwise, with the nation's attention focused on the war effort and so many troops overseas, the park system languished in relative obscurity with few visitors.

This period of quietude ended abruptly after the war. By the early 1950s, new pressures beset the parks, threatening to eliminate any semblance of remaining wilderness. A suite of development proposals designed to promote and accommodate population growth swept across the western states. In one notorious instance, the Bureau of Reclamation proposed a massive dam at Echo Park where the Green and Yampa Rivers merge inside Dinosaur National Monument, ostensibly to ensure water and power for the growing upper basin states. Originally, the Park Service chose not to oppose the dam. That role fell to the Sierra Club and other conservation groups that viewed the dam as an unwarranted intrusion into the remote reaches of a national park and completely inconsistent with wilderness preservation objectives.

A youthful David Brower, recently named the Sierra Club's executive director, conceived a brilliant nationwide publicity campaign to save Echo Park by bringing political pressure to bear on Congress. Brower and his allies sought public attention for the cause through books, photos, films, and media coverage. Brower personally enlisted award-winning writer Wallace Stegner to help produce a large-format photo book titled *This Is Dinosaur* designed to acquaint Americans with this remote yet remarkable wilderness setting. Stegner knew the country well and did his part:

> To this moment, at least, the Green and Yampa canyons have been saved intact, a wilderness that is the property of all Americans, a 325 mile preserve that is part schoolroom and part playground and part—the best part—sanctuary from a world paved with concrete, jet-propelled, smog-blanketed, sterilized, over-insured, aseptic . . . with every natural beautiful thing endangered by the raw engineering power of the twentieth century.[13]

In the end, Brower's efforts succeeded, the dam was blocked, and Dinosaur's defenders were emboldened to pursue a much broader wilderness legislative campaign to protect the remaining undisturbed landscapes across the federal public lands. The Park Service's reaction, though, was quite different: it constructed new roads into Echo Park to expose more people to the area's natural beauty—a move that effectively compromised the area's wilderness qualities.[14]

In fact, the Park Service seized the postwar period as a new opportunity to open the national parks to new visitors by launching a construction campaign that transformed park landscapes across the system. To accommodate a war-weary citizenry in the midst of a baby boom and eager for new recreational outlets, the director of the Park Service, Conrad Wirth, conceived Mission 66—so named to coincide with the agency's upcoming fiftieth anniversary in 1966—to expand and upgrade the deteriorated national park infrastructure to ensure visitors a quality experience.[15] By the time Mission 66 concluded, the Park Service had built more than twenty-five hundred miles of new or improved roads, 114 new visitor centers, and dozens of new lodges while also adding a wide array of new trails, bridges, and campgrounds. The Hurricane Ridge Road was designed to open the mountainous heart of Olympic National Park to auto traffic, and Stevens Canyon Road was built into an undisturbed corner of Mount Rainier National Park, just two examples of the effect Mission 66 construction projects had on the remaining intact park wilderness lands.

Indeed, these new roads and facilities plainly imperiled the very wilderness qualities that the Park Service extolled as an important part of the national park experience. Director Wirth argued that the Mission 66 developments represented "zones of civilization in a wilderness setting" and that the new roads were "corridors through the wilderness linking these zones." And there is evidence that the Park Service was becoming more sensitive to the effect roads were having on its landscapes. The early pattern of national park road projects was to construct loop or bisecting roads through the heartland of a park. In Yellowstone, the loop road system, built from 1883 to 1918, was designed to connect such key features as Old Faithful, Yellowstone Lake, Norris Geyser Basin, and the Upper and Lower Falls on the Yellowstone River, and it cut through the core of the park. The Park Service's newer roads, however, were designed merely to provide visitors access to previously inaccessible areas and thus usually affected less acreage. The road to Wonder Lake in Denali National Park and the road to Flamingo in Everglades National Park—both new Mission 66 roads—were built to allow some (but not too much) visitor access into these remote regions while otherwise leaving the surrounding lands undisturbed. That did not stop Brower and other conservationists from lambasting the agency's penchant for new roads into pristine areas, however. Park visitors, according to Brower, were being relegated to mere "roadside wilderness," hardly an encounter with the untamed nature that had dominated these landscapes little more than half a century earlier.[16]

The Wilderness Act and Its Impact

Meanwhile, buoyed by its Echo Park success, a growing wilderness movement was poised to write the notion of wilderness preservation into federal law. Aiming to build on the Forest Service's early primitive area designations yet convinced that such administrative designations did not adequately safeguard wilderness values against rapacious timber companies, The Wilderness Society spearheaded a determined legislative campaign to bring formal legal protection to select roadless areas scattered across the public lands. In doing so, the society and its allies were acknowledging that the national parks, as administered under the Organic Act, were not sufficiently protective of wilderness values. Their goal was to take conservation to another level of protection.

After eight long years, the effort finally succeeded when Congress passed the Wilderness Act of 1964 and instructed the Park Service, U.S. Forest Service, and U.S. Fish and Wildlife Service to inventory their roadless lands for potential wilderness protection. The agencies were to make wilderness recommendations to the president, who would then make his own recommendations to Congress, which retained ultimate legal designation authority for itself (a process roughly mirroring how new national parks were ordinarily created). At the same time, Congress designated more than nine million acres of "instant wilderness" on national forests lands, thus giving the rival Forest Service significant preservation responsibilities similar to those the Park Service had long pursued. There was one key difference, however. The Wilderness Act, which defined *wilderness* as "an area where the earth and its community of life are untrammeled by man, where man himself is a visitor who does not remain," plainly did not contemplate any tourism- or recreation-related development within a protected wilderness area. Thus, the Park Service's recent road-building binge would not be repeated in designated wilderness areas.[17]

Curiously, the Park Service initially opposed the Wilderness Act and argued against extending it to the national parks. Agency officials asserted that the Organic Act provided them with sufficient legal authority to protect wilderness values on their own lands, thus making congressional wilderness designations unnecessary. They worried, too, that the new wilderness legislation would undercut their own management prerogatives under the Organic Act, not only making it more difficult (if not impossible) to continue building roads and other visitor facilities within the parks, but also limiting their backcountry management options. In addition, there was a sense among Park Service officials that the rival Forest Service would now be in a much stronger position to resist any proposal that would transfer national forest lands to the Park Service for safekeeping, something that had happened regularly in the years preceding the Wilderness Act. That was the pattern surrounding the creation of Olympic National Park, Kings Canyon National Park, and Isle Royale National Park, all to the lasting chagrin of Forest Service officials. But wilderness advocates no longer trusted the Park Service, fearing that the agency's protective backcountry designations could be dropped with the speed of an administrator's pen. They believed, in short, that legal wilderness designations were necessary to protect the national parks from the Park Service itself.[18]

Resistance from Within

Reluctantly reconciled to the Wilderness Act, the Park Service responded timidly with its original wilderness recommendations, not making any formal recommendations until 1970, more than five years after the act's passage. Given the opportunity to overlay the national parks with additional legal protection, agency officials recommended only remote portions of its major parks for wilderness protection, carefully leaving out lands near developed facilities, roads, and scenic vistas. For example, intent on constructing a transmountain highway through Great Smoky Mountains National Park, the Park Service proposed to protect less than half of the park as wilderness, with that acreage broken into six different blocks, some as small as five thousand acres.Displeased congressional wilderness champions chastised Park Service leaders for their minimalist approach and then pointedly added more wilderness acreage to several early national park wilderness bills, including Bandelier and Cumberland Island.Moreover, the Park Service proved reluctant to aggressively pursue its original recommendations, which has enabled powerful local congressional delegations to effectively block formal wilderness protection for such parks as Yellowstone, Grand Teton, Glacier, and Canyonlands.[19]

It is no surprise, then, that the first national park statutory wilderness designations did not occur until 1970 and covered only modest acreage at

Petrified Forest and Craters of the Moon. Since then, Congress has steadily enlarged the national park wilderness inventory, which has risen from 3.2 million acres in 1978 to more than 43 million acres today. In California, for example, nearly 7 million acres of national park lands are protected as wilderness; Death Valley National Park, with 3.15 million protected acres, is now the largest wilderness area in the contiguous United States. Similarly, the three major national parks in Washington state—Mount Rainer, Olympic, and North Cascades—boast more than 1.7 million acres of designated wilderness. But Yellowstone, Glacier, Grand Canyon, and other major parks, although often surrounded by protected wilderness lands, still lack any official wilderness within their borders.

In fact, the Park Service has been seriously laggard in seeking wilderness protection for its lands and has yet to complete the wilderness review process required by the 1964 Wilderness Act. Charged with submitting wilderness recommendations to the president within ten years of the act's passage, the Park Service still had not conducted wilderness reviews for thirty-nine units of the national park system by the year 2000. Also, several of its wilderness recommendations were yet to even be forwarded to the secretary of the Interior, including those for Big Bend National Park and Glen Canyon National Recreation Area, both of which contained large blocks of undisturbed lands. Following his retirement in 2004, Jim Walter, the Park Service's own national wilderness coordinator, harshly criticized the agency for neglecting its Wilderness Act responsibilities. His litany of its transgressions included these inexplicable delays in completing wilderness inventory recommendations and the absence of wilderness management plans.[20] Citing these same repeated failures, The Wilderness Society sought a court order to force the Park Service into taking action, but a federal appellate court ruled in 2006 that the agency was not legally obligated to complete its wilderness review obligations or to develop wilderness management plans.[21] Since then, however, the agency has begun completing additional wilderness assessments and management plans, and now recognizes that more than twenty-six million acres of national park lands are eligible for wilderness designation.[22]

Moreover, the Park Service has shown a disturbing tendency to discount statutory wilderness management standards in deference to its own mission priorities. At the Cumberland Island National Seashore, much of which is designated wilderness, park officials regularly ferried visitors in park vehicles (including a fifteen-person van) across the established wilderness area to popular historical sites on the northern part of the island, when park employees were otherwise traveling there for maintenance purposes. A federal court, however, concluded that the agency's visitor shuttle violated the Wilderness Act's prohibition against using motorized vehicles in designated wilderness areas, also finding that the arrangement was inconsistent with the "primitive and unconfined type of recreation" that Congress had in mind for wilderness areas.[23] At Olympic National Park, a Park Service proposal to helicopter prefabricated trail shelters into a designated wilderness area was blocked by a federal court, which ruled that the Wilderness Act's limitations took precedence over the Organic Act and related recreational, historic preservation, and safety concerns.[24] In addition, although the Park Service's management policies have long required each park to develop wilderness management plans,[25] nearly three-fourths of the agency's wilderness areas do not have these plans in place.

Wilderness in the Parks

It is thus quite ironic that the Park Service, despite its reticence toward legal wilderness, is now the largest wilderness manager among the federal land management agencies. This distinction is directly attributable to the landmark 1980 Alaska National Interest Lands Conservation Act, which created ten new national park units in the state and expanded three others while attaching a wilderness designation to more than thirty-two million acres of national park lands. As a further irony, shortly after the Wilderness Act authorized the establishment of wilderness areas in the parks, the Park Service radically altered its own resource management policies, bringing them into alignment with the notion of unmanaged wild nature. Beginning in 1968, on the heels of the groundbreaking Leopold report, the Park Service began to allow nature to take its course without significant human intervention. Backcountry wildfires were allowed to burn unchecked, bear feeding displays were halted, previously despised predators were tolerated, a wolf restoration program was conceived, and wildlife were no longer managed intensively. Despite its antipathy toward formal wilderness and wilderness management plans, the Park Service has nonetheless entered the business of wilderness management, both on its own officially designated lands and those that were not so designated.[26]

Of course, the concept and reality of a national park as a wilderness setting does not depend on a formal congressional wilderness designation.

Even though Congress has not passed wilderness legislation for Yellowstone or Glacier, no one familiar with these parks would suggest that their expansive backcountry terrain does not offer as authentic a wilderness experience as can be found in formally designated wilderness areas. In fact, with the grizzly bear and wolf—two iconic symbols of wild nature—now roaming these two parks, most visitors would see them as a closer representation of the wilderness setting our forebearers encountered than can be found in many official wilderness areas, where these creatures have long been missing. Yet there is something about the contemporary national park setting that belies the notion of wilderness. Perhaps it is the road networks or the hoards of visitors (most of whom will never alight in the backcountry) or the ubiquitous presence of automobiles. Perhaps it is the Park Service's continued ambivalence toward official wilderness or the relentless development pressures from beyond park boundaries that are incrementally chipping away at the landscape. Whatever it is, controversy still haunts the image and reality of wilderness in the national parks.

THE GRAND CANYON WILDERNESS CONTROVERSY

Ever since Congress first authorized wilderness protection for national park lands, the rugged Grand Canyon has seemed an obvious candidate. Apart from a few roads and hotels on the canyon's rims, the rest of the park within the canyon walls appears much as it did when Major Powell first explored this little-known corner of the Southwest in the nineteenth century. Dig a little deeper, though, and you encounter a flourishing white-water rafting industry on the Colorado River, a major upstream dam at Glen Canyon, noisy air tour overflights, and pollution-tainted skies, all of which belie the notion of a pristine, undisturbed environment. Wilderness advocates, with occasional support from the Park Service, have mounted a thirty-five-year campaign to secure a congressional wilderness designation for much of the park, but without any success. The Grand Canyon wilderness campaign proffers several lessons on the role and reality of wilderness in today's national park.

An Untamed Landscape

In 1869, during his first exploratory expedition down the Colorado River, Powell did not doubt that he was engulfed in an untamed landscape. After eleven weeks on the river and poised to enter yet another massive unknown canyon, Powell penned the following journal entry on August 14:

"The walls now are more than a mile in height—a vertical distance difficult to appreciate. . . . A thousand feet of this is up through granite crags; then steep slopes and perpendicular cliffs rise one above another to the summit. The gorge is black and narrow below, red and gray and flaring above, with crags and angular projections on the walls, which, cut in many places by side canyons, seem to be a vast wilderness of rocks."[27] Once through this stretch of the river, its sights and sensations so impressed Powell that he named it the "Grand Canyon."

Bisected by the mighty Colorado River, the Grand Canyon is cut a mile deep and stretches along 277 river miles. The distance across the gorge averages ten miles; the North Rim sits at more than eight thousand feet in elevation and is snowbound more than half the year, whereas the South Rim sits at seven thousand feet and is less prone to winter storms. The heart of this spectacular landscape is the Colorado River, which has carved the canyon and its myriad geologic features over the centuries and is still at work shaping the surrounding terrain. The dry desert landscape that covers much of the canyon gives way to thick pine forests and a cooler climate on the canyon rim. Today, this geologic wonder attracts more than 4.5 million visitors annually, most of whom are content to marvel at the gaping chasm from atop one of the rims.

Less than two decades after Powell's epic journey, efforts to protect the Grand Canyon were under way. During the mid-1880s, well aware that Congress had recently safeguarded the picturesque Yellowstone and Yosemite country, Indiana senator Benjamin Harrison sought legislative protection for the canyon as a "public park." Although those efforts failed, a few years later, in 1893, now President Harrison acted to create Grand Canyon Forest Reserve, thus preserving these lands in federal ownership but still leaving them open to mining, logging, livestock grazing, and other such activities. After a 1903 visit to the canyon, President Theodore Roosevelt first designated part of it a federal game preserve; then, in 1908, he protected more than 800,000 acres stretching rim to rim as Grand Canyon National Monument, declaring it "an object of unusual scientific interest, being the greatest eroded canyon within the United States." In doing so, Roosevelt invoked his newly delegated powers under the Antiquities Act of 1906, which empowered the president to protect "objects of historic or scientific interest" as national monuments.[28] Fearing that the president's sweeping declaration could ruin his thriving canyon rim business, a local entrepreneur named Ralph Cameron, who later served as U.S. senator from Arizona, challenged the legality of Roosevelt's actions, arguing that the president could not unilaterally transfer so much acreage from the public domain to the Park Service. The U.S. Supreme Court, however, eventually sustained Roosevelt's new monument, concluding that the Antiquities Act vested broad authority in the president to safeguard federal land as national monuments and to define their size.[29]

Grand Canyon National Park

Finally, in 1919, Congress created Grand Canyon National Park as the seventeenth park in the nascent national park system. By then, the local business community, which had originally linked its fortune to the area's minerals and other raw materials, realized that the spectacular canyon country might prove more lucrative as a tourist attraction, particularly once the Santa Fe Railroad opened the area to ready access. The enabling legislation transferred management responsibility for the park from the Forest Service to the fledgling National Park Service, which soon joined in making the area safe and attractive for visitors, a commitment that involved exterminating predators to protect more attractive wildlife (the highly visible deer, elk, and big horn sheep) and suppressing all wildfires that might blacken the surrounding scenery. (Not surprisingly, elimination of the area predators soon led to an explosion in the local deer population, forcing the federal government to initiate a controversial deer reduction campaign.) But evidently unwilling to forgo entirely future development possibilities, Congress inserted provisions that would allow access roads, mineral development, and dam construction "whenever consistent with the primary purposes of said park."[30]

Creation of Grand Canyon National Park set off a chain of events over the ensuing decades that has dramatically changed the surrounding area as well as the park itself. The new park—marketed early on both nationally and internationally by the railroads for its unparalleled scenic splendor—quickly became the anchor for a burgeoning regional tourism industry that now includes once-in-a-lifetime commercial raft trips down the Colorado River and $200-an-hour scenic air tours over the rugged canyons as well as the more traditional family sojourn to the South Rim to digest the canyon's majestic vistas. With tourism flourishing, few people objected when Congress, in 1975, passed the Grand Canyon Enlargement Act and added more than four hundred thousand acres to the park. Today, the Grand Canyon is promoted around the world as a must-see natural wonder, the park is a featured part of a popular circuit tour through the American Southwest, and Las Vegas promoters link the park to their own advertising efforts.

Even when Powell first entered the canyon, however, it was not an uninhabited wilderness, at least not to the Native Americans who had long made their home along the river and its sundry side canyons. By most estimates, the canyon area has been continuously occupied for more than twelve thousand years, through the Basketmaker and Ancestral Puebloan time periods and by such well-known ethic groups as the Zuni, Hopi, and Navajo. In the original park enabling legislation, Congress expressly recognized the Havasupai tribe's "use and occupancy" rights on its lands within the park. These rights were made even more explicit in the 1975 Grand Canyon Enlargement Act, which transferred 185,000 acres to the Havasupai reservation and opened them to traditional uses, hunting, and even small business opportunities, but otherwise mandated that the lands "shall remain forever wild." In addition, the Hualapai tribe occupies a one-million-acre reservation on the South Rim adjacent to the park, and Navajo reservation lands abut the park on its eastern flank. Like their Anglo counterparts, the tribes have now entered into the tourism business. The Havasupai reservation, with its blue waters and sparkling waterfalls, has become a popular destination; the Hualapai tribe offers pontoon boat trips into the lower canyon and has constructed a most unnatural sky bridge that extends over the canyon rim.[31]

For most of its first fifty years as a national park, the Grand Canyon was still not widely traveled, leaving its wilderness appearance largely unsullied. When World War II drew to a close, only a few intrepid souls were known to have hiked the canyon; twenty-five years later, fewer than twenty-five thousand visitors had entered the park's backcountry below the canyon rim. The same held true on the Colorado River; during the 1950s, fewer than a hundred individuals annually ventured down the white-water-filled river, and most of them relied on a budding but still quite primitive rafting industry. All these numbers soon shot upward, however, propelled by the baby-boom population bulge, sturdy new hiking and rafting equipment, and a surging national economy that permitted more leisure time. By the mid-1980s, the Park Service recorded more than eighty-two thousand annual overnight backcountry visits and estimated that eight hundred thousand more people ventured there as day visitors. A decade later, the overnight visitor number was hovering around one hundred thousand annually, with no letup in the growing number of day hikers either.[32]

An even greater explosion occurred among the canyon's river rafters. In 1965, only 547 people floated the river, but by 1972, a total of 16,428 people made the same trip, spending more than a hundred thousand visitor nights on the river corridor's fragile beaches. Increasingly concerned about overcrowding and environmental damage, the Park Service instituted a rafting permit system that allocated ninety-seven thousand user days between twenty-one commercial outfitters (92 percent) and private groups (8 percent). No distinction was drawn between motor-driven rafts, which had helped shorten the traditional two-week journey to just over one week, and oar-powered rafts. Not only did these escalating visitation numbers call into question the nature of a canyon wilderness experience, but the commercial-private split set up a conflict that has also undermined formal wilderness designation efforts.[33]

Wilderness Designation for the Canyon?

Once the Wilderness Act was passed in 1964, an effort was launched to designate the park's backcountry as a protected wilderness area. In the Wilderness Act, as noted, Congress had reserved for itself the power to create official wilderness, but the Park Service was required to review its undeveloped lands for possible inclusion in the new national wilderness preservation system and to make its recommendations to the President. After some initial hesitation, the Park Service set about evaluating the Grand Canyon, realizing immediately that the key question would be whether to include the Colorado River corridor in its wilderness recommendations, which would preclude the increasingly popular motorized raft trips through the canyon. In 1977, after further prompting from Congress,[34] the Park Service director submitted a final wilderness recommendation covering 1.1 million acres (approximately 94 percent of the park) that included the river corridor and the more remote North Rim area. Two years later, Grand Canyon officials completed a Colorado River management plan that called for phasing out motorized boats to "perpetuate a wilderness river-running experience in which the natural sounds and silence of the canyon can be experienced . . . and the river is experienced on its own terms." During the phase-out period, the river corridor would be managed as "potential wilderness," the agency's lawyers having concluded that temporary motorized use would not permanently compromise the area's wilderness character. The stage was set for a rim-to-rim wilderness designation that would ensure a quiet and less crowded experience on the river.[35]

The dream of a canyon-wide wilderness has not come to pass, however. The newly energized rafting concessioners—fearful that the recommended wilderness designation and the prohibition on motors would force a sizable reduction in visitor numbers and hence their profits—protested the wilderness recommendation and enlisted their clients to contact prominent politicians. In 1980, responding to these entreaties, Senator Orrin Hatch from Utah attached a rider to the Department of the Interior appropriations bill that forbid the Park Service from reducing the number of visitors using motorized rafts below 1978 levels.[36] Although only effective for a year, the Hatch rider sent agency officials scrambling back to the drawing board, where they promptly abandoned the no-motors plan and significantly increased overall usage limits, which grew by 30 percent for commercial trips and by 600 percent for private trips. Not only would motorized rafts continue to disturb the canyon's natural quietude, but even more people would be allowed on the river, further undermining any semblance of wilderness solitude and ensuring continued environmental degradation.[37]

Since then, the Park Service has revisited the wilderness question as well as its Colorado River management plan, but without any serious effort to merge the two issues. The park's 1995 General Management Plan seeks to "protect and preserve the [Colorado River corridor] in a wild and primitive condition," but it reflects no change in actual management. A 1998 draft wilderness management plan proposed rigorous standards for administering the recommended wilderness acreage, requiring rehabilitation of heavily used campsites and trails, revegetation of primitive roads on the park's North Rim, and use of a permit system to better control visitors. Although the plan continued to treat the Colorado River corridor as merely "potential wilderness," it nonetheless rekindled old antagonisms and prompted yet another effort to extend wilderness protection to the river corridor, not only eliminating motorized rafts but also opening more permits for private rafting groups. When the Park Service abruptly canceled the planning process, another lawsuit forced the issue back on the table.[38]

In 2006, following another lengthy planning process, the Park Service issued another revised river management plan that once again permitted motorized rafting to continue, but not from November to March, a time set aside mostly for private, oar-powered trips. In addition, rafting permits were reallocated between commercial and private users: total allowed user days were set at 219,000 per year, with roughly half allocated to commercial trips, which are about two-thirds motorized. The plan also permitted helicopter access into the lower canyon to remove commercial raft passengers after they completed their journey, further increasing noise levels and degrading the experience. Moreover, to address tribal economic development concerns, the plan granted the neighboring Hualapai tribe a motorized pontoon boat concession in the Lower Gorge, enabling them to ferry 480 passengers daily into the park, an arrangement the Park Service characterized as having "impacts [that] will be adverse and of major intensity during peak use period" for visitors seeking a wilderness experience. The courts rejected a lawsuit challenging the plan's motorized rafting provisions as inconsistent with national park and wilderness management requirements, deferring to the Park Service's judgment to continue allocating significant periods to commercial motorized trips. Plainly, the commercial outfitters have succeeded in protecting their interests; getting more people down the river in a shorter period of time takes precedence over the natural setting, leaving the canyon's wilderness values compromised in the absence of stronger legal protection.[39]

The Park Service's Colorado River Management Plan illustrates both the legal and practical differences between the National Parks Organic Act and the Wilderness Act. Under the Organic Act, the Park Service must maintain its lands and resources in an unimpaired condition while also accommodating visitors; under the Wilderness Act, motorized uses are expressly forbidden so as to provide outstanding opportunities for primitive recreation and solitude.[40] Absent an official wilderness designation for Grand Canyon National Park that includes the river corridor, the Park Service need only meet the less rigorous Organic Act standard. And according to the agency, the river management plan "will not constitute an impairment to [the park's] resources and values," and it will "provide for high quality visitor experiences." The plan nonetheless acknowledges that "impacts to wilderness character as described by natural, undeveloped, and recreation opportunity characteristics will be detectable and measurable during most of the year, but more apparent during the higher mixed-use period. . . . Impacts to natural conditions (except soundscape) and undeveloped character will be of minor intensity."[41] Stated more bluntly, the Park Service takes the view "that the services provided by commercial outfitters, which enable thousands of people to experience the river in a relatively primitive and unconfined manner and setting are necessary to realize the recreational or other wilderness purposes of the park."[42] As a result, the Grand Canyon's recommended wilderness lands are effectively split in half by the unprotected river corridor, the park's heart and soul that define its history, appearance, and character.

The Park Service's wilderness recommendations also cover the remote Kaibab Plateau area on the park's North Rim and would help restore more natural conditions there, too. Although the North Rim remains lightly visited, that would not be true had Horace Albright's idea to construct a tramway across the canyon come to fruition during the park's early years, but it did not. The North Rim therefore now provides the Park Service with the opportunity to restore wilderness characteristics to an important ecosystem. This area constitutes part of the larger Kaibab Plateau, which spreads across several million acres and extends northward into southern Utah. Logging roads crisscross the plateau, and extensive clear-cutting has fragmented and reduced wildlife habitat, leaving the national park lands as the last remnant of the plateau's ancient forest ecosystem and an important refuge for displaced species. The North Rim park lands are also marred by old logging roads that provide recreational opportunities for park visitors, but the park's wilderness recommendations and management plan would close most of these roads. To no one's surprise, the motorized recreation community has fought these plans in yet another instance in which motors and wilderness have come into conflict within the park. If the wilderness recommendation ultimately carries the day, then these North Rim lands would not only help sustain the park itself but the larger Kaibab Plateau ecosystem as well.[43]

Wilderness Protection from an Ecosystem Perspective

Even before the Park Service's wilderness deliberations had begun, the Grand Canyon's natural integrity was already seriously compromised if not permanently altered. In 1963, following a political trade-off that scuttled the proposed Echo Park Dam in Dinosaur National Monument, the Bureau of Reclamation completed the Glen Canyon Dam on the Colorado River, less than forty miles upstream from the original park boundary. This dam is a multipurpose facility designed for water storage, hydropower, and recreation; its completion prompted Congress to establish the Glen Canyon National Recreation Area, which encompasses and surrounds the resulting reservoir and represents another protected area in the Colorado River watershed.[44] Although neither the dam nor the reservoir (now known as Lake Powell) intruded into Grand Canyon National Park, the dam dramatically changed downstream river flow patterns and hence the canyon ecosystem.

Where warmer, silt-laden water had historically surged down the river during regular springtime floods, much colder, clear water now flowed from the base of the dam at a regular and predictable rate. These unnatural cold-water discharges from the dam created a new blue-ribbon trout fishery that supports a multimillion-dollar local sport fishing industry. Meanwhile, native warm-water fish species, like the humpback chub and razorback sucker, were soon added to the endangered species list, victims of the severely modified river system. Without flood-borne silt, the park's sand beaches gradually began to erode, both altering the riverbank ecology and reducing the number of suitable campsites for river rafters.[45]

As these negative environmental effects mounted, the dam's mere presence disclosed just how closely attached the park is to the river and to the larger Colorado River watershed. In an effort to restore the park's deteriorating river corridor ecology, Congress adopted the Grand Canyon Protection Act of 1992, explicitly acknowledging the key role the river environment plays in the overall health of the park. The act authorized, following an environmental study, modified operation of the Glen Canyon Dam to release additional water during the springtime to simulate prior flooding events through the Grand Canyon. Although several organizations would prefer to remove the dam entirely and restore the river to its former majesty through the canyon, currently there is not sufficient political support for this more radical restoration strategy. The three flooding experiments that have occurred to date nonetheless suggest that the water releases can help restore sand beaches, native vegetation, and other historical conditions within the canyon, which is a modest step toward a more natural, wilderness-like canyon environment.[46]

Besides the upriver Glen Canyon Dam, Grand Canyon National Park has faced a phalanx of potentially destructive dam proposals over the years. Most notably, during the 1960s, the Bureau of Reclamation sought congressional approval for two dams that would have flooded portions of the park, one at Bridge Canyon and the other at Marble Canyon. Although unnerved by the dam proposals, the Park Service was constrained from officially opposing them by Interior secretary Stewart Udall, who viewed the idea as essential to regional development. That did not stop Brower, however. Incensed at the prospect of losing more wilderness-quality canyon country, Brower once again rallied the Sierra Club and other allies, caustically observing, "If we can't save the Grand Canyon, then what the hell can we save." They unleashed another nationwide publicity campaign in the *New York Times* and other newspapers patterned after the earlier Echo Park effort, but this time with even more alluring messages: "Should we also flood the Sistine Chapel so tourists can get nearer the ceiling?" In the end, the park was spared, leaving the river corridor through the canyon without any permanent structures and thus still a suitable candidate for wilderness protection.[47] Whether the dam proposals might have been more readily defeated were the river corridor an official wilderness area is open to speculation. At least the Wilderness Act—unlike the National Parks Organic Act—explicitly prohibits construction of dams inside designated wilderness areas, absent an express presidential finding that the project is "needed in the public interest."[48]

The park's wilderness atmosphere has also been compromised by commercial air tour overflights that skim low enough to effectively eliminate any sense of solitude or natural quiet. Just as the park was being formally created in 1919, the first air tour flight took wing over its canyons, followed eight years later by establishment of the first commercial air tour business. By 1975, the air tour industry had become such a problem that Congress added a provision to the Grand Canyon Enlargement Act explicitly identifying natural quiet as a park resource and requiring agency officials to control "significant adverse effect[s] on the natural quiet and experience of the park." A 1986 midair collision between two tour planes precipitated the 1987 National Parks Overflights Act, giving the Park Service authority to establish noise limitations to "substantially restore natural quiet in the park." Since then, a succession of proposed rules and lawsuits challenging these limitations has done little to achieve meaningful progress. Although the Federal Aviation Administration has set ninety thousand air tour flights as an annual ceiling and adopted quiet technology standards, the number of overflights has still doubled since 2000, and aircraft noise continues to penetrate into the deep canyons, diminishing the visitor experience in several locations. The air tour companies, much like their Colorado River rafting concessioner counterparts, insist that their customers are entitled to view the park on their own terms, regardless of the effect low-flying aircraft have on the natural setting or other park visitors. Although progress toward restoring natural quiet is painfully slow, the Park Service Organic Act and the related overflight legislation seem to provide as much protection as would be available under the Wilderness Act for this occasionally noisy and less wilderness-like park landscape.[49]

Besides Las Vegas–based air tour operators, other external activities have further eroded the park's wilderness appearance and character. Regional haze—traced to coal-fired power plants, some located hundreds of miles away, and to automobile exhaust fumes from as far away as the Los Angeles area—regularly befouls the park sky and obscures its classic vistas. Not only do these polluted skies violate the Grand Canyon's protected class one airshed status under the Clean Air Act, but the problem prompted Congress to create a special multistate Grand Canyon Visibility Transport Commission in an effort to restore the park's crystalline air quality. A growing assortment of Clean Air Act lawsuits has shut down the dirtiest of the regional power plants, whereas others have been forced to upgrade their smokestack scrubber technologies. A recent upsurge in the price of uranium, responding to renewed global interest in nuclear power, has prompted new exploratory drilling activity on the national forest lands adjacent to the park. Arizona congressman Raul Grijalva, fearing that this type of creeping industrialization could further degrade park resources, has proposed legislation to create a no-mining buffer zone around the park, but his bill has gone nowhere. Meanwhile, Interior secretary Ken Salazar has used his administrative power to withdraw the area from new mining to protect the local aquifer as well as nearby park lands, prompting several industry-driven lawsuits challenging his authority.[50]

The striking reality is that the Grand Canyon's wilderness character is under assault from an array of forces linked to commercial activities that have long capitalized on the park itself as well as more distant activities that are inexorably altering the park's natural appearance and condition. Whether the ostensible goal is to enable more visitors to enjoy the park or to accommodate everyday industrial activity, the cumulative effect of motorized rafts, upstream dams, air tour overflights, coal-fired power plants, uranium mining, and regional growth has endangered the very wilderness-like qualities that initially prompted creation of Grand Canyon National Park in 1919. Perhaps the reality of pristine wilderness even in this remote setting is ultimately untenable in today's ever more populated and demanding world. But it has not deterred the park's stalwart defenders from seeking formal wilderness designation to add another layer of legal protection, nor has it deterred them from challenging—one after another—these profuse threats to the park's natural integrity. There is little likelihood, however, that these visitor pressures or other external forces will abate in the years ahead.

Any meaningful effort to protect and restore wilderness qualities to the Grand Canyon must acknowledge the park's location at the center of a larger regional ecosystem and wilderness complex. Surrounded by mostly undeveloped public lands, the park serves as a critical ecological core essential to safeguarding regional water supplies, migratory wildlife populations, and the area's overall natural integrity. Whether aware or not of these ecological connections, Congress acted in 1975 to expand the park's boundaries, not only bringing additional acreage under protection but also expressly noting the park's wilderness potential. In 2000, President Bill Clinton built on this legacy of protection by creating two new national monuments abutting the park: the 1,014,000-acre Grand Canyon–Parashant National Monument on the park's northwestern flank and the 293,000-acre Vermilion Cliffs National Monument on its northeastern flank.[51] Congressman Grijalva's proposed buffer-zone legislation would effectively expand the park's influence yet again beyond its formal boundaries, which is effectively what Salazar's withdrawal order has done. Although these legal and de facto expansions reveal a willingness to protect the Grand Canyon as a national park, they do not ensure that its wilderness qualities are similarly protected.

THE VALUE OF WILDERNESS PROTECTION

Today, few people would confuse our national parks with the untamed wilderness that Lewis and Clark encountered on their epic voyage of discovery. Lodges, roads, trails, and warning signs all belie the notion of a wilderness setting, at least as the parks are encountered by most visitors. But beyond the asphalt and traffic, the larger parks retain sufficient wilderness characteristics to thrust intrepid visitors back to another era when nature—not humankind—dominated the landscape. John Wesley Powell's primitive wooden dories may no longer ply the Colorado River's tumultuous rapids, but a raft trip through the Grand Canyon still offers vestiges of a wilderness experience. Similar experiences await those who venture into Yellowstone's remote backcountry or the expansive territory blanketed by the Alaskan national parks. Even for those who never journey into these settings, just the opportunity to gaze into the Grand Canyon's multihued folds or to peer across Glacier's forest-covered mountainsides will conjure up vivid images of a wilderness landscape still run on its own terms.

Indeed, the values traditionally attached to wilderness mirror many of those associated with the national parks. By protecting wild nature, both designations seek to preserve an important dimension of our cultural heritage: the wilderness experience that shaped the early American character on the frontier and helped instill a shared sense of self-reliance. A journey into either place can be therapeutic, an opportunity to escape the daily pressures of modern urban life and to learn "the trick of quiet."[52] As Professor Joe Sax put it in his influential book *Mountains without Handrails*, a visit to a "national park provides an opportunity for respite, contrast, contemplation, and affirmation of values for those who live most of their lives in the workaday world."[53] National parks and wilderness areas not only provide an important venue for scientific research, but they are also educational settings where visitors can learn about nature and nature conservation. They also serve as wildlife sanctuaries and play a critical role in conserving biodiversity. And they offer manifold recreational opportunities where outdoor skills can be tested and honed. Exposure to either a national park or wilderness area can help implant an ecological consciousness—a "land ethic" in Aldo Leopold's words—that is an essential first step in safeguarding our natural heritage.[54]

Does it really matter, then, whether a national park is also formally denominated a wilderness area? Is there any real difference between legal and de facto wilderness? The National Parks Organic Act contains rigorous protective standards designed to safeguard park resources, most notably the nonimpairment mandate, and these standards seem to offer as much protection to the parks as the Wilderness Act would afford. Both laws contemplate preserving designated areas in their natural or undisturbed condition, and both laws clearly prohibit commercial mining, logging, and other industrial activities. The Park Service takes justifiable pride in its historical commitment to protecting its lands from any such intrusions and regularly notes that sweeping backcountry venues are available just beyond park road corridors. And the agency is correct: most national parks offer de facto wilderness settings where visitors can encounter wild nature in a largely unaltered state. Wilderness advocates, plainly cognizant that most national park backcountry areas are not at great risk, have thus focused their political efforts on securing wilderness protection for roadless national forest and Bureau of Land Management (BLM) lands, where the threat of new roads and industrial incursion is much more real and imminent.

There are, however, fundamental differences between the two areas, making formal wilderness protection appropriate—perhaps even essential—in the national park setting. National park status simply does not offer the same level of protection that wilderness designation affords. The National Parks Organic Act expressly provides for public enjoyment of the parks, and the Park Service historically has been in the business of serving visitors, not only by constructing lodges and visitor centers, but also by building roads into remote wild areas. As we have seen, the quest for formal wilderness designation at Grand Canyon National Park has stalled over the Park Service's unwillingness or inability to curb motorized rafting on the Colorado River. Whether the sight and sound of raft motors, not to mention hovering helicopters, are compatible with the Organic Act's conservation mandate and nonimpairment standard may be open to debate, but there is no debate about their compatibility with the terms of the Wilderness Act: they would be prohibited. The Wilderness Act defines "wilderness" as an area "retaining its primeval character and influence . . . [with] outstanding opportunities for solitude or a primitive and unconfined type of recreation," and it also prohibits the "use of motor vehicles, motorized equipment or motorboats . . . [or] other form of mechanical transport."[55] Once lost, the sense of raw nature that wilderness conveys cannot be readily reclaimed.

Pragmatic considerations support extending wilderness protection to eligible national park lands. The political nature of the national parks means that local congressional delegations routinely exert considerable influence over park officials, usually in response to constituent pressures and commercial interests that depend on the parks for their livelihood. Hatch's intervention on behalf of the Grand Canyon commercial rafters and their clients to block the Park Service's wilderness recommendation during the early 1980s serves as one such example. When faced with these pressures, the Park Service has frequently acquiesced, taking refuge behind the Organic Act's visitor provisions while downplaying its conservation obligations. Where applicable, however, the Wilderness Act's more rigorous management standards provide agency officials with a legal justification for preservation-based decisions as well as an important layer of political insulation, which should better enable them to give park resource protection first priority when confronted with such demands.

The point is not to suggest that the national parks should be converted wholesale into wilderness areas. The national park idea, from the beginning, has embraced a broader view of nature conservation, one that engages the public generally in this lofty endeavor. As a result, few people conceive of a national park as solely a wilderness sanctuary. An attractive feature of the national parks is that they offer different levels of experience with nature, something that is not available in designated wilderness areas where the trappings of civilization are absent, a reality that discourages many families and novices from ever entering a wilderness area. Another alluring feature is the educational role the national parks play in introducing people to nature and instilling important environmental values. The parks—with their existing roads, lodges, visitor centers, and interpretive programs as well as undeveloped backcountry areas—represent an elegant compromise: wilderness is one of several available experiences, but it is not the only one.

The ultimate question is whether national park wilderness values—whether formal or de facto—are at risk from other competing uses or external forces. Has the Park Service's commitment to accommodating visitors and their autos undermined its willingness to pursue appropriate wilderness legislation or to maintain its undeveloped landscapes in a roadless and untrammeled condition? Wilderness is not compatible with commercialism or motorized recreation or hordes of noisy tourists, but it is fully compatible with key park values: nature research, contemplative recreation, outdoor education, wildlife protection, and ecological conservation. Further, wilderness protection creates an additional buffer against the diverse pressures mounting from within and without the parks. Beyond the existing roads and lodges, wilderness management provides the highest level of protection available, where the natural world is exalted over the humdrum of commercialism and mass recreation. With this wilderness backdrop, the unique and timeless qualities of our national park settings are more secure over the long term.

CHAPTER 3

"A Pleasuring Ground"

Tourism in the Wild

Tourism has long occupied a central role in the national parks. Even before the term *national park* was coined, Congress directed that Yellowstone and other early parks were set aside as "pleasuring grounds" or "public parks." The underlying principle was obvious: America's most spectacular landscapes were open to the general public; they were not preserved for the benefit of a privileged few, unlike the elitist tradition that prevailed in Europe. The National Park Service's early leaders eagerly embraced the visitation theme as well as automobile tourism. They not only vigorously promoted the growing national park system, but also set about making the parks accessible and attractive to everyone. Once World War II ended, park visitation exploded, calling into question the agency's commitment to its basic preservation mission in the face of mounting visitation and commercial pressures. Whether these sensitive natural settings can accommodate a steadily rising flow of visitors and their ubiquitous cars and still retain their wilderness attributes, ecological integrity, and scenic beauty is a question that goes to the very heart of the national park concept.

TOURISM IN THE NATIONAL PARKS

The national park idea first gained official recognition in 1864 in the guise of the Yosemite Valley Act, legislation designed to protect the scenically spectacular valley from falling into private hands. Concerned over the prospect of commercial exploitation, Congress conveyed the valley, along with the nearby Mariposa grove of ancient giant sequoia trees, to the state of California, stipulating that this land must be "held for public use, resort, and recreation." Following the transfer, Frederick Law Olmstead, already the nation's premier landscape architect, was appointed chair of the new Yosemite Park Commission, which was charged with determining how the state should manage these breathtaking landscapes to meet the congressional mandate. As chair, Olmstead drafted a report that firmly endorsed the idea of a protected yet publicly accessible park: "The first point to be kept in mind then is the preservation and maintenance as exactly as is possible of the natural scenery; the restriction, that is to say, within the narrowest limits consistent with the necessary accommodations of visitors, of all artificial constructions and the prevention of all constructions . . . which would unnecessarily obscure, distort or detract from the dignity of the scenery." Convinced that visitation would number "in the millions" within a hundred years, Olmstead concluded that "the first necessity is a road" to provide access to the valley and its scenic grandeur. Olmstead's report, however, although now regarded as a classic in national park lore, was initially suppressed by his fellow commissioners to advance their own more aggressive development agenda.[1]

Railroads, Roads, and Autos

Indeed, commercial interests played an important role in promoting establishment of the early national parks. Yellowstone's origins are linked to the Northern Pacific railroad; Grand Canyon's to the Santa Fe Pacific; Glacier's to the Great Northern, and so on. Recognizing the potential tourism market, Northern Pacific executives helped persuade Congress to adopt the Yellowstone Act in 1872, setting this unique landscape aside as a protected "pleasuring ground." To accommodate the expected visitors, Congress authorized the secretary of the Interior to grant short-term leases to private commercial interests, with the revenues to be used to build roads and bridle paths into the park. Soon the Northern Pacific and other railroad companies were constructing large European-style lodges in Yellowstone, Glacier, and other parks to provide their wealthy customers with attractive accommodations. Although allied with the railroads in the tourist trade, however, park officials resisted running rail lines into the early parks. In 1885, when the Northern Pacific extended its rail line from Livingston, Montana, through Paradise Valley toward Yellowstone's northern entrance—representing the first rail line specifically constructed to a national park—the park superintendent steadfastly insisted that the track stop outside the park boundary. His refusal incurred the wrath of magnate Jay Cooke, who envisioned the railroad essentially capturing national park visitors for the entire duration of their stay.[2]

Given the expense and time involved in late-nineteenth-century rail travel, only the wealthy elite could afford to vacation in the new western national parks. Well-heeled easterners, prompted by Congress's 1905 "See America First" campaign and other corporate promotional efforts, headed westward to behold for themselves the awe inspiring scenery and learn about these recently protected areas. Early local entrepreneurs vied with the railroads and other large national firms to provide them with services. It was, according to one historian, an era of heritage tourism, characterized by a limited and exclusive clientele focused on viewing, understanding, and appreciating the nation's natural and archeological wonders, but the era proved short-lived. As park visitation grew, the tourism business became more organized across the region, reflecting its growing economic importance and reach.[3]

Along with the railroads, roads were an early key to promoting park visitation. Wagon roads constructed outside the parks initially opened them to a wider audience, and similar roads constructed inside the parks allowed visitors to reach special attractions once there. At Yellowstone, to make the expansive park accessible to early visitors and their carriages and other conveyances, park officials built a road system that looped through the interior and authorized the construction of hotels near the park's famous geyser basins and other natural features. Even John Muir, an early national park proponent and wilderness lover, in his classic *Our National Parks* tome, observed that "all the western mountains are still rich in wilderness, and by means of good roads are being brought nearer civilization every year."[4] As his struggle against the Hetch Hetchy Valley dam proposal in Yosemite intensified, Muir supported opening the embattled park to new roads, believing that better access would bring more visitors who could then be enlisted as allies in his battle against the dam. Muir even wrote an early Yosemite guidebook, part of a larger series designed to attract new visitors to the parks, including those who might not be able to afford the luxury of train travel.[5]

As the twentieth century dawned, the first automobiles appeared, soon changing the appearance of the country as well as national park visitation patterns. In 1902, the first car arrived at the Grand Canyon rim, foreshadowing a substantial new automobile trade that would alter forever the national park visitor profile and behavior. Within a decade, the automobile was becoming a mainstay of American society, presenting the question of whether autos should be admitted to the national parks. Muir seemed to capture the prevailing ambivalence during a 1912 conference on Yosemite's future: "Under certain precautionary restrictions these useful, progressive, blunt-nosed mechanical beetles will hereafter be allowed to puff their way into all the parks and mingle their gas-breath with the breath of the pines and waterfalls, and, from the mountaineers standpoint, with but little harm or good." Interior secretary Walter Fisher, at the same conference, echoed similar sentiments: "Some automobiles make a great deal of noise . . . emit very obnoxious odors . . . drop their oil and gasoline all over the face of the earth . . . [and] are sold by people who regard it as a hardship to be excluded from any particular road." That said, however, Fisher concluded that the national parks should allow automotive travel once safety concerns were addressed. In 1913, Yosemite opened to cars, and Yellowstone followed two years later, presaging a new era in park visitation and a new set of management challenges.[6]

Attracting the Public

By adopting the 1916 National Parks Organic Act, Congress gave the newly created National Park Service a dual mission that combined nature preservation with public use. No longer employing the earlier "pleasuring grounds" terminology, the legislation instructed the Park Service to "promote and regulate the use of the . . . national parks" so as "to conserve the scenery and . . . wild life therein" and "to provide for the enjoyment of [them] . . . by such means as will leave them unimpaired for the enjoyment of future generations." Congress was not explicit as to how the new agency was to accomplish this dual mission, which included promoting the new national parks. It did, however, empower the agency to enter into limited term leases and permits with private entities to provide visitors with accommodations and other services, but outlawed any arrangements that would deny them free access to natural wonders or curiosities. The Organic Act legislation, when viewed through the prism of early park history, delivered a powerful message endorsing visitation in these protected areas, a view that was confirmed once the Park Service began to make its policy priorities explicit.[7]

The agency's inaugural director, Stephen Mather, placed park visitation at the top of his agenda, as we have seen. Mather subscribed to the views earlier expressed by Olmstead, who asserted in his report on the Yosemite Valley that "the enjoyment of scenery employs the mind without fatigue and yet exercises it; tranquilizes it and yet enlivens it; and thus, through the influence of mind over the body, gives the effect of refreshing rest and reinvigoration of the whole system." Mather feared that the national parks and the new Park Service were vulnerable not only to the whims of Congress but to the rival Forest Service, which had opposed creating the new agency. According to Forest Service chief Gifford Pinchot, whose deep-seated utilitarian philosophy clashed with Mather's strong preservationist views, it made no sense to set public lands aside for merely scenic protection. Mather's response was to invite the public to visit these wondrous settings, believing that park visitors would develop a special appreciation for these special places and could then be rallied to defend them.[8]

Mather's views were incorporated into the 1918 Lane letter, which cemented the Park Service's commitment to tourism as well as an early alliance with the railroads and automobile associations. Affirming that the national parks "are set apart for the use, observation, health, and pleasure of the people," the Lane letter instructed the Park Service to work with the railroads, chambers of commerce, tourist bureaus, and automobile associations to inform the public about how to reach the parks. It acknowledged an ongoing role for national park concessioners in providing a range of accommodations to visitors "at carefully regulated rates" that do not "impose a burden upon the visitor." It also admonished that "automobile fees in the park should be reduced as the volume of motor travel increases." These automobile provisions proved prescient. By the early 1920s, the auto had displaced the railroad as the primary means of visiting the national parks, signaling the end of the heritage tourism era and the advent of recreational tourism. In the spirit of the Progressive Era values that then prevailed across the nation, no longer would the national parks be the primary domain of the wealthy elite; rather, they were rapidly becoming everyone's playground.[9]

To promote visitation, Mather set about marketing the national parks, much as he had once marketed Borax soap products. Mather's immodest goal was to make the national parks "an American trademark in the competition for the world's travel." His strategy was straightforward: build and upgrade park roads, construct more hotels and other accommodations, and establish alliances with an emerging tourism industry. Fully recognizing the transformative potential of automotive travel, Mather conceived and promoted an adventuresome park-to-park auto tour that looped across the West, connecting all the major national parks. As part of his visitation campaign,

Mather instructed all his superintendents to caravan in their autos to the 1925 park superintendents' conference in Mesa Verde National Park. He also regularly toured the parks in his own roadster, prominently bearing the license plate "NPS-1." Individual parks also engaged in their own efforts to attract and entertain visitors, using such gimmicks as the nightly firefall at Yosemite and evening bear viewings at Yellowstone and providing such services as swimming pools, golf courses, tennis courts, and downhill ski areas. To further enhance the visitor experience, the Park Service's resource management policies called for removing harmful predators and extinguishing all wildfires to protect its "good" animals and the scenic splendor.[10]

These policies, predictably, altered the national park setting, transforming it from a remote and uninviting wilderness to a democratized and commercialized nature reserve. The automobile opened the parks to a new type of visitor, one who could not afford luxurious railroad travel but instead came seeking adventure. These new visitors demanded more and different services within the parks: better roads, affordable accommodations, accessible campgrounds, gasoline stations, grocery stores, and the like. Mather was more than willing to meet these new demands. During his tenure, the Park Service constructed 1,298 miles of new roads along with numerous new trails, campgrounds, telephone lines, sewer and water systems, and other buildings. Visitation numbers reflected the increasing accessibility of the major western parks: In 1906, slightly more than 2,500 people entered them, while in 1920 this number stood at nearly 250,000 people. But by 1922, Mather's park superintendents, concerned with the escalating level of construction, were sounding a cautionary note: "Over-development of any national park, or any portion of a national park is undesirable and should be avoided. Certain areas should be reserved in each park, with a minimum amount of development, in order that animals, forest, flowers, and all native life shall be preserved under natural conditions."[11]

The Great Depression years brought major changes to the national park system, ones that would further transform the parks into even more visitor-friendly destinations. As the nation's economy spiraled downward, park visitation numbers dropped to 3.4 million in 1933, but then bounced back to nearly 12 million in 1936. In part, the increase can be explained by President Franklin D. Roosevelt's 1933 decision to transfer more than fifty historic sites, battlefields, and national monuments from the War Department and other federal agencies to the Park Service, a move that greatly enlarged and fundamentally reshaped the national park system. Moreover, once Roosevelt established the Civilian Conservation Corps (CCC) as part of his New Deal program, the Park Service promptly signed on, seizing the opportunity to improve and expand its facilities. By the time the CCC was done, it had added 2,186 miles of road, 188 new water lines, 5,310 new campground acres, and various other building projects to the national parks. The 1930s also saw Congress add three major eastern parks to the system—Great Smoky Mountains, Shenandoah, and Mammoth Cave—a move that brought the prospect of a national park visit much closer to where most people lived. In 1936, the Park Service took its first step into the business of recreation management when it agreed with the Bureau of Reclamation to manage its new Boulder Dam National Recreation Area in southern Nevada. Congress later placed the reservoir area in the national park system as the renamed Lake Mead National Recreation Area, thus giving the Park Service a major new role in managing national recreation areas and meeting the growing public demand in outdoor recreation opportunities.[12]

Mission 66 and Industrial Tourism

World War II intervened, however, pushing the national parks to the back of the nation's domestic agenda and curtailing visitation,[13] but once the war ended, the national parks confronted a mounting flood of visitors as the nation entered a postwar period of relative peace and prosperity. By 1955, annual visitation had ballooned to more than fifty-six million, compared with the seventeen million who had visited the parks in 1940, just before the war. Several forces were at work. The postwar baby boom represented an unprecedented population explosion that soon meant that more families were seeking affordable vacation options, which were readily available at the national parks. The postwar economic boom and the advent of paid vacation periods meant more money and time for American families, enabling them to contemplate a distant national park visit. Automobile use and ownership soared, and the growing federal interstate highway system put the parks within a reasonable driving distance. The availability of military surplus camping equipment, along with the eventual development of new lightweight camping gear, made family camping trips an attractive and affordable vacation option. National park concessioners responded to this surge in visitation by pushing to expand their operations, arguing that more hotels, restaurants, and other facilities were necessary to accommodate the new visitors.[14]

The Park Service used the occasion of its upcoming fiftieth anniversary to launch its ambitious Mission 66 project, as previously noted. The initiative got a jump-start in 1953, when prominent historian Bernard DeVoto penned a scathing *Harper's Magazine* article that proposed closing the national parks, citing the deplorable condition of park roads, campgrounds, and other facilities. Faced with a deteriorating system and escalating visitor pressures, the Park Service director, Conrad Wirth, promoted Mission 66 as an opportunity to rebuild and expand park facilities and infrastructure. The construction effort was essential, according to Wirth, because the American public was "loving the parks to death," and park visitation was only expected to skyrocket in the years ahead. Wirth's solution was to build new facilities, roads, and trails to channel visitors into more developed areas, thus sparing other undeveloped areas from overuse and degradation. To do so, Wirth astutely saw to it that Mission 66 construction projects touched most congressional districts, thereby ensuring Congress's support and the needed funding.[15]

Once the money was in hand, the Park Service set off on a massive, decade-long building spree that stretched across the system. By the time Mission 66 was brought to a close, the list of construction projects was impressive: 1,200 miles of new roads; 1,500 miles of repaired roads; 900 miles of new or upgraded trails; 1,800 new or rehabilitated parking areas; 575 new campgrounds; 220 new administrative buildings, and the list goes on. More facilities meant more visitors, of course, and the tide just kept rising. Park visits climbed from 56 million in 1955 to 121 million in 1965. Other beneficiaries of the building boom were the park concessioners, including such mainstays as the Fred Harvey Company and the Yosemite Park and Curry Company, as well as nearby communities, all of which were eager to welcome more park visitors demanding their services. Never one to mince words, writer Edward Abbey branded it "industrial tourism" and accused the Park Service of abetting an unseemly commercially driven onslaught on the natural world it was pledged to protect. In many respects, the national park visit was becoming just another commodity to be acquired, part of the nation's growing mass consumerism culture.[16]

As Mission 66 unfolded, the National Parks Association and other conservation organizations soon grew disenchanted with the Park Service's road-building and construction spree, believing it was irreparably damaging park landscapes solely to cram more people and their cars, trailers, and boats into these delicate areas. Pointing to new roads in Everglades, Olympic, and Yosemite, along with poorly located and unsightly new developments in Yellowstone, Grand Teton, and Everglades, they accused the Park Service of ignoring its conservation responsibilities with its obsession for promoting park visitation. At the same time, as we have seen, a budding national wilderness movement emerged to champion the newly proposed Wilderness Act, which would extend nature preservation responsibilities across the federal land management agencies and thus provide an alternative to the development-oriented national parks. That the Park Service initially opposed the wilderness bill added support to this argument, even as the agency's leaders responded that they were trying to protect the parks by keeping visitors away from sensitive backcountry areas.[17]

Taking Nature into Account

Fixated on visitation, the Park Service continued to view its resource management responsibilities primarily in scenic preservation terms. This approach ignored, however, the agency's fundamental conservation responsibilities as well as accumulating lessons from the ecological sciences. The 1963 Leopold report not only exposed the problem, but called for a new emphasis on the biological sciences to safeguard park resources and ecosystems from degradation. Rather than just protect park scenery and wildlife for the visitor's pleasure, the Leopold report urged park managers to understand how ecological systems operated and to emulate them when managing park wildlife, forests, grasslands, and even natural processes such as wildfires and floods. The report's authors expressed their deep concern over the Park Service's current hyperdevelopment direction: "Perhaps the most dangerous tool of all is the roadgrader. Although the American public demands automotive access to the parks, road systems must be rigidly prescribed as to extent and design. . . . The goal . . . is to maintain or create the mood of wild America." In response, Interior secretary Stewart Udall admonished that physical structures must be "limited to those that are necessary and appropriate . . . under carefully controlled safeguards . . . so that the least damage to park values will be caused."[18]

Other changes were afoot that would redefine the national park visitor experience, even as visitation numbers continued to increase. During the 1960s, Congress began approaching new national park designations differently, establishing Canyonlands, King Canyon, and North Cascades as primarily wilderness parks with little or no infrastructure development. During the early 1970s, Congress passed several new environmental laws, including the National Environmental Policy Act and the Endangered Species Act, which forced the Park Service to begin preparing an environmental analysis for its development projects and to consult with the U.S. Fish and Wildlife Service over endangered species matters, new legal requirements that made its construction and other decisions more transparent and also subject to legal challenge. In 1972, the Conservation Foundation, in its seminal *National Parks for the Future* report, strongly condemned the Park Service's overemphasis on development and visitor facilities. The report's recommendations were bold: "New hotels and similar elaborate visitor facilities should be located outside park boundaries, and present facilities and developments inappropriate to the natural setting should be relocated outside park boundaries as well," while "in-park automobile use should be phased out [and] off-road use of vehicles should be prohibited." A few years later, the Park Service's own 1980 *State of the Parks* report acknowledged that important park resources, values, and experiences were imperiled system-wide by heavy visitor use, excessive vehicle noise, and extensive soil erosion.[19]

Reform moved slowly within the agency, however, partly a reflection of the political climate that prevailed during the 1980s. In fact, Interior secretary James Watt gave priority to upgrading visitor facilities and park infrastructure, in essence endorsing the agency's discredited visitor-centered agenda and ignoring the Leopold report's environmental concerns. Under Watt, as we shall see, the Park Service reversed its decision to close Yellowstone's Fishing Bridge campground to protect important grizzly bear habitat, instead leaving it open to accommodate more park visitors and appease politically influential merchants in nearby Cody, Wyoming. By the end of the decade, even the Park Service's most ardent supporters were bemoaning an increasingly impoverished park visitor experience.[20]

Recasting the Visitor Experience

The so-called "Vail Agenda" report, released in 1993 as part of the agency's seventy-fifth anniversary celebration, observed that "the Park Service is in danger of becoming merely a provider of 'drive through' tourism or, perhaps, merely a traffic cop stationed at scenic, interesting or old places." Echoing earlier reports, it recommended that the "Park Service should minimize the development of facilities within the park boundaries" and that "facilities . . . purely for the convenience of visitors should be provided by the private sector in gateway communities." To enrich what was becoming a humdrum visitor experience, it envisioned the Park Service assuming an ever more prominent educational role to "interpret and convey each park unit's and the park system's contributions to the nation's values, character, and experience." The challenge was clear: it was time for the Park Service to rethink and reorient its priorities, particularly as related to unlimited park visitation.[21]

During the 1990s, while noting shifts in park visitation, agency officials began to seriously address visitor management problems. Despite an early recession, the decade saw visitation continue to increase in the national parks, jumping from 256 million in 1990 to 286 million in 2000, but these numbers obscured other disturbing trends. For example, visitation actually declined when measured against gross domestic product, suggesting less public engagement with nature and outdoor activities, and the profile of the park visitor remained mostly white and middle class, hardly reflective of the nation's burgeoning ethnic and racial diversity. The Clinton administration, to better protect park resources and to enhance the visitor experience, sought to reduce automobile use and facilities inside the parks. It introduced a new shuttle bus system at Zion and Bryce Canyon National Parks in southern Utah and at Acadia National Park in Maine, proposed a new rail line to carry visitors to the Grand Canyon's south rim, and issued a new plan for Yosemite Valley, long regarded as the prime example of overdevelopment in a national park.[22]

At the same time, the concept of ecotourism was taking hold, casting the national park visit as an important educational opportunity for visitors hungry to understand and experience the parks as a dynamic natural environment, not as just a pretty postcard setting. An array of nonprofit entities associated with the parks—the Yellowstone Institute, Canyonlands Field Institute, and others—stood ready to meet the demands of this new tourism market. Although incremental, the evidence suggests a discernible policy shift away from catering to every tourist's needs or desires—whether in the form of more lodges, roads, stores, or other amenities—within the national parks and toward a more nature-focused visitor experience attuned to the natural setting and the ecological processes that account for it.

The national park visitor experience, however, is not simply confined to the parks. From the beginning, adjacent communities have played an influential role in how visitors approach and encounter the parks. Local sentiments in the towns adjoining Yellowstone, Grand Canyon, Rocky Mountain,

Yosemite, Great Smoky Mountains, and elsewhere have helped shape park visitor policies and facilities. Even communities located far from the national parks can affect how visitors encounter the parks. As we have seen, air tour operators based in Las Vegas daily shuttle tourists over the Grand Canyon as a sideshow in their Vegas entertainment package, an extraordinarily noisy intrusion on this serene landscape that hardly brings people into contact with the park itself. Given the Park Service's recent push to locate visitor facilities outside park boundaries, the role nearby communities play in park tourism will only intensify. But towns and businesses located outside the national parks, faced with a short tourist season, are driven primarily by their own economic concerns and desire for greater profits, not the nature or shape of their customer's national park experience. Moreover, when it comes to air tours and other issues arising outside park boundaries, the Park Service has no immediate jurisdiction over these matters and is mostly left to rely on its persuasive powers to influence them. Thus, although the visitor's role and experience has evolved inside the parks, it is also heavily influenced and shaped by powerful forces external to the parks.[23]

THE YOSEMITE VALLEY CONTROVERSY

Few national park controversies have stirred emotions like the acrimonious struggle over Yosemite Valley, which has long served as a magnet for park visitors. From the time Europeans first encountered the valley, it was hailed as a scenic marvel, an exemplary specimen of nature's handiwork. Sheer granite cliffs soared thousands of feet above the tranquil valley floor, waterfalls cascaded down from atop the cliff faces, and the light appeared ethereal to even casual visitors. The Merced River courses through the narrow valley, which is never more than a mile wide and extends seven short miles into this natural cathedral. Deer, bear, cougar, and other wildlife originally teemed across the valley floor, itself a mix of conifer and oak forests merging with bucolic meadows. Its patron saint, John Muir, was ecstatic about the place that he called the "sanctum sanctorum of the Sierra" and "the grandest of all the special temples of Nature."[24]

A Tourist Attraction

Despite Frederick Law Olmstead's early pleas to put scenic preservation first, Yosemite Valley's initial state caretakers viewed the setting as an unrivaled tourist attraction. To entice visitors into the new park, they set about promoting it with a guidebook and photo album extolling its natural beauty. Roads were built for the visitors, and local residents were encouraged to provide them with accommodations and other services. People soon began coming: from 1864 to 1870, almost 5,000 visitors arrived, a dramatic increase from the 653 who visited from 1855 to 1863. New hotels and other facilities gradually spread across the valley floor, but the building activity was haphazard. Concerned over the number of trees being felled, Muir and other critics complained that the valley's beauty and charm were being "despoiled by commercialism and exploitation." The matter first came to a head in 1890 when Congress, citing uncontrolled logging and overgrazing, designated the high-country lands surrounding the valley as a new national park. By then distrustful of the state's management priorities, Congress decided to retain these lands in federal control. Fifteen years later, California returned Yosemite Valley to the federal government, a decision hailed by both the Sierra Club and Southern Pacific, each seeing it as clearing the way for better management and improved tourist services.[25]

As the twentieth century got under way, transportation improvements brought the new park closer to the visiting public. The growing local federal presence helped eliminate the bothersome private toll roads that had long hampered access to this out-of-the-way destination. In 1907, the Yosemite Valley railroad reached Merced, putting the park within easy striking distance by stagecoach or horseback. Six years later, a road was completed into the valley, opening it to automobiles. By 1916, when the newly created National Park Service assumed control, the annual number of tourists arriving by car equaled the number arriving by rail, just over twenty-eight thousand in total. Within ten years, following completion of an all-season highway connecting Merced with the valley, most visitors were arriving in their own cars, relegating the railroad to history. With improved access, annual visitation numbers approached five hundred thousand by the end of the 1920s. The presence of so many people inevitably took a toll on the park, as reflected in this 1927 account: "Yosemite Valley is getting to be an awful place. We have had crowds all season and right now the camps are very much crowded. The air is filled with smoke, dust, and the smell of gasoline."[26]

Of course, the Park Service and its concessioners viewed the mounting crowds through only slightly different lenses, one measured success in terms of visitor numbers and the other in monetary profits. Consistent with Mather's priorities, the new agency worked in tandem with its concessioners to provide decent facilities and attract more people to the fledgling national park system. During these early years, this work involved constructing more accommodations, upgrading the roads and bridges, and improving park trails. Still, the unusually aggressive Yosemite Park and Curry Company also regularly pushed new proposals designed to lure even more visitors into the park, which the Park Service only rarely rejected: a renewed nightly firefall at Glacier Point, a new bear-viewing platform on the Merced River (which attracted more than two thousand visitors on just one summer evening during 1929), and a new midsummer Indian Field Days rodeo event. To invigorate visitation during the slow winter season, the Park Service joined with the Yosemite Park and Curry Company to construct new sports facilities—a toboggan run, ice-skating rink, and a downhill ski area—and to inaugurate a Winter Sports Carnival. In 1927, the Ahwahnee Hotel was completed, thus providing wealthier visitors with more luxurious accommodations than were available at the rustic Camp Curry tents, long the main overnight option in the valley.[27]

The Great Depression and World War II years in Yosemite followed the same pattern evident throughout the national park system. As the Depression deepened, visitor numbers initially fell before rebounding for a few years and then plummeted during the wartime years. The Depression brought the CCC to Yosemite, where it maintained several camps and actively built new roads, bridges, shelters, picnic sites, trails, and the like, all designed to help improve the park visitor experience. Seizing the moment, the park's concessioners pushed for even more lodging, parking, shops, and other facilities, even arguing that bars and dance halls were necessary to entertain the park's visitors. As the war was winding down, Yosemite officials, with support from Park Service director Newton Drury, expressed growing concern over the "carnival atmosphere" that was taking hold in the valley. The problem, they suggested, might be addressed by relocating some facilities outside the valley, but their efforts were thwarted by the concessioners' strenuous opposition.[28]

The Crowds Arrive

Once the war ended, giving way to a postwar boom period, the Park Service predicted that it would face an onslaught of visitors, and it was correct: between 1946 and 1966, Yosemite's annual visitation soared from 640,000 to more than 1.6 million people. With visitor numbers rocketing upward across the national park system, the Park Service initiated its Mission 66 program, which called for more facilities and infrastructure with a heavy emphasis on automobile tourism and recreation. In Yosemite, the Mission 66 building boom included a new Valley Visitor Center, several new campgrounds, and additional motel units.

Critics, however, soon began decrying the ongoing transformation of the valley from an inspirational natural setting into just another commercial enterprise. According to photographer Ansel Adams, one of the valley's most famous residents and stoutest defenders, incremental development pressures were converting the park into a "resort" and putting it on "the brink of disaster." Incensed by the presence of so many "typical urban installations," Adams argued that "people, things, buildings, events, and evidence of occupation and use simply will have to go out of Yosemite if it is to function as a great inspirational natural shrine for all our people." The Mission 66 program nonetheless brought further changes to the valley, including more cars and people who stayed overnight and demanded additional amenities. It also improved access elsewhere in the park by, for example, widening the historic Tioga Pass road to speed up auto travel, an improvement widely championed by the park's east-side communities that believed that more traffic would bring more visitors to their businesses. The Sierra Club, however, condemned the road project, flaying the Park Service for blasting away pristine granite outcroppings with no regard for its preservationist responsibilities.[29]

The valley's overcrowding problems came to an unexpected head in 1970 when riots erupted over the Fourth of July weekend. By then, annual visitation stood at 2.2 million people, a significant increase over the 820,000 who had visited in 1950. Even though the riots—mainly attributed to youthful recklessness and anti–Vietnam War sentiment—were not aimed at the park, the event lent credence to the park's critics that development and crowding had converted the valley into a problem-ridden urban setting. Any sense of a national park experience amidst overflow crowds, expansive parking lots, assorted commercial establishments, routine traffic jams, and mounting air-quality concerns was increasingly difficult, if not impossible, to maintain. Although the Park Service had eliminated the firefall and bear-viewing spectacles, it was plainly time to address the larger question of how many people and autos could be accommodated in Yosemite Valley. While the agency focused on developing a new master plan for the valley, it closed the eastern third to motor vehicles and installed propane-fueled shuttle buses to reduce auto use and bring visitors into more direct contact with the natural setting.[30]

In 1980, the Park Service completed its planning process by adopting a new general management plan that staunched any further development in the valley, but it postponed the fundamental question of visitation limits. Now committed in the aftermath of the Leopold report to restoring natural processes in the park, the plan acknowledged the need to reduce crowding and traffic congestion, but only called for additional studies. An earlier draft version had actually suggested closing the valley to autos as one alternative, but the park's major new concessioner, the politically well-connected Music Corporation of America (MCA), had objected to the closure so as to maintain its customer base. Instead, the plan merely observed that "the ultimate goal of the National Park Service is to remove all private vehicles from Yosemite Valley," citing the need to reduce air and noise pollution. Most visitors continued to put the spectacular valley at the top of their itinerary, and park concessioners, abetted by the Reagan administration's visitor-first policies, continued working hard to attract them.[31]

By the 1990s, park visitation had swelled to more than three million annually and even once exceeded four million. The automobile still reigned supreme, an array of commercial establishments still dotted the valley floor, and traffic jams were more commonplace than ever. Despite contrary admonitions in the park's general management plan, concession operations in the valley actually increased during the 1980s. The MCA, for example, established a new rafting concession on the Merced River that was putting more than thirteen hundred people a day on the river, and the number of parking spaces in the valley was approaching 5,000 rather than the 1,270 originally called for in the plan. On some summer weekends, park visitors waited more than an hour for admission at park entrance stations. Moreover, as a silent testament to these unrelenting visitor pressures, the Park Service completed a new jail in the valley that would house twenty-two prisoners and a new courthouse that was expected to handle a thousand cases annually.[32]

Confronting the Overcrowding Problem

The condition of the Merced River—one of the valley's most popular and environmentally sensitive features—was becoming a major concern. In 1987, citing the river's outstandingly remarkable scenic, recreational, and other values, Congress had designated a 122-mile stretch of the Merced as a "Wild and Scenic River," thus giving it an added layer of legal protection and setting the stage for a protracted legal battle aimed at forcing the Park Service to finally confront the valley's overcrowding problem. Under the Wild and Scenic Rivers Act, the Park Service was required to prepare a comprehensive management plan for the 81-mile river corridor running through the park. Because the river bisected the valley, the plan would necessarily include most of the valley's roads, developed areas, and campgrounds. In addition, it was supposed to address the river's "user capacities," which is legalese for determining and establishing limits on the number of visitors to protect environmental values. When completed, the Merced River comprehensive management plan would hold equal stature with the park's general management plan.[33]

But in January 1997, before the Park Service had finished its Merced plan, the river flooded, severely damaging nearby roadbeds, campgrounds, and motel units. With the river cresting at eight feet above flood stage, the damage was serious enough to force an unprecedented three-month park closure. Viewing the flood as a potential opportunity, the Park Service announced that it would reassess the park's persistent overcrowding and overdevelopment problems by preparing a new Yosemite Valley plan. At the same time, the agency found itself in federal court for failing to complete the Merced River plan in a timely fashion. Either separately or together, the two pending plans provided park officials an opportunity to establish visitation limits and reduce vehicle traffic.[34]

When released, however, neither draft plan confronted these questions head-on. The Merced River plan, although acknowledging the potential for environmental damage, relied on an experimental new Visitor Experience and Resource Protection (VERP) analysis to address any overuse problems that might arise. Under VERP, the Park Service indicated that it would monitor environmental conditions in the river corridor and take action to limit visitor numbers if there was evidence of deterioration, but only after it had validated the VERP process over a five-year testing period. Park officials asserted that the existing wilderness quota system as well as limited parking and other facilities in the valley effectively served to constrain visitor numbers. The Yosemite Valley plan, when released in 2000, likewise demurred on the park visitation question. Instead, the agency proposed consolidating day-visit parking on the eastern valley floor while adding new parking areas outside the valley and new shuttle buses to transport visitors into the valley. Although the proposal reduced the number of parking spaces available in the valley, it actually only prevented autos from entering the valley during the extended summer peak season. It also would remove a few motel buildings at Yosemite Lodge, reduce the number of campsites to below pre-1997 levels, add a new visitor center near the new parking lot, relocate the equestrian concession outside the valley, and remove a bridge to safeguard the fragile river corridor.[35]

Not surprisingly, the plans failed to placate the agency's critics, who saw little real change and the likelihood of more environmental degradation. The problem, in their view, was twofold—no proposed limitations on visitor or vehicle numbers and the prospect of more development, including a new visitor center, on the valley floor—but even the mention of limiting access to the valley had drawn strong resistance from the park's concessioners as well as the nearby gateway communities, who feared a loss of business. Many local park visitors also objected, viewing the valley as a traditional playground setting. Their concerns were echoed by the local congressmen, who repeatedly warned Park Service officials at congressional hearings against any thought of "locking out" park visitors. Although agency officials were talking with nearby communities about a regional transportation strategy, the projected costs—roughly $400 million for a rail system—dwarfed the park's annual $24 million congressional appropriation. Clearly, the question of how to manage the valley and whether to control visitor access resonated well beyond park boundaries, highlighting the regional implications that would accompany any agency imposed visitor limitations.[36]

Time to Limit Visitation?

Perhaps predictably, litigation ensued over the Merced River plan, which the Park Service consistently lost over the course of six federal court rulings. The key question, under the Wild and Scenic Rivers Act, was whether the Park Service's VERP analysis complied with the legal requirement that it determine "user capacities" for the Merced River corridor, much of which traverses Yosemite Valley. The courts, in short, concluded that the agency must "deal with or discuss the maximum number of people" at the Merced River, which seemingly will require the Park Service to set a numerical limit on visitor use.[37] The rulings rejected the agency's efforts to rely on monitoring of environmental conditions as a touchstone for later visitor limitations as well as its argument that parking and overnight accommodation limits were sufficient to address visitor capacity concerns. In fact, citing the litany of existing facilities in the valley river corridor, the court of appeals exclaimed: "The multitude of facilities and services provided at the Merced certainly do not meet the mandatory criteria for inclusion as a remarkably outstanding value" in the statute.[38] Having found the Merced River plan legally inadequate, the courts then blocked several Park Service construction projects, effectively curtailing further development in the valley.[39]

The fundamental question—one that the Park Service and its constituencies have wrestled with for nearly a hundred years—is how to reconcile visitation and preservation in the stunning but limited confines of Yosemite Valley. With annual park visitation regularly pushing 3.5 million people and with most visitors arriving in their own personal automobiles, a new sense of urgency surrounds the question. As the "heart of the park," Yosemite Valley merits much greater protection from the onslaught of visitors than it has received. It should not be treated as a mere sacrifice zone, a place where multitudes are invited to congregate, thus leaving the rest of the park less disturbed in fact and more wilderness-like in appearance. Although the Park Service has acknowledged the problem, imposed modest limits on automobile access, and taken other incremental steps, it has not placed a cap on the number of visitors or cars permitted in the valley. Despite its Organic Act responsibility to preserve park resources in an unimpaired condition, the agency's historical impulse to welcome all visitors continues to prevail. Indeed, fearing possible political repercussions, the Park Service has only rarely placed a hard limit on the number of visitors allowed in any park, most notably at Denali where the number of daily buses allowed on the park road is limited. If the nonimpairment mandate does not force its hand in Yosemite, then the Wild and Scenic Rivers Act may compel reconsideration of its perpetual open-door policy in this age of mass tourism.[40]

THE MASS TOURISM CHALLENGE

Tourism pressures on the national parks have steadily escalated to the point where park resources are sometimes endangered and the visitor experience can take on an urban veneer. Although Park Service officials, who have consciously abetted the growth in visitation since the agency's inception, showed little initial concern over tourism's environmental effects, agency policy has shifted over the years and now reflects a much greater ecological sensitivity. Visitation numbers, however, still count within the Park Service and Congress, especially when the agency is seeking additional funding support for the parks. And the automobile continues to dominate most park visits, strongly affecting how people encounter the parks and thus presenting the Park Service with one of its greatest challenges. What responsibility, in this era of mass tourism and the automobile culture, does the Park Service have to manage visitors and shape their park experience?

Tourism in the national parks is much different today than in Stephen Mather's time. The initial visitors to Yosemite, Yellowstone, and other early parks encountered a largely undeveloped, wilderness-like setting with primitive accommodations, few roads or trails, and little oversight from park officials. Most park visitors who had traveled any distance represented the upper strata of society and were there to view the stunning scenery. Since those early days, however, the parks have built an extensive infrastructure that includes miles of paved roads, diverse overnight accommodations, massive parking lots, various commercial establishments, and the like, all designed to meet visitor needs and even create new ones. Annual visitation at several major parks—including Yosemite, Grand Canyon, Great Smoky Mountains, Olympic, and Yellowstone—regularly exceeds three million people, most of whom tend to cluster around the developed areas and popular natural attractions. Some have come seeking outdoor recreational opportunities, others for solitude and self-renewal in a wilderness setting, and yet others are attracted to the parks by clever marketing. What they find, in too many cases, are hordes of like-minded people, the sounds of civilization, and growing evidence of cumulative environmental degradation. Yosemite Valley, as one prominent example, often seems as much a smog-shrouded urban venue as it does a peaceful natural one.

Nothing has affected the national park visit as much as the automobile. Early on, Mather astutely perceived the auto's potential effect on national park visitation, and the Park Service proceeded to make the parks into an auto-friendly, if not auto-dominated, setting. The result was a national park system soon accessible to most middle-class Americans, converting the parks into a much more democratic institution than was true at their inception. As the automobile became more widely available to the American populace, so, too, did the national parks, and so, too, did the demand for more roads, accommodations, and services. The Park Service obliged, as reflected in the CCC infrastructure improvements and the Mission 66 construction frenzy. Although these developments helped attract more people to the parks, they also confronted the Park Service with the twin challenges of controlling the auto's effect on the natural setting and enticing people out of their cars to actually experience nature. If a park visit was merely a "drive-through"—or "windshield"—experience, then the national park idea might well remain one dimensional.

As the national park system has evolved, however, the Park Service and its allies have begun addressing the automobile problem. In designing and managing several of the newer parks, including Olympic, Canyonlands, Everglades, and North Cascades National Parks, the Park Service has consciously abstained from constructing roads into every prominent feature or alluring site, choosing instead to limit park road miles and thus contain the automobile. More recently, the agency has employed shuttle bus systems to bring visitors into Yosemite Valley, Zion, Bryce Canyon, and Acadia to alleviate traffic jams, parking problems, and air pollution concerns.[41] Roads in some parks remain unpaved, while new road proposals are now routinely resisted, accounting for the Park Service's decision, after more than four decades of contention, to finally mothball the Fontana Lake "road to nowhere" proposal in Great Smoky Mountains.[42] Where implemented, the new shuttle bus systems provide visitors with an alternate way of experiencing parks, one that usually brings them into more intimate contact with the natural surroundings. Moreover, the bus systems enable the Park Service to reach more visitors directly, educate them about the park and its attributes, and explain the broader ecological context, all of which can only expand the public's environmental knowledge and awareness.

The sheer presence of so many people, autos, and other trappings of civilization inevitably has an effect on national park environmental conditions. Although the Park Service's early efforts to make the parks friendly to tourists by exterminating predators, extinguishing wildfires, and building modern facilities have now changed, the agency still faces difficult challenges managing park wildlife and sustaining ecological processes in the presence of so many visitors. Too often, park visitors do not understand the unpredictable nature of wild creatures, the dangers of habituating them to the human presence, or the role that wildfires, floods, and other natural processes play in sustaining healthy ecosystems. And though the Park Service now strives to minimize its construction activities, it still periodically turns a blind eye when replacing run-down facilities and deteriorated roads.

The obvious antidote to mass tourism is better management and more education. It is thus no surprise that the Park Service has become extremely adept at managing people. It has realized that it is more efficient and effective to manage people rather than animals: by restricting visitor activities, educating them as to proper behavior, or relocating facilities. Park officials clearly have the authority to limit access to particular areas or resources, for example, by placing denning sites or other sensitive locations off-limits to protect wildlife during critical breeding and rearing periods. In Everglades National Park, the courts sustained a park rule that closed certain areas to all public entry to protect at-risk fisheries. On the educational front, to prevent injurious wildlife encounters and harassment incidents, the agency regularly supplies visitors with brochures explaining proper behavior, and it posts warning signs to convey the same message. Simply put, park officials have the authority and experience to control the multitudes that regularly visit as long as they also have the foresight and fortitude to intervene before environmental damage occurs.[43]

Perhaps the most difficult visitation issue looming over the national parks is the question of limiting visitor numbers, either to protect park resources or to ensure a quality experience. At Yosemite, park officials have steadfastly avoided any decision that would cap the number of visitors allowed in the namesake valley. Instead, they have addressed the overcrowding problem indirectly, limiting parking and camping spaces, deploying shuttle buses, trying to entice visitors elsewhere, and even relocating facilities outside the park. Whether or not the Organic Act's nonimpairment standard obligates them to do more, the courts have invoked the Wild and Scenic Rivers Act to suggest that hard limits may be necessary to protect the Merced River's fragile riparian corridor. In fact, park officials have long used quota systems to control backcountry usage, popular white-water venues, and fishing opportunities, sensitive to important environmental and experiential values in these settings.

There are no obvious legal impediments to imposing limits on visitor numbers as long as they are designed to protect park resources or the visitor experience. The Park Service's management policies instruct park superintendents to assess visitor carrying capacities and, before ordering any closure or restriction, to make a written determination that the measure is necessary to protect park resources or values, an onerous but not impossible burden.[44] However, even the suggestion of limiting the number of visitors allowed into a beloved national park or scenic park locale invariably generates counter pressures from the visiting public as well as concessioners and gateway communities. As a result, the Park Service has never imposed visitor limits for attractive frontcountry venues, such as Yosemite Valley, the Grand Canyon's South Rim, or Cade's Cove in Great Smoky Mountains, which draw so much tourist interest. Sensitive to the likely public uproar and potential political consequences that could accompany any decision denying access to popular park features, the agency has persisted in giving visitation priority, even when it may degrade natural conditions and the experience for others. Viewing the national park as a tourist destination is, after all, a deeply embedded institutional tradition, one that cannot be divorced from a park's larger surroundings, where powerful vested interests have long been committed to keeping the numbers growing. In short, the national park as a tourist destination is a concept that endures and continues to shape the national park idea.

CHAPTER 4

"The Nation's Playground"

Recreating in Paradise

National parks offer a prime recreation setting, and millions of visitors would seem to agree. From the outset, the Park Service has encouraged visitors to avail themselves of the hiking, camping, fishing, and other outdoor opportunities available in the parks. Along the way, beset by diverse new recreational requests backed by powerful businesses and constituencies, the Park Service has opened some areas to snowmobiles, off-highway vehicles, personal watercraft, and similar activities. It has also denied or limited access for other users, including hunters, mountain bikers, white-water kayakers, and aerial enthusiasts, asserting that these recreational pursuits are incompatible with national park purposes. At the same time, Congress has significantly expanded the park system to meet escalating recreational demands, creating new national recreation areas, gateway parks, and related designations. In addition, the private sector—concessioners, outfitters, gateway businesses, and others—has come to play an important role in promoting and providing recreational opportunities for park visitors. Underlying this evolution in recreation policy is the fundamental question of whether the national parks are appropriately regarded as playgrounds or whether they should aspire to more lofty goals.

RECREATION IN THE NATIONAL PARKS

From the beginning, national park advocates viewed recreation as an important dimension of the national park experience and a means to introduce Americans to the parks. In his classic tome, *Our National Parks*, John Muir opined that a visit to the parks could refresh the spirit of an overcivilized populace, giving visitors an opportunity for "briskly venturing and roaming . . . [and] jumping from rock to rock, feeling the life of them, learning the songs of them, panting in whole-souled exercise, and rejoicing in deep, long-drawn breaths of pure wildness." The railroads, early advocates for new park designations, not only promoted the parks as scenic attractions, but also extolled their recreational potential, advertising mountain climbing at Glacier as a "glorious sport" and casting Crater Lake as "Oregon's mountain playground." Enos Mills, principal proponent for Rocky Mountain National Park, believed that "every one needs to play, and to play out of doors. Without parks and outdoor life all that is best in civilization will be smothered." Having been lured to the parks to view their spectacular scenic attractions, the public was also encouraged to experience them firsthand by tramping through the forests, climbing the rugged mountains, and wading into the pristine waters.[1]

An Early Promotional Impulse

Congressional passage of the 1916 National Parks Organic Act confirmed that public recreation was part of the new Park Service's mission. Although framed simply in terms of "enjoyment," the legislative history underlying the Organic Act indicates that Congress saw the new national park system as an important recreational outlet. Frequent references were made to parks as "playgrounds" throughout the multiyear debates that proceeded passage of the bill. Secretary of the Interior Franklin Lane, in his policy-setting 1918 letter to the Park Service, pointedly described the national parks as "this national playground system" and stated that "the recreational use of the national parks should be encouraged in every practicable way," thus affirming the principle that the parks were "set apart for the use, observation, health, and pleasure of the people." The Lane letter also held that "all outdoor sports which may be maintained consistently with the observation of safeguards thrown around the national parks by law will be heartily endorsed and aided wherever possible," further elaborating that "mountain climbing, horseback riding, walking, motoring, swimming, boating, and fishing will ever be the favorite sports." From the outset, however, hunting in the national parks was forbidden; park animals—at least the good ones—were more valuable as a visitor attraction than as a trophy or meat.[2]

The Park Service, under the early leadership of Stephen Mather and Horace Albright, was intent on enticing people to visit the parks and building a supportive constituency. Providing attractive recreational opportunities was one way to do that. Singling out motoring as a favored form of recreation also helped, especially with the automobile growing in popularity around the country, making the national parks more accessible to more people. As we have seen, the Park Service strongly endorsed the construction of new roads, lodging, and other visitor accommodations, all designed to make the parks more accessible to the general public. But Mather, ever the marketing executive, went much further and supported the building of golf courses, ski areas, and tennis courts in the parks, all with the hope of luring even more people to visit. There was little overt opposition to these early recreation policies; park visitation was relatively light, few conflicts surfaced between different recreational users, and motor vehicles were not equipped for off-road forays into the backcountry or to traverse snow-covered roads.[3]

Conceiving a National Recreation Policy

During the early twentieth century, recreation began to emerge as a national concern, driven by rising prosperity, greater leisure time, and continued growth in automobile ownership. The Park Service soon discovered that it was not the only federal agency interested in outdoor recreation. The Forest Service, with its expansive portfolio of scenic and game-rich lands, also laid claim to a recreation mission. In 1920, chief forester Henry Graves penned a paper titled "A Crisis in National Recreation" that argued for a new nationally comprehensive recreation policy and, more ominously, for transferring the fledgling Park Service to the Department of Agriculture, where the Forest Service was already ensconced. By then, of course, an interagency rivalry between the utilitarian-oriented Forest Service and the preservation-oriented Park Service was quite apparent, with its origins tracing to the high-profile Hetch Hetchy controversy. Not only had the Forest Service initially opposed creation of the national park system, but Graves's article decried the new Park Service's acquisitive tendencies, namely its penchant for targeting scenic national forest lands for new national parks. Recreation policy differences between the two agencies soon became even more evident once the Forest Service endorsed the wilderness concept as an area to be left untouched, an idea that ran counter to the Park Service's efforts to open its own lands to visitors with roads, lodges, and other conveniences. This early interagency rivalry still endures, reflected in federal recreation policies that are not always harmonious or coordinated.[4]

By the 1930s, with onset of the Great Depression and creation of the Civilian Conservation Corps (CCC), the national parks were soon transformed into more-visitor-friendly settings. The CCC, as noted, constructed hundreds of miles of new roads and trails, dozens of new campgrounds and bridges, and other improvements, which encouraged more visitation and increased recreational demands. Although the nation was gripped by the Depression and travel to most national parks was still not easy, public recreational demands continued to mount, prompting yet more reports that called for a new national recreation policy to better integrate federal, state, and local programs. Congress responded by passing the Parks, Parkway, and Recreation Act of 1936 despite resistance from the Forest Service, which feared that more of its lands would be lost to the Park Service. The legislation instructed the Park Service to inventory federal, state, and municipal lands (excluding national forests as a concession to the Forest Service) for their recreation potential and to develop recommendations for a comprehensive national recreation policy. The report, released in 1941, offered a blueprint for improving and expanding recreational opportunities across the nation; it specifically encouraged coordination at a regional level, endorsing the use of interstate compacts to support such efforts as establishment of the Appalachian Trail. By all appearances, the Park Service was now the primary federal recreation agency.[5]

At the same time, Congress began to expand significantly the federal role in recreation, designating the Park Service to oversee an array of new recreational sites and programs. Once it recognized the recreation potential associated with new federal multipurpose water projects, Congress authorized the first National Recreation Area at Lake Mead on the Colorado River in 1936; the Park Service soon assumed management responsibility for the lake and adjacent lands, but not for the dam operations. Soon after, in an effort to bring outdoor recreational opportunities nearer to the coasts where most people lived, Congress created the first National Seashore at Cape Hatteras (renamed the Cape Hatteras National Seashore Recreational Area in 1940), also vesting the Park Service with its management. The legislation establishing these new recreation-oriented sites typically instructed that they should be administered "for general purposes of public recreation, benefit, and use" and for preservation, but only as long as consistent with competing water, power, recreational, and other uses.[6] Some Park Service officials, fearing that the agency had become too enamored with its growing recreation leadership role, originally resisted taking over these new areas, but their concerns were overridden with the argument that the agency had an existing legal obligation—derived from the Organic Act—to meet public recreational demands.[7] Over the ensuing years, Congress has created seventeen more Park Service–administered national recreation areas as well as various other special areas—national lakeshores, national parkways, national rivers, and the like—where recreation was given explicit priority.[8]

The Rise of Mass Recreation

Although the nation's interest in recreation understandably dwindled during World War II, the immediate postwar period witnessed an unprecedented spurt in the public demand for outdoor recreational outlets. A rapidly growing and more affluent populace with yet more leisure time, combined with construction of the interstate highway system, soon put even greater pressures on the national parks and other outdoor settings. The availability of postwar surplus, including rugged jeeps, sturdy river rafts, and new camping equipment, opened previously unreachable areas—backcountry lands, whitewater rivers, and other such venues—to ready public access. Visitor numbers tell part of the story: by the mid-1950s, national park visits had reached sixty million, a five-fold increase from 1945 figures, and national forest recreational visits experienced a similar postwar surge. As visitation escalated, so, too, did the calls for additional recreational opportunities, ranging from backcountry hiking and river rafting to off-road jeeping and snowmobiling. Simply put, the postwar period marked the rise of mass recreation, which placed ever-expanding demands on all the federal land management agencies.[9]

The initial congressional response, in 1958, was to create the first Outdoor Recreation Review Commission (ORRC). Chaired by Laurance S. Rockefeller, this high-powered group was charged with completing another inventory of recreational resources and recommending ways to meet surging demands over the next several decades. To its dismay, the Park Service was not given a prominent role in this new study and soon found itself displaced as the preeminent federal recreation agency, in part because park visitation figures had slipped behind the national forest figures. The 1962 ORRC report contained some key recommendations: one was to create a federal Outdoor Recreation Bureau separate from existing agencies, and another was to establish a federal funding mechanism to support land acquisitions for recreational purposes. Congress responded, and both of these recommendations soon became law. Although the Bureau of Outdoor Recreation proved short-lived, the Land and

Water Conservation Fund Act of 1964 has underwritten the acquisition of millions of acres of new federal, state, and municipal lands primarily for recreational uses, including important additions to the national park system and existing park inholdings. The ORRC also introduced the notion of charging user fees for recreational activities, an idea that has proven controversial over the years, but one that has gradually been incorporated into federal policy.[10]

The National Park Service responded to the post–World War II visitation surge by launching the aforementioned Mission 66 program. Under the leadership of Conrad Wirth, who had burnished his executive credentials overseeing earlier recreation studies, park facilities were expanded and upgraded to better serve the mounting number of visitors. The resulting new roads, bridges, trails, and campgrounds significantly enhanced access to the parks and recreational opportunities, including backcountry areas, and with greater access, the push for more diverse recreational uses only intensified. During this same period, Congress continued to expand the national park system, adding ten new natural areas, eleven new lakeshores and seashores, and eight new recreational reservoirs. Congress also further diversified the federal portfolio of protected recreation sites to include wilderness areas, wild and scenic rivers, and national trails. Further, in the General Authorities Act of 1970, Congress clarified that the national park system was one unified system subject to uniform management standards, rejecting the Park Service's effort to distinguish between natural, recreational, and historical areas for management purposes.[11]

Addressing the Mass Recreation Challenge

Along with the Park Service's growing commitment to meeting the new mass recreation demands came a rising chorus of criticism, even from within the agency. As early as 1936, Sequoia National Park's outspoken Superintendent Colonel John R. White bluntly raised the issue in a speech to his fellow superintendents: "We should boldly ask ourselves whether we want the national parks to duplicate features and entertainments of other resorts, or whether we want them to stand for something distinct, and we hope better, in our national life." Conservation groups publicly accused the Park Service of ignoring national park standards in its rush to promote recreation. Newton Drury, who then headed the Redwood League and who would shortly become National Park Service director, disparaged the agency as the "Super-Department of Recreation" and a "glorified playground commission." Some groups were so incensed that they opposed Park Service management of the Kings Canyon area in California, arguing that it would be in better hands under the Forest Service. By the early 1970s, the Conservation Foundation was moved to conclude that "the national park was fast becoming a playground, a bland experience little different from what the visitor can and does find at a thousand other areas." The ever-present—yet regularly ignored—question in this new era of mass recreation was how recreational opportunities should be attuned to the national park setting.[12]

The postwar upsurge in park visitation and the expanding interest in outdoor recreation soon confronted the Park Service with the need to control and regulate visitor activities, both to protect the environment and to reduce user conflicts. These concerns eventually came to a head, first at Sequoia and Kings Canyon during the 1960s and then at the Grand Canyon a decade later. Although it was apparent during the 1940s that the Sequoia and Kings Canyon backcountry meadows were being severely damaged from excessive livestock use—the Sierra Club regularly brought up to 250 people and 100 head of livestock into the fragile alpine terrain—the Park Service did not begin to regulate livestock usage until the late 1950s, when it closed some sensitive areas and limited group sizes. The problem worsened, however, once backpackers started appearing in increased numbers, prompting a proposal to eliminate pack stock from the high country. In 1960, park officials responded with a groundbreaking backcountry management plan, detailing the growing environmental damage and recommending new recreation carrying-capacity limits designed to preserve the wilderness experience. The document, which soon circulated throughout the Park Service, represented an early agency effort to actively manage growing recreational pressures.[13]

Not long thereafter, Grand Canyon National Park officials faced a twenty-fold increase in the number of visitors floating the Colorado River and mounting ecological damage within the famed canyon. In 1972, as earlier noted, they responded by freezing the park's previously open-ended permit system to limit the number of river trips; then, seven years later, they allocated the permits between commercial rafting companies (92 percent) and private trips (8 percent) and announced the phaseout of motorized rafts. No one was happy, and the commercial operators promptly convinced Congress to overturn the ban on motorized trips. Dissatisfied noncommercial private boaters sued, claiming that they enjoyed a "free access" right to the river and accusing the Park Service of "commercializing" the park. The federal courts, however, sustained the permit system, finding that park officials were justified in protecting the river and had reasonably divided the permits between the two classes of river users, most of whom required the assistance of professional guides. Subsequent river management plans have modified these original permit allocations, but the Park Service's long-standing efforts to develop a workable plan that protects river corridor resources and the park experience still have not satisfied all the multifarious constituencies seeking this grandest of recreational experiences. Despite persistent complaints that motors are incompatible with the river experience, motorized rafts continue to ply the canyon waters, proof of how difficult it can be to outlaw established recreational uses from a park.[14]

The contentious issue of appropriate recreational activities in a national park has evolved over time. Early on, Park Service officials, often joined by local concessioners and businesses, viewed recreation as a primary means of attracting visitors to the parks. In Yosemite, when Mather unabashedly promoted golf, tennis, and swimming pools, the park's major concessioner—the Curry Company—lobbied vigorously for a year-round suite of recreational opportunities so as to increase its off-season revenues. The Park Service obliged, first by approving the Curry Company's ski area, ice rink, and toboggan run construction projects and then by mounting an unsuccessful campaign to bring the 1932 Winter Olympics to Yosemite National Park. Dismayed by the agency's growing commitment to commercialized recreation, Superintendent John R. White from neighboring Sequoia National Park spoke out against the construction of mechanized ski lifts or toboggan elevators, fearing that they would "make a hurly burly of the park in winter."[15] These sentiments were soon echoed by others who admonished the Park Service to eliminate resort-like facilities and focus instead on recreational activities appropriate to the parks' wilderness-like attributes. Gradually, the Park Service has phased out most—but not all—swimming pools, tennis courts, golf courses, and ski areas in Yosemite, Rocky Mountain, and elsewhere, concluding that these activities really did not fit the national park setting and were readily available elsewhere.[16]

More recently, national park recreation controversies have involved the use of motorized equipment and the propriety of new thrill-seeking activities. Having initially sanctioned motoring as a preferred form of recreation, the Park Service has encountered sustained resistance when it has moved to eliminate or regulate the new generation of motorized equipment, namely off-highway vehicles (OHVs), snowmobiles, and personal watercraft. The agency has also faced similar resistance when addressing new nonmotorized forms of recreation, many of which involve thrill-centered or speed-oriented activities, such as mountain biking, base jumping, and adventure racing. In addition, an old issue that continues to resurface is access to the national parks for hunting and trapping. The Park Service has steadfastly opposed those activities, although sport fishing has long been allowed in park waters, often with catch-and-release restrictions. Drawing consistent and workable lines between these diverse and ever-growing outdoor activities has been a real challenge, one that has forced park managers to reconsider the role that recreation should play in these unique and fragile scenic venues.[17]

The two principal questions for the Park Service, in the face of these mass recreation demands, are how should it determine when recreational use crosses the line to threaten park resources and values and what limitations are appropriate to protect them. The agency's answer is found in its official *Management Policies*, which invoke the Organic Act's nonimpairment standard as the basis for regulating recreational uses. The standard approach has been to identify "appropriate uses," which are those uses "appropriate for the purposes for which the park was established and that can be sustained without causing unacceptable impacts."[18] Even when deemed "appropriate," however, the use still cannot cause "unacceptable impacts," which must be mitigated through carrying-capacity limitations enforced through educational efforts, permit requirements, timing restrictions, technology standards, and the like. To set limits on the number of visitors, the carrying-capacity strategy first identifies desired resource and visitor experience conditions through agency planning processes and then monitors the situation over time.[19] To enforce these limitations, the Park Service has drawn on its general regulatory power to adopt rules that limit, for example, the number and type of OHVs, snowmobiles, and commercial raft trips allowed in the parks. Individual park superintendents are responsible for promulgating these rules and for documenting them in resource protection, scenic values, user conflicts and other terms.[20]

The courts have consistently ruled that the Park Service has sufficient power under the Organic Act to prohibit and restrict different types of recreational activities to avoid impairing park resources. In one case, the National Rifle Association's effort to overturn the Park Service's prohibition on hunting in national recreation areas, absent explicit congressional authorization, was rejected, with the court observing: "In the Organic Act, Congress speaks of but a single purpose, namely, conservation." Another federal court overturned a park rule allowing snowmobiles in Yellowstone, stating that "the National Park Service is bound by a conservation mandate, and that mandate trumps all other considerations," although this ban was subsequently reversed. Yet another case sustained the Park Service's decision to curtail mountain biking in national recreation areas, finding that "resource protection [is] the overarching concern." And in one case challenging the Voyageurs National Park superintendent's decision to temporarily close snowmobile trails to protect wolves from harassment, the courts found that the Park Service had properly given priority to safeguarding the wolves from possible injury. Under these precedents, Park Service officials have the authority as well as guiding standards to address recreation-related problems, but that does not ensure that they will exercise their power, given the other pressures they face.[21]

Dealing with Outside Pressures

The unrelenting demands of mass recreation have been driven by forces well outside the Park Service's immediate control and the national park setting. Just as the Curry Company early on promoted winter recreational activities in Yosemite, a growing assortment of concessioners, gateway communities, product manufacturers, and organized user groups have regularly besieged the Park Service with new recreation demands, arguing that the agency is obliged to ensure public enjoyment of the parks. Whether the issue involves snowmobiles, personal watercraft, OHVs, or even mountain bikes, the question is whether each type of recreational activity is compatible with the national park setting and the preservation ethic. In response to these pressures, which regularly ignore competing environmental and compatibility concerns, the Park Service's answers have not always been consistent. Snowmobiles, for example, have been banned in Glacier National Park since 1977, whereas they were permitted in Yellowstone with little regulation until 2000, when they were briefly banned and then reinstated subject to additional limitations. Given the amount of money involved in mass recreation today and the attractive playground venue that the parks represent, nearly every effort made to curb particular forms of recreation in the parks has been resisted politically and has frequently been litigated. With many nearby communities embracing new tourism and amenity-based economies linked to the presence of national parks, there is little likelihood that these recreation pressures will dissipate soon, especially given general population growth trends.[22]

In fact, a major controversy erupted in 2005 over the Park Service's *Management Policies,* driven principally by the role of recreation in the parks. Although the Park Service, following a comprehensive three-year review, had just finished revising its all-important *Management Policies* in early 2001, a politically appointed deputy assistant secretary in the Interior Department secretly penned wholesale revisions to the policies following the 2004 election. That the deputy assistant secretary was Paul Hoffman, who had previously been director of the Cody, Wyoming, Chamber of Commerce, where he led opposition to the Clinton administration's efforts to ban snowmobiling in Yellowstone, sparked immediate suspicion about his motives, suspicions that proved well-founded. The Hoffman revisions sought to water down the agency's definition of "impairment" by requiring that any impairment finding be judged not just by the proposal's effect on park resources but also by its effect on visitor enjoyment. The revisions also equated visitor enjoyment with recreational use and reduced the protection afforded natural soundscapes, opening the door for more motorized recreation, air tour overflights, and the like. As a whole, the proposed revisions would have rewritten the long-accepted interpretation of the Organic Act that prioritized resource conservation as the Park Service's primary responsibility.

Once Hoffman's proposed changes were made public, park supporters erupted in fury, accusing the Interior Department of reversing a half-century of consistent policy for political purposes. The charge had much validity, given the ongoing contacts between high-level Interior political appointees and recreation vehicle industry representatives who were promoting a "National Outdoor Recreation Policy" initiative designed to open the parks to more motorized recreation. An array of groups, spearheaded by the National Parks Conservation Association and the Coalition of National Park Service Retirees, collectively mobilized public opposition to the revisions, converting this somewhat abstruse policy controversy into a matter of national concern. Critics of the revisions abounded, within the Park Service and outside it. The agency's Pacific West regional director, Jon Jarvis (who was appointed National Park Service director in 2010 by President Barack Obama), spoke for many when he characterized the proposed changes as "the largest departure from the core values of the National Park System in its history, posing a threat to the integrity of the entire system." Following a well-publicized Senate hearing and massive editorial criticism in the nation's newspapers, the Park Service quietly adopted a few modest revisions to the document that preserved its essential integrity and priorities. The management policies controversy nonetheless serves as a stark reminder that the struggle over national park recreation policy is ongoing in the form of persistent pressure from motorized groups, industry organizations, and their political allies.[23]

Plainly, national park recreation policy issues cannot be separated from the surrounding landscape. Not only do adjacent communities often face strong local political pressure—driven by main street businesses and concessioners—to expand recreational activities and opportunities, but they also have powerful economic incentives—tax revenues, local jobs, and the like—to support an ever-expanding repertoire of activities in nearby parks. National parks are not the only suitable venue for snowmobiling, backcountry jeeping, or similar activities, however; other nearby federal, state, and private lands may offer an alternative motorized recreation venue, albeit perhaps a less scenic one. The important point—one recognized by the first federal ORCC and repeatedly reconfirmed since—is the need to coordinate recreation policy at the federal, state, and municipal levels to meet these growing demands in an efficient and environmentally acceptable manner. Put simply, national park recreation policy should not be formulated in a vacuum, but must take account of other venues and options that might help ameliorate the relentless demands associated with mass recreation today.[24]

SNOWMOBILES, OHVS, AND OTHER CONFLICTS

A seemingly endless series of recreation controversies threaten to reshape the true meaning of the national park experience. As these conflicts have grown in number and intensity, the Park Service has repeatedly found itself having to reconcile mass recreation demands with its nonimpairment mandate. While charged with promoting the parks and providing public "enjoyment" opportunities, the agency must also minimize environmental impacts and visitor conflicts as well as address less apparent aesthetic and symbolic concerns. Just as cumulative environmental effects can eventually unravel an ecosystem, so, too, can incremental concessions to intensive new recreation demands recast the park experience. Not everyone can play in paradise on their own terms.

Snowmobiles in Yellowstone

Few national park recreation controversies have evoked the passionate response that has driven the Yellowstone snowmobile imbroglio. In 1948, with

World War II receding and tourist traffic picking up again, the town of Cody, Wyoming, situated near the park's east entrance, unsuccessfully importuned the Park Service to begin plowing the park's snow-covered roads and thus create a new winter visitation season. Six years later, two West Yellowstone entrepreneurs started hauling winter visitors into the park on new but quite primitive snow coaches. In 1963, the first snowmobiles entered the park; their number quickly grew from one thousand to more than five thousand in three short years. Faced with renewed pressure from Wyoming politicians and businesses to plow the park roads, Yellowstone officials struck a compromise that officially sanctioned snowmobiles in the park. The plan, finalized in 1968, left the roads unplowed, but allowed over-snow access on groomed roads and opened winter lodging at Old Faithful.[25]

Park superintendent Jack Anderson, in the spirit of Mather, set about actively promoting park snowmobiling, even arranging a winter visit for popular radio personality Lowell Thomas, who then shared his experience with his weekly audience. By the early 1970s, more than twenty-five thousand winter visitors were annually coming to the park, many of them on machines rented from burgeoning new businesses in West Yellowstone. The town promptly rebranded itself as the "snowmobile capitol of the world" and began paving its streets from the new wintertime revenues. For his support, Anderson received a special recognition award from the International Snowmobile Industry Association.

By the mid-1970s, however, complaints about increased air pollution, noise levels, and wildlife harassment began to mount. Yellowstone officials, faced with two presidential executive orders requiring all public land agencies to reassess off-road vehicle usage on their lands, simply designated all park roads open to snowmobiles, ignoring the regional office's request to undertake a formal environmental analysis of the situation. To the north, Glacier National Park officials seized the same opportunity and proceeded to reevaluate the park's growing snowmobile usage. After preparing a formal environmental assessment, they banned snowmobiles from Glacier, finding that they not only caused environmental damage and shattered the silent landscape, but were also disfavored by more than 90 percent of those who commented on the matter. At the same time, Yosemite, Sequoia, and Lassen likewise banned snowmobiles, while Rocky Mountain decided to only allow them on its remote west side. The opposite view prevailed at Yellowstone, however, where park officials expanded snowmobile and other winter opportunities, erecting new warming huts and opening another hotel to winter visitors. In response, winter visitation surged yet again, growing from 70,000 in 1983 to more than 143,000 in 1992.[26]

Confronted with this unprecedented growth in winter usage, Yellowstone officials were finally forced to address the effect that snowmobilers and other winter visitors were having on the park setting. Under new leadership, the park issued its first winter use plan in 1990, but it mostly endorsed the status quo, and its decade-long visitation and pollution predictions proved outdated in just three years as more and more snowmobiles flooded into the park. Next, Yellowstone officials initiated a visitor use management planning process designed to finally establish limitations on the number and types of winter visitors, but the severe winter of 1996–1997 disrupted these plans. Faced with starvation, the park's free-roaming bison herd traversed the hard-packed roads and exited the park in record numbers in search of accessible forage, only to be shot by Montana state officials, who viewed the animals as a disease vector that could imperil the state's livestock industry if they transmitted the brucellosis bacteria to domestic cattle. After more than 1,000 bison had been killed, disgusted observers led by the Fund for Animals filed a lawsuit challenging the park's plainly inadequate winter use program. The case was soon settled when the Park Service agreed to prepare an environmental impact statement examining its open-access and road-grooming winter use policy.[27]

This initial lawsuit soon triggered a series of winter use policy shifts, competing court cases, and congressional bills that has kept the Yellowstone snowmobile issue unsettled. Once forced to examine the environmental effects of its snowmobile policy, the Park Service was compelled to admit that they violated its fundamental resource management responsibilities: "Continued [snowmobile] use hinders the enjoyment of resources and values for which the parks were created, most notably natural soundscapes, clean and clear air, and undisturbed wildlife in a natural setting."[28] Not only were park rangers regularly donning respirators at the West Yellowstone entrance station and elsewhere to protect against harmful snowmobile exhaust fumes, but the high-pitched whine of two-cycle engines was penetrating far into the park backcountry, and wildlife harassment incidents were growing in number. Invoking the Organic Act's nonimpairment standard, its *Management Policies*, and other legal authorities, the Park Service concluded that snowmobiles would be phased out over a three-year period, while less intrusive snow-coach access would be encouraged so that visitors could continue using the park in winter. Whatever economic effect might be felt by West Yellowstone and other nearby communities by eliminating snowmobiles from the park would be ameliorated by licensing local businesses to operate the new snow coaches and by phasing in the snowmobile shutdown. For the first time in more than thirty years, Yellowstone was poised to return to a more quiet and placid winter landscape.[29]

This effort to eliminate an established motorized use from the park proved short-lived, however. Once the George W. Bush administration replaced the Clinton administration, which had overseen the adoption of this new no-snowmobile policy, the Park Service announced that it would reevaluate the closure decision, noting that new technological advances might quiet snowmobiles and reduce exhaust emissions to within legal limits. In early 2003, armed with a new supplemental environmental analysis, the Park Service reversed course and, rejecting its own environmentally preferred alternative, adopted a new policy that increased the number of snowmobiles permitted into the park, imposed new emissions requirements on entering snowmobiles, and adopted a guide requirement for most trips.[30] Within days, a Washington, D.C., federal court blocked the new rule, ruling that the Park Service had not explained this policy turnaround in view of the agency's "clear conservation mandate and the previous conclusion that snowmobile use amounted to unlawful impairment." Further, the judge found that the Park Service's policy shift "was completely politically driven and result-oriented" and had disregarded the public input, which by more than 90 percent favored no more snowmobiles in the park. The court denied a request to stay its ruling due to its potential effect on local communities and park visitors, concluding that "any economic or emotional harm to those who made plans to visit the Park falls squarely on the [agency's] shoulders."[31]

The effect of the court's ruling was to reinstate the Clinton-era no-snowmobile rule, but the International Snowmobile Industry Association had other ideas. Reviving an earlier lawsuit, the association convinced a Wyoming federal judge that the Clinton-era rule was illegal and politically driven, too; it had not adequately examined the environmental and safety effects of snow coaches, and the public had not been afforded an adequate opportunity to comment on the proposed rule. The court also found that the Clinton-era rule would have a substantial economic effect on local snowmobile-related businesses that outweighed any environmental harm snowmobiles might cause to the park. With both the Clinton and Bush plans now enjoined, the Park Service was ordered to draft a new temporary winter use plan. It did so and issued new rules permitting nearly eight hundred snowmobiles into the park daily, with new pollution-control equipment standards and a commercial guide requirement. Caught between dueling federal courts, the Park Service was now on record as favoring an increase in the number of snowmobiles allowed into Yellowstone, even in the face of overwhelming scientific evidence that showed an array of adverse environmental effects on park resources and consistently strong public sentiment against the machines.[32]

The courts, though, were not yet finished, nor was Congress content merely to stand on the sidelines. In 2008, when the Park Service issued another revised winter use plan that would allow up to 540 snowmobiles daily into the park, the D.C. federal court again found the agency in violation of the Organic Act, observing that "the fundamental purpose of the national park system is to conserve park resources and values" and that "conservation is to be predominant." After reviewing the agency's own environmental data, the court found that "the Winter Use Plan will increase air pollution, exceed the use levels recommended by NPS [National Park Service] biologists to protect wildlife, and cause major adverse impacts to the natural soundscape in Yellowstone."[33] Not to be outflanked, the Wyoming federal district court, lamenting the potential economic effect on local businesses if snowmobiling were shut down in the park, reinstated the 2004 temporary rule, observing that it would "provide businesses and tourists with the certainty that is needed in this confusing litigation."[34] Moreover, Congress has taken up the matter and considered, but not adopted, several legislative proposals: some designed to eliminate snowmobile use in national parks, and others designed to ensure snowmobile access into the parks.[35] The Obama administration has taken the next step in this seemingly interminable struggle, proposing a plan that would impose a daily limit on the total number of snowmobiles and snow coaches allowed into the park, but would nonetheless permit as many as 350 snowmobiles per day.[36]

The critical question, of course, is whether high-powered snowmobiles represent an acceptable recreational activity in the national park setting. The 20-horsepower snowmobiles that initially ventured into Yellowstone during the mid-1960s are a pale shadow of today's machines, which boast 145-horsepower engines that are more powerful than a Honda Civic auto engine.[37] Industry efforts to quell snowmobile engine noise have largely proven unsuccessful, ensuring that conflicts will persist between motorized and non-motorized winter visitors. That Cody, Wyoming, politicians have pressured the Park Service to keep Sylvan Pass open at the park's eastern entrance during the winter season to foster snowmobile recreation as a local business opportunity, despite the area's extreme avalanche danger and the $1 million annual cost, stands as further proof that the snowmobile controversy is less about park recreation values and more about promoting local commerce. Remove these concerns from the equation and the snow-coach option would keep the park accessible for winter visitors, reduce wildlife conflicts, improve air quality, and generally enhance the winter visitation experience. Besides, there are manifold alternative snowmobiling opportunities on the national forest lands bordering the park, where the sharp whine of high-speed motors does not conflict with management policy and where silence and solitude are less cherished values.[38]

OHVs at Canyonlands

Snowmobiles are not the only motorized recreational activity to affect the national parks. From the national park system's inception, the Park Service has welcomed automobiles into the parks. After World War II, it was just a short step from the conventional automobile to the all-terrain jeep and more recently to OHVs, which have grown enormously in popularity. Between the 1999 and 2008, the number of four-wheeled all-terrain vehicles in use almost tripled, increasing from 3.6 million to 10.2 million, and their performance standards improved, too. Many OHVs can now reach speeds of sixty miles per hour or more, and they can access remote backcountry areas that were previously unreachable by standard motorized transport. OHV advocacy organizations, such as the Blue Ribbon Coalition and the Shared Access Alliance, have emerged and mounted effective political campaigns, along with OHV manufacturers, to secure access across the federal public lands. As OHVs have proliferated, concern and conflict have mounted within the national park system, where major battles have been waged at the desert and seashore parks.[39]

Canyonlands National Park in southeastern Utah has witnessed a particularly bitter struggle over motorized access to Angel Arch, a remote natural landmark long accessible to OHV users by driving up the Salt Creek streambed. Jeeps are allowed in the park backcountry on designated routes like the White Rim Trail, so it was not surprising that park officials initially balked at closing Salt Creek to OHV traffic. After a court decision forced them to reconsider, however, they finally closed the route to motorized access, acknowledging that the fifty required stream crossings had polluted this rare desert watercourse with motor oil and that the OHVs were damaging the fragile, ecologically important desert soil crust. While environmental groups applauded the decision, local OHV enthusiasts challenged the closure order in court, arguing that it violated the Organic Act's injunction that the national parks were open for public enjoyment. A Utah federal court rejected that position, however, ruling that the Park Service had correctly interpreted congressional intent that "conservation be the predominant conclusion in making management decisions where there is a conflict between conserving resources and providing for the enjoyment of them." The decision reinforces that the agency, when confronted with OHV-caused damage to park resources, has the duty and the power to take corrective action, even to the point of reversing a long-standing policy that allowed OHVs into the park backcountry.[40]

Elsewhere, however, the Park Service has chosen not to prohibit OHVs, despite claims of environmental damage and visitor conflicts. At Cape Cod National Seashore, faced with a tradition of dune-buggy use, park officials allowed these vehicles to continue plying an eight-mile stretch of beach, concluding that they were an "appropriate public use" that was not causing any ecological harm. The federal courts rejected a lawsuit challenging these conclusions, holding that the park's enabling legislation permitted this form of recreation, at least on a portion of the beaches, and that the management plan effectively protected the beaches from damage. The Cape Cod motorized recreation access controversy, which has been duplicated at Cape Hatteras, illustrates just how much latitude park superintendents have in addressing recreational issues, particularly when the park-enabling legislation does not prohibit OHVs or other potentially damaging recreational activities.[41]

Significantly, motorized recreation has become a pervasive problem that extends beyond the national parks. Across the public lands, as OHV use has grown, agency officials have documented mounting soil erosion, decreased air quality, wildlife disturbance, and visitor conflict problems, not unlike the problems confronting the national parks. In response, the Forest Service and the Bureau of Land Management (BLM) have enacted national rules closing their lands to OHVs except on designated routes and trails.[42] Trail designation decisions then occur at the local level through a public planning process, which delineates the routes that are open or closed to motorized recreation.

This travel management planning process roughly mirrors the Park Service's park-by-park approach to OHV use, an approach that leaves the agencies open to pressure from local politicians, businesses, and residents who tend to support motorized recreation and the commerce it generates. But even with the Forest Service's and the BLM's newfound willingness to limit OHVs to designated routes, extensive road and trail systems still crisscross these public lands, providing OHV enthusiasts with diverse riding challenges and experiences. More coordinated motorized recreation planning among the public land management agencies, including the Park Service, could help take some pressure off the national parks while still ensuring that OHV riders have a place for their sport.[43]

Other Recreation Conflicts

New nonmotorized recreational activities also pose significant management challenges in the national parks. The once unknown sport of mountain biking has soared in popularity, offering a relatively quiet, human-powered means of transport over backcountry roads and trails as well as the physical challenge of maneuvering through difficult terrain. Under a Reagan-era rule, the Park Service prohibited bikes outside of developed roads, parking areas, and designated routes, which required a "written determination that such use is consistent with the protection of a park area's natural, scenic and aesthetic values, safety considerations and management objectives and will not disturb wildlife or park resources." Although the rule focused on environmental concerns, many park officials and visitors believed that mountain biking, given the speed and thrill-seeking spirit many riders bring to the sport, simply did not fit the traditional profile of a national park recreational activity. Indeed, the Park Service prohibits other thrill-seeking recreational activities, such as base jumping and hang gliding, on the grounds that these sports are incompatible with the national park setting, pose inherent safety risks, and create possible conflicts with other visitors.[44] When challenged, mountain biking and other such prohibitions have not been disturbed by the courts, which have agreed that the Park Service's governing legislation gives priority to resource protection over recreational interests and sustained the agency's determinations of possible environmental damage. But a recent policy change allows park managers, following a formal environmental review process, to open park trails to bicycle use if it will not impair resource values or create safety problems.[45]

The more traditional adventure sports of rock climbing and mountain climbing are permitted in the national parks, however. In the case of rock climbing, the Park Service regularly imposes permit requirements to control climber numbers and ensure minimal competency, and it can also close areas to protect environmental resources, such as cliff-side vegetation, cliff-dwelling animals, and overused access routes. Whether to allow or outlaw fixed anchors—bolts that are permanently affixed to the rock surface for safety purposes on popular climbing routes—has generated considerable controversy within the ranks of climbers and environmentalists, resulting in detailed regulations governing their use at individual parks.[46] In the case of mountain climbing, parks like Denali and Mount Rainier offer world-class mountaineering experiences, luring climbers from near and far. To protect against overuse on popular mountains, the Park Service uses permit systems to control the number of climbing parties and to allocate access between commercial and noncommercial trips, much like its white-water permit system. Park officials also have the authority to close areas and routes to avoid environmental damage. As climbing-related recreation increases, the Park Service will undoubtedly face even greater user pressures (as it already does at the Black Canyon of Gunnison National Park and Yosemite's Half Dome), testing its resolve to limit access when park resources are potentially at risk or the climbing experience might be compromised.[47]

Because Park Service policy gives individual superintendents broad authority to manage recreational pursuits in the parks, local controversies have regularly flared when prohibitions are selectively imposed on some activities. Yellowstone's boating restrictions are a case in point. Since the mid-1980s, park officials have prohibited canoeing or kayaking the park's rivers, which offer an attractive and challenging white-water venue. These craft are permitted, however, to ply the park's lakes, and white-water rafting and kayaking are commonplace in the Grand Canyon, Dinosaur, and other parks. In 1987, responding to an initial petition from white-water enthusiasts for access to park rivers, Yellowstone's superintendent cited wildlife displacement, conflict with other backcountry users, and maintaining the natural river corridor setting to justify the ban. Public comments on the matter widely supported his position. When confronted with subsequent petitions, Yellowstone officials have grown increasingly concerned about the daredevil nature of white-water kayaking—plunging over steep waterfalls and the like—that also raise potential liability issues and rescue costs. The disappointed paddlers have responded by questioning why the park permits motorboats on its lakes and snowmobiles on its roads, but denies them access, even when they agree to strict limits on numbers, location, and timing.[48]

Some national park recreational issues have an undeniable commercial flavor. In 2011, organizers of the 2012 Quiznos Pro Challenge bicycle race sought access to Colorado National Monument for one stage of this week-long statewide event. The proposal would have closed the monument for a day, bringing six hundred riders and thousands of spectators to the site as well as support vehicles, aircraft, and helicopters. Promotional materials for the event extolled "the high speeds, danger and adrenaline of professional cycling." Although pressured intensely by state political officials and the local business community, who believed that the event would be good for local businesses and pay broad promotional dividends, park superintendent Joan Anzelmo steadfastly denied the requested permit. Citing the Park Service's *Management Policies*, she explained that such "a mega sporting event" was inconsistent with the monument's purpose; it would close the site to visitors for a twelve-hour period, disturb nesting raptors, disrupt desert bighorn sheep mating behavior, and generate unwanted litter. Park supporters lamented that the event would commercialize the park, setting a bad precedent system-wide. Earlier, the Park Service had denied a similar permit request in Yosemite, also citing the commercial nature of the event along with its anticipated negative effects. Whether viewed as a recreational or commercial activity, such a mass event is plainly incongruent with national park values. A few months later, the National Parks Conservation Association publicly recognized Anzelmo with its coveted Stephen T. Mather award for her courageous decision.[49]

It is difficult to reconcile these diverse recreation management decisions, which often vary across the national park system. OHVs are permitted in several parks, at least on designated routes, but mountain bikes are prohibited off-road in most parks; base jumping is permitted once a year at the New River Gorge National River site, but outlawed elsewhere; and white-water boating is an accepted fact in Grand Canyon and elsewhere, but prohibited on Yellowstone's rivers. The rationale—both expressed and implied—is as varied as the park settings where these issues have arisen: environmental degradation, visitor conflicts, fear of liability, the weight of tradition, aesthetic impingement, and the value of silence. Thus far, when the Park Service has linked its regulatory reasoning to its resource protection responsibilities, the courts have usually sustained its recreation management decisions, holding the line against more motors and other very intensive activities. These controversies, however, mask more fundamental questions about the nature of the national park experience and the appropriate role of recreation in these paradise-like settings.

BEYOND A PLAYGROUND MENTALITY

There is little reason to expect that national park recreation conflicts will soon abate. If anything, these conflicts will likely intensify in this era of mass recreation as new sporting activities appear on the scene, commercial and political pressures continue to mount, technological advances open even more ways to experience the outdoors, and the growing populace asserts a right to pursue individual activities of preference. Given the opportunity, who would not choose to recreate in a scenic national park playground setting? But the Park Service, although long committed to recreational activity on its lands, has increasingly sought to control, prohibit, and even eliminate some forms of recreation, drawing lines between different types of activities that may not immediately appear logical or defensible. Whether environmental damage is imminent or not, the agency has adopted the view that some forms of recreation are simply not appropriate in the national park setting, invoking such intangible qualities as natural quietude, self-reliance, and contemplative reflection as key dimensions of the national park experience.

For the most part, the Organic Act's nonimpairment mandate provides sufficient justification to prohibit or control potentially harmful recreational activities. The havoc that unregulated snowmobile use in Yellowstone can wreak on the park's wildlife and its crystalline winter air quality is evident and thus provides ready justification for prohibiting or rigorously regulating these mechanical intruders. The real problem is not that one or a few snowmobiles will harm the park but, rather, with the cumulative environmental effect that legions of snowmobiles visit on the park ecosystem. The same holds true for OHVs, mountain bikes, and other mechanical devices, all of which can, like the automobile before them, facilitate greater access and demonstrably impair park lands and resources. According to the courts, the Organic Act is crystal clear: the Park Service's first obligation is to resource protection, and recreational uses may be precluded if they could impair park resources or values. In addition, if the Organic Act's nonimpairment mandate does not apply, then other powerful laws—such as the Endangered Species Act, Wilderness Act, and National Environmental Policy Act—might be invoked to prevent or regulate potentially harmful recreational activities.[50]

National park recreation controversies extend beyond visible environmental damage, however, touching on the aesthetic and intangible dimensions of the park visitor experience. Some forms of recreation—base jumping and hang gliding, for example—are not inherently harmful, but just seem out of place in the national park setting. As Professor Joe Sax insightfully opined, the national parks are special places that offer visitors a unique opportunity to engage with nature that is not readily available elsewhere. Few enough such places are left in our increasingly urban and clamorous world, making the national park setting unique for its quiet spaces and uncluttered vistas. Whether the park visitor's experience is spiritually transcendent or just a momentary respite, the mere fact of being able to set aside daily concerns and concentrate on the fundamentals of existence serves both to purge and energize one's self. As noted in the Park Service's *Management Policies*, park resources and values include scenic features as well as natural landscapes, natural visibility, and natural soundscapes, and the agency is obligated to protect these key attributes.[51] So, include the intangible values of silence, visibility, and scenic integrity as an additional rationale for controlling or channeling the recreational impulses of park visitors.

The unremitting demands of mass recreation have also prompted increased visitor conflicts and created significant oversight problems. One person's thrill-a-minute jeep ride up an imposing mountain slope is another's worst nightmare as she labors up the same slope with a loaded pack, expecting to enjoy the backcountry in silence and solitude. But even in the busier and noisier frontcountry, where most park visitors spend their time and hold lower expectations, motorized sports are problematic. These locations may be the only opportunity that many visitors have to encounter nature, which argues strongly against spoiling the experience by converting it into the commonplace. Indeed, the acceptance of incompatible recreational activities anywhere in a national park establishes an unsettling precedent for acceptable visitor activities, one that can spill over onto other public lands where the same activities may also be unwelcome. In addition, it inevitably raises the question that if OHVs and the like are permissible where visitors congregate, then why not permit them in the backcountry where fewer people will be disturbed. Dispersed recreation is not easily policed, however, especially in expansive backcountry areas, where park rangers are only occasionally present and violations often go undetected.

In fact, the Park Service can credibly distinguish between various types of recreational activity to justify its extensive regulatory limitations. The argument that such distinctions are inherently discriminatory has an answer, albeit a complex one tied to the agency's dual focus on resource protection and park values. It may not be easy to distinguish between jumping from the top of El Capitan with a parachute and free climbing the face of El Cap: both can be classified as thrill-seeking sports, both have a real potential for injury, and the likelihood of environmental damage may be greater with rock climbing, especially if bolts are permanently affixed to the granite face. It may also be hard to distinguish between mountain bikes and horses on a backcountry trail: both can cause serious erosion problems, neither causes undue noise (unlike OHVs), and both can create conflicts with hikers and other trail users. As a policy matter, however, when environmental harm is not determinative, the answer turns on a subtle but crucial judgment about the type of recreational experience that is appropriate in national parks, one that involves nonmechanized physical challenges and the opportunity for contemplative experiences that draw on the unique natural setting. The answer is also influenced by tradition: horse travel and rock climbing have long been associated with the national parks, which is not the case with hang gliders and mountain bikes. No doubt these judgments are tough to make, but they are consistent with the aspirations and values that have animated our conception of a national park and the experiences available there.

Of course, recreation policy in the national parks cannot be divorced from the surrounding landscape. Because the national parks are such a powerful magnet for visitors, the mere presence of a park creates commercial opportunities for nearby communities and businesses, many of which will have been attracted by the park itself. Because most national park visitation seasons are relatively short, the concessioners, outfitters, guides, and gateway businesses have a powerful incentive to pack as many customers and activities as possible into the park, sometimes with little regard for the consequences. Indeed, recreation-based marketplace pressures have produced unintended consequences across the federal estate, often triggering growth and development that have radically altered the landscape. After World War II, for example, the Forest Service actively promoted downhill skiing on the national forests and leased entire mountainsides for ski areas, effectively creating the modern alpine ski industry. Few people, however, anticipated the resulting real estate development frenzy that accompanied these new ski areas, repeatedly transforming placid mountain valleys into urban archipelagoes and creating new fire protection and other forest management problems as reflected in Colorado's overcrowded Vail Valley. A similar phenomenon transpired in Moab, Utah, once the town was identified with mountain biking on the nearby public lands, and the same pattern is evident outside national parks in such locations as Estes Park adjacent to Rocky Mountain National Park and the Flathead Valley that adjoins Glacier National Park.[52]

Local political and commercial pressures can be quite intense, as is evident in the Grand Canyon rafting, Yellowstone snowmobile, and Colorado National Monument bike race conflicts. Because individual park superintendents are responsible for making recreation use decisions in their parks, they are often in the political cross-hairs for congressional delegations as well as state and local officials intent on protecting their constituents' individual interests. Whenever an important local business—such as West Yellowstone's snowmobile operators—cry foul over restrictive park recreation policies, the argument is inevitably framed in economic terms, decrying the local jobs at risk and lost revenue sources. For the Park Service, a response framed in terms of anticipated environmental damage or the more intangible benefits of solitude and a contemplative visitor experience can be a difficult position to sell locally, but the Lane letter was clear from the beginning: "the national interest must dictate all decisions affecting public or private enterprise in the parks." Moreover, Congress has flatly rejected the notion that parochialism should drive park recreation policy; its 1970 and 1978 amendments to the Organic Act eliminated any distinction between national parks, national recreation areas, and other designations, reaffirming nonimpairment as the universal standard governing park recreation policy as well as other resource management concerns. The Park Service, citing this single standard, might thus better insulate itself from these local political pressures by establishing more uniform recreation regulations across the national park system, much like the Forest Service and BLM have done by setting a national OHV policy for their lands.[53]

That said, the national parks cannot—and should not—isolate themselves from the surrounding landscape when it comes to recreation policy. An island mentality has no place here. The public lands adjacent to many national parks offer potential venues for an array of recreational activities—motorized sports, mountain biking, white-water kayaking, and the like—that are not always appropriate in the national park setting. Other than being in a splendid natural setting, it is difficult to understand why a tantamount recreational experience cannot be realized on other public lands outside the parks, where visitor conflicts are less likely and where environmental effects are more readily tolerated. Conversely, national park policies can inadvertently affect recreation policy on other public lands. Whenever the Park Service opens a park to particular types of recreation—snowmobiling in Yellowstone, for example—it establishes a precedent for allowing that same activity on other public lands because the potential environmental and other consequences are acceptable in the restrictive national park setting. The obvious answer to these dilemmas is to coordinate recreation management among the different public land agencies, which is also essential to sustain shared ecosystems and resources.

In sum, national parks are not mere playgrounds. The parks embody ideals that reflect the American public at its best, not at its most commonplace or self-indulgent. The Organic Act directive to conserve unimpaired for future generations captures that ideal, importuning us to enjoy the parks on nature's terms, not just on our own. Whether we want that engagement to take the form of snowmobiles rather than cross-country skiing is a policy choice only partially answered by the Park Service's nonimpairment mandate, however. It provides an important starting point—but not always the end point—in the national park recreation debate. The array of intangible values integral to the national park experience—silence, solitude, self-reliance, and personal reflection—are equally important in determining appropriate recreation policies. Moreover, a broad vision that extends beyond national park borders to address recreation issues at a landscape scale enlarges the management options and reduces the pressures on park lands. With these fundamental values firmly in mind and a broad strategic vision, park managers have the authority to ensure that the national parks do not become just another playground.

CHAPTER 5

"A Commercial Commodity"

Putting Nature on Sale

As tourist destinations and recreational meccas, the national parks are imbued with a distinct commercial overtone that can be traced to park concessioners and nearby communities. From the beginning, the national park designation has served as a beacon for entrepreneurs who have viewed park visitors as a resource to be exploited. Initially, local residents stepped forward to provide intrepid, nineteenth-century wilderness adventurers with accommodations and food, only to see their efforts soon displaced by outside corporate entities eager to exploit a burgeoning captive market. Whether their commercial activities were centered inside the parks as concessioners or outside them in gateway communities, each recognized the park as the proverbial "goose with the golden eggs," and each has long been deeply entwined with the making of park policy, yet another instance in which the parks cannot be divorced from their larger surroundings. The Park Service, unschooled in the commercial marketplace, has long looked to private business to help meet its visitors' needs, creating sometimes unholy relationships with adverse ramifications for park resources and values. The public interest in nature preservation and the private interest in profit are rarely in harmony.

CONCESSIONERS AND GATEWAY COMMUNITIES

The remote early national parks, with all their wilderness attributes, were not initially hospitable settings for tourists. Most early park visitors, having traveled long distances by rail, coach, and horseback to view these splendid places, required lodging, sustenance, and other services once they arrived.

The few residents who were on the scene usually obliged; they constructed often-primitive accommodations, livery stables, and the like, eager to capitalize on the nascent tourist trade. In the case of Yosemite, once the federal government transferred ownership of the valley to the state of California in 1864, local residents not only laid claim to choice parcels but also constructed a disparate array of buildings, roads, irrigation ditches, agricultural fields, and other facilities so that they could offer lodging and services to arriving visitors. Congress, having decreed that Yosemite Valley would be used for "public use, resort, and recreation," ultimately denied their land ownership claims, but did authorize ten-year leases to private entities. The newly created state park commission readily issued leases, disregarding Frederick Law Olmstead's recommendations, which urged limiting construction in the valley "within the narrowest limits consistent with the necessary accommodation of visitors" to ensure "the preservation and maintenance as exactly as is possible of the natural scenery." Yosemite Valley soon took on the tawdry appearance of Niagara Falls, Arkansas's Hot Springs spa, and other early national tourist destinations, becoming a ramshackle collection of lodges, stores, eateries, and other crassly commercial establishments.[1]

Early National Park Concessions

With the designation of Yellowstone as the first national park in 1872, Congress confirmed that private enterprise was welcome in these new nature reserves. The Yellowstone legislation included a provision authorizing the secretary of the Interior to "grant leases for building purposes for terms not exceeding ten years, of small parcels of ground, at such places in said park as shall require the erection of buildings for the accommodation of visitors." Although local merchants already offered rudimentary accommodations, financial backers of the Northern Pacific Railroad Company—an important proponent of the seminal Yellowstone legislation—took full advantage of this provision and promptly established a virtual monopoly over lodging facilities within the park. Although denied the right to extend their rail line into the park, Northern Pacific's promoters, with strong political support from their own senators, secured a concession contract from the secretary of the Interior allowing them to build hotels at Mammoth Hot Springs, Old Faithful, and elsewhere for the nominal annual rent of $2 per acre. The arrangement, decried even then as an unwise monopoly by Yellowstone's superintendent, heralded the advent of the Yellowstone Park Improvement

Company, which served as the park's major concessioner for nearly a hundred years. It also established the principle that private enterprise, in the form of park concessioners, would play an important role in visitor services in the national parks and thus in shaping national park policy.[2]

A similar pattern played out in other early national parks. At Glacier, despite objections from local businesses that were only offering lackluster accommodations on the park's periphery, the Great Northern Railway Company secured the right to provide lodging within the park. Under this concession contract, the railway proceeded to construct the classic lodges at East Glacier, Many Glacier, and Waterton and to purchase Lake McDonald Lodge, providing park visitors (who were primarily its rail customers) rather luxurious accommodations in this wilderness setting. At the Grand Canyon, even before it achieved national park status, the local entrepreneurs serving the canyon's growing number of visitors found themselves increasingly marginalized once the Atchison, Topeka and Santa Fe Railroad completed a rail spur to the canyon rim. In alliance with the Fred Harvey Company, the railroad used its financial and political clout to secure the land needed to build El Tovar lodge on the South Rim in 1905. With the Forest Service's support, these two corporate giants soon brought the handful of local businesses under their wing, transforming the canyon rim into an enticing but ever more commercialized tourist destination. The Park Service, once it took over the canyon, confirmed the monopoly arrangement with new concession contracts, giving these powerful national corporations effective control over most visitor services.[3]

As their national park connections grew, the railroads aggressively pursued their tourism-based financial interests in various forums. In a display of their influence and importance, railroad executives regularly attended the annual national park conferences that preceded formal creation of the National Park Service, unabashedly promoting an expansive tourism agenda for the new parks. At the first conference, convened in 1911, the president of the Northern Pacific Railroad Company noted that his company had invested literally millions of dollars in Yellowstone constructing lodges and other tourist facilities, whereas the government had yet to make any significant financial investment in the park. The Southern Pacific Railroad Company, seeking to attract visitors to the new California national parks that it had helped create, founded *Sunset* magazine as a vehicle to promote park visitation. The railroads also helped conceive the "See America First" campaign, designed to lure wealthy tourists away from Europe by extolling the new parks' awesome scenery. And during the seven-year campaign for the comprehensive Organic Act legislation, the railroads played a key role advocating for the proposed park system and oversight agency.[4]

Mather and the Private Sector

After Stephen Mather was appointed to oversee the fledgling national parks, the railroads and other national corporate interests were practically assured an expanded role in the parks. A successful businessman, Mather had an abiding faith in private enterprise, which he was prepared to enroll on behalf of the national parks to correct the shoddy conditions he had encountered during earlier park visits. In fact, Interior secretary Franklin Lane offered Mather the Park Service directorship after he complained about conditions in the national parks: "Dear Steve, If you don't like the way the national parks are being run, come on down to Washington and run them yourself." As one of his first acts, Mather hired newspaper editor Robert Sterling Yard to churn out publicity about the parks to promote visitation, a decision that complemented the railroads' already considerable advertising efforts. When Congress finally acted in 1916 to adopt the National Parks Organic Act, it employed language from earlier park statutes to empower the Interior secretary to "grant privileges, leases, and permits for the use of land for the accommodation of visitors in the various parks . . . for periods not exceeding thirty years," but enjoined lessees from "interfere[ing] with free access to [the natural wonders] by the public." The organic legislation also directed the secretary to "promote" use of the new parks, an admonition that aligned well with the corporate interests that had already staked their claim to visitor service opportunities in the existing parks.[5]

Once Congress established the National Park Service to oversee the new park system, park concession arrangements soon shifted from ad hoc, locally negotiated agreements to more standardized contracts governed by system-wide rules. Viewing tourist services as vital to attracting Americans to their national parks, Mather was intent on ensuring reliable, affordable, and quality facilities that catered to a range of visitors. Because neither the new Park Service nor the federal government had any experience running hotels or other tourist services, the private sector—already well established in the existing parks—was the obvious choice to meet these needs. This arrangement coincided neatly with Mather's view that the national parks would serve as engines of growth for western communities. Or, stated differently, that the parks constituted an essential commodity that could be exploited for local economic gain.

The 1918 Lane letter sought to reconcile these apparent conflicting perspectives, instructing the Park Service that resource protection was its first priority and that "the national interest must dictate all decisions affecting public or private enterprise in the parks." Calling for accommodations ranging from low-priced camps to luxury hotels, the Lane letter endorsed a concession system for the parks, but noted that concessioners, faced with large initial investments and regulated rates, "must be given a large measure of protection" from competition. It also admonished Park Service officials to work with local chambers of commerce and tourist bureaus to help increase park visitation, thus giving the new agency a promotional role not unlike the one already assumed by the railroads, its concessioners, and nearby community businesses.[6]

To ensure reliable facilities and services, Mather concluded that park concessions should be treated as regulated monopolies, an idea that Fred Harvey had proposed at the initial 1911 national park conference. Mather thus aligned himself with the emergent corporate concession interests and entrusted park tourist accommodations to the private sector, which was protected against rampant competition and practically guaranteed a return on its investment. Indeed, Mather was convinced that corporate concessions, rather than the hit-and-miss local operators he had encountered during earlier park visits, would provide higher quality and more dependable services to the visiting public. The Park Service's concession policy was not transparent, however; contracts were negotiated in secret, where political connections could be as important as a proven track record, and neither the contract terms nor the resulting revenues were disclosed. Although the agency expected some financial return from these concession arrangements, its primary goal was to provide visitors reliable and affordable accommodations.[7]

Not surprisingly, local entrepreneurs who had initially capitalized on the new national parks were gradually displaced, with the Park Service's blessing, but not without controversy. At Rocky Mountain National Park, after the Park Service granted exclusive in-park transport rights to one company, other local businesses offering this same service challenged its decision. Hotel owner Enos Mills, who had played a critical role in helping establish the park, commenced a lengthy court battle over the concession arrangement, supported by other outraged local entrepreneurs. One civic group condemned "the present transportation concession . . . as monopolistic, unnecessary, unjustifiable, unlawful, unjust, unreasonable, undemocratic, un-American, corrupt, vicious, and iniquitous . . . autocratic favoritism . . . an alliance of bureaucratic politicians and profit-grabbing special interests." And one local businessman voiced his opinion by posting a notice in his shop window warning about "the possibility of the Prussianized control of National playgrounds." The Park Service prevailed in the end, however. Local resistance to the arrangement gradually faded, although hard feelings lingered among local business owners who felt entitled to benefit from the park's presence. At Mount Rainier National Park, Mather arranged for a group of Seattle and Tacoma business owners to get the concession contract for the park's hotels, disregarding objections from more-local businesses that already provided services in the park and coveted the contract.[8]

The Effect of the Automobile

Once the automobile appeared and park visitors began arriving in their own private autos, the railroads soon waned in importance. Not only did the automobile undermine the railroad monopoly, but it changed the relationship between the national park and nearby communities. Whereas early visitors, after enduring long train rides, spent their time and money at railroad-owned park hotels, the new auto tourists had much more flexibility in where they stayed and how they saw the parks. Rather than encamping at a concessioner-run park lodge, they could choose a local hotel (or motel) and frequent local eateries and shops, or they could stay at one of the Park Service's auto-friendly campgrounds and buy their supplies outside the park. These new auto-driven realities, according to historian Hal Rothman, broke the hegemonic hold that the railroads and allied national corporations held over national park tourism.[9]

As a result, communities situated near the national parks—eventually dubbed "gateway communities"—discovered that they could compete with park concessioners for the new auto tourism business. But just as the early national park concessioners were driven by a profit motive, the same held true in the gateway communities. For local businesses, the key was to capture the visitors to ensure that their money was spent in town rather than in the park. Because most towns could not compete with the natural splendor or recreational allure of the parks, they offered visitors other attractions, ranging from reptile farms and wax museums to Old West–style shoot-outs and souvenir shopping opportunities that often clashed with the nature-based experience that the parks offered.

The advent of gateway communities, however, created new challenges for the nearby national parks, which had little say over what happened outside their borders. Rothman chronicles the rise of White's City, New Mexico, situated on the road to Carlsbad Caverns National Park, where local entrepreneur Charlie White developed a three-hundred-room motel, the Million Dollar Museum, and an assortment of other commercial establishments. To promote his enterprise, White planted thirty billboards along the highway to the park. Though White's commercial blandishments were a jarring contrast to the Park Service's natural history message, White's City regularly attracted hordes of visitors during the summer tourist season. Because White owned the land beneath his buildings, the Park Service had no control over how he conducted his business, highlighting a recurrent problem the agency has confronted with its gateway neighbors. Outside park boundaries, the agency's jurisdictional authority is limited if not nonexistent. Although the Park Service can essentially dictate how its concessioners run their businesses, its neighbors enjoy near carte blanche in how they conduct themselves and their businesses.[10]

Perhaps predictably, Mather's commitment to promoting park visitation fostered a strong—and sometimes unhealthy—relationship between the agency and many of its concessioners. Motivated by the financial bottom line, park concessioners regularly sought to increase their customer base, frequently cajoling park officials to approve new facilities or activities designed to entice more visitors. Yosemite's concessioner-driven firefall and evening bear-viewing spectacles were two notorious examples, rivaled by its decision approving a concession-driven winter sports complex in an effort to secure the 1932 Winter Olympics. Most parks supported on-site conventions as well as souvenir sales, many involving Native Americans, even though the reality of these commercial ventures was often quite different than the image. In a presentation to his fellow park superintendents, the outspoken Sequoia superintendent, Colonel John White, warned: "In their desire . . . to make proper returns on their investments, the [concession] operators may easily damage park atmosphere . . . [with] cheap vaudeville shows designed merely to entertain the average visitor unappreciative of nature." He also condemned the trend toward in-park conventions, pointedly noting that the "park is not in competition with other resorts." And he deplored in-park curio sales, asserting "that much sold in Sequoia is atrocious."[11]

These mounting commercialization concerns found a friendly ear with Interior secretary Harold Ickes, who favored a government takeover of park concessions. At the 1934 park superintendents' conference, Ickes observed that the parks face "pressure from the concessionaires" and then continued: "I do not want any Coney Island. . . . We must forget the idea that there is competition between the parks and the seaside or mountain resorts or that we must have share of the trade. . . . I wish we had the statutory power and the money to take over all of these concessions and run them ourselves." Consistent with these sentiments, the new eastern national parks created during Ickes' tenure—Great Smoky Mountains, Shenandoah, and Mammoth Cave—were each designed without in-park lodging; instead, the plan was for the surrounding communities to supply visitor accommodations with only daytime visitor services in the parks. The arrangement, while diminishing the park concessioner's role, greatly enhanced the connection between the parks and gateway communities, which has not always proven to be a positive one.[12]

Early Reform Efforts

By the time World War II had drawn to a close, concession reform efforts were under way. During the prewar years, according to historian Paul Sutter, the advent of automobile travel had succeeded in both democratizing and commercializing nature, a development reflected in the escalating numbers of middle-class auto tourists visiting the parks and needing accommodations. In 1950, concerned about these trends, Interior secretary Oscar Chapman issued a policy statement that "permit[ted] the development of accommodations within the [parks] . . . only to the extent such accommodations are necessary and appropriate for the public use and enjoyment of the areas, consistent with their preservation and conservation. Where adequate accommodations exist or can be developed by private enterprise outside of such area, accommodations shall not be provided within the area." Six years later, however, faced with an unremitting onslaught of visitors and more on the horizon, the Park Service shifted course to pursue its new Mission 66 agenda, which director Conrad Wirth conceived to rehabilitate and expand obsolete park lodging while also "encourag[ing] private business to build more accommodations in the gateway communities near the parks."[13]

The gateway communities generally welcomed the Mission 66 program and the idea of locating visitor accommodations outside the national parks. With park visitation surging in the postwar period, gateway businesses could expect real economic returns along with new business opportunities. At Yosemite, for example, the east-side towns of Lee Vining and Bishop strongly endorsed the Park Service's decision to upgrade Tioga Pass Road, believing that improved travel conditions would not only bring more visitors to their communities but also expand the visitor season. At Yellowstone, the Wyoming legislature, motivated by a study revealing the financial benefit of visitor spending, passed a law empowering the state to buy and operate the park's concessions. At Everglades National Park, local business interests succeeded, over the Park Service's objection, in promoting development of tourist facilities at Flamingo inside the park, believing that it would attract more visitors who would also patronize their outside-the-park businesses. The national parks were coming to be seen as "a great regional cash factory."[14]

The Mission 66 program helped uncover serious problems with the Park Service's concession program, prompting a major congressional overhaul. As Mission 66 unfolded, park concessioners pushed to expand and upgrade their facilities to accommodate even more visitors, particularly wealthy ones. Critics complained that additional development would imperil park resources and that high-end accommodations were pricing middle-class visitors out of the parks, problems exacerbated by the lack of a coherent national concession policy. In response, Congress adopted the Concessions Policy Act of 1965, intended to provide concessioners with sufficient security to enable them to finance needed improvements to their properties while enabling the Park Service to ensure that they were providing adequate and affordable services while protecting park values. As a policy matter, the legislation called for "carefully controlled safeguards against unregulated and indiscriminate use, so that the heavy visitation will not unduly impair these values and so that development of such facilities can best be limited to locations where the least damage to park values will be caused." It also cautioned that new "development shall be limited to those that are necessary and appropriate for public use and enjoyment . . . consistent to the highest practicable degree with the preservation and conservation of the areas." To achieve these goals, Congress granted the Park Service clear regulatory authority over concession activities. In addition, the legislation instructed that revenue production was subordinate to protecting park values and providing visitor services at reasonable rates.[15]

Congress, though, acutely aware of concessioner concerns, also confirmed the regulated monopoly features of the existing concession system, giving concessioners a "possessory interest" in their properties and preferential renewal rights. These provisions effectively safeguarded existing concessioners against termination and precluded any meaningful competition. Although the courts consistently acknowledged that the Park Service had broad discretion in overseeing its concessioners, agency officials only rarely employed that authority to limit their operations. Pointing to instances of excessive profiteering and undue influence over park policy, critics were soon regularly complaining about the cozy relationship concessioners enjoyed with the agency. As evidence, they noted that concession contracts were secretive bilateral arrangements negotiated without any opportunity for public input. With these legal protections in place, however, the Park Service was not about to incur potentially crippling financial liability by ousting concessioners, absent abject neglect or unbridled exploitation. That situation occurred during the late 1970s in Yellowstone, prompting a $20 million taxpayer funded buyout of the Yellowstone Park Company shortly after it was acquired by General Host, a national corporate conglomerate. The episode revealed just how entrenched the concessioners were, ensuring private enterprise an ongoing role in the development and management of the parks.[16]

The Park Service's Mission 66 building binge, strongly abetted by its corporate concessioners, crystallized the view that the national parks were at risk from the unabated onslaught of park visitors. At Yosemite, the Music Corporation of America, which became the park's principal concessioner during the early 1970s after acquiring the Yosemite Park and Curry Company, unashamedly pressured the Park Service to expand its luxury lodgings and parking areas and to build an aerial tram from the valley floor to Glacier Point, all to attract more potential customers. A 1972 Conservation Foundation report highlighted the growing problem: "The concessioner has a disproportionate influence on planning and policy-making for the national parks. His objective is to generate as much demand for the services he provides as is possible. . . . [which] too often brings people to the parks for the wrong reasons. The predictable result is that the concessioner makes a case for further facilities to accommodate a market that he—not the parks—has created." A 1974 General Accounting Office report confirmed the problem, finding that the national parks, at concessioners' behest, had hosted 174 conventions, more than a third of them during the peak season. The answer, according to the Conservation Foundation, was to eliminate private concessions from the parks in favor of "non-profit, quasi-public corporations whose primary allegiance is to appropriate public use of the parks." To avoid a resort environment, the foundation recommended locating visitor facilities outside any new or smaller parks while limiting in-park accommodations at the larger and older parks to those necessary to enjoy the park itself. In addition, to further reduce visitor impacts, the report advocated eliminating automobiles from the parks. The challenge, simply put, was how to deal with "industrial-strength tourism" promoted by "conglomerate concessionaires."[17]

Economics and Ecology

By 1980, however, in-park management concerns were giving way to external concerns, bringing the relationship between national parks and their neighbors into sharper focus. The Conservation Foundation's 1972 report had not only observed that external activities were adversely affecting park resources, but also urged more coordinated regional planning to help protect the parks. The mid-1970s Redwood National Park controversy, precipitated by extensive upstream logging that was damaging the park's namesake Redwood groves, further focused the problem. In 1980, the Park Service released its seminal *State of the Parks* report, describing the inherent ecological connections between the parks and neighboring lands as well as an incredible array of threats the parks faced from these lands. The report served as a call to action for the Park Service and its allies to strengthen the agency's hand in dealing with adjacent communities and landowners. Confronted with mounting energy development, clear-cut logging, and subdivision pressures, the national parks could no longer look on their neighbors as benign presences; rather, park officials would have to begin asserting their preservation obligations more aggressively. Whereas local communities had long understood that economic concerns connected them to the nearby national parks, it was now evident that the two were also conjoined by ecological ties, which were as essential to the parks' welfare as the tourist trade was to the gateway communities.[18]

These emergent external concerns were soon relegated to the back burner, however, as park concessioners and gateway businesses saw their stock rise. The 1980 election that brought Ronald Reagan to the White House also brought his free-market economic philosophy to the fore. Reagan appointed James Watt as his secretary of the Interior, and Watt promptly moved to refocus Park Service policy on meeting visitor needs while downplaying environmental problems. In his inaugural missive to the Park Service, Watt baldly stated that "the concessioner is essential to the national park experience," admonishing park managers to work with them to ensure reasonable profits and to upgrade park visitor facilities. Watt also instructed park managers to focus on in-park problems and not matters involving local communities outside park boundaries. During Watt's tenure, the Park Service regularly bent to gateway community pressures, as in the case of Yellowstone's reversal of its Fishing Bridge campground closure decision in deference to nearby Cody, Wyoming, business interests who feared that the closure would deter visitors from using the park's eastern entrance. Deeply committed to expanding visitor services and the private-sector role in the national parks, Watt disregarded the corrupting influence of creeping commercialism, and he used his political clout to promote these goals.[19]

Watt's priorities squarely raised the question of whether—and how—the Park Service could reconcile its resource protection and visitor use mandates. Although the Organic Act authorized concession leases to meet visitor needs and the Park Service had long entered into these arrangements, it was not required to provide lodging or other resort-style amenities for its visitors. In 1985, increasingly concerned about park resource conditions, the Conservation Foundation issued another report further cataloguing the negative effects of growing commercialization within the parks and lamenting the concessioners' influence on park policy. According to the report, gross concessioner receipts nearly doubled, rising from $189 million in 1977 to $340 million in 1983, while pretax income grew by 65 percent, going from $18.5 million to $30.6 million. It also disclosed how concessioners had pressured park officials at Yosemite, Crater Lake, and Zion to reverse key management decisions. At Zion, for instance, a politically well-connected concessioner actually stopped the Park Service from removing deteriorated visitor cabins from the scenic canyon area, even though the agency owned the cabins. The report, while acknowledging that it would be impractical to eliminate all accommodations from the parks, urged removing structures inappropriately located near major features and strengthening the Park Service's oversight role. Notwithstanding the Park Service's considerable authority over concession operations, buying out concessioner interests, given the property rights that obtained under the 1965 Concessions Policy Act, was financially infeasible as well as politically sensitive. If, however, the national parks were not to become mere commodities, then it was necessary to address the twin threats of unbridled commercialism and unbounded development, threats emanating from park concessioners and their gateway counterparts.[20]

Reform and Retrenchment

By the early 1990s, the idea of relocating lodging and other visitor facilities outside park boundaries was gathering momentum. The Park Service's "Vail Agenda" report not only called for "use and enjoyment on the *park's* terms," but recommended that "facilities . . . purely for the convenience of visitors should be provided by the private sector in gateway communities." Citing the growing litany of threatening activities originating outside the parks, the report urged that "the prevention of external and transboundary impairment of park resources and their attendant values should be a central objective of Park System policy." At the same time, environmental groups, having long highlighted the perils posed by logging and energy development on the periphery of the national parks, put forth the argument that adjacent communities could profit by reorienting their economies away from the traditional extractive industries to new amenity-based economic opportunities—tourism, recreation, retirement living options, and the like—tied to the presence of national parks. The idea of relocating lodging and other visitor facilities outside the parks was consistent with this approach, although it presaged new tensions between national concessioners and local business owners who resented any outside corporate intrusion into the local tourism market.[21]

But this approach, even though it might reduce concessioner influence and pressure inside the parks, still conceived of the national parks as a commodity—or "anchor tenant"—to be exploited for profit. Moreover, towns with local economies built on recreation and tourism—as was plainly evident in places like Jackson Hole, the Flathead Valley, and Estes Park—were beset by rampant development pressures in the form of new roads, housing, subdivisions, and the like, all of which put additional environmental pressures on nearby national parks. Because these pressures emanated from private lands, the Park Service and its allies had fewer legal tools to address them than would be true if adjacent public lands were involved, where an assortment of federal environmental laws applied.

In the face of continued criticism, Congress finally enacted concession reforms in 1998 designed to promote real competition among concessioners while better protecting park resources. The National Parks Omnibus Management Act, which repealed the 1965 Concessions Policy Act, plainly states that park resource values take priority over visitor uses and limits park accommodations and services to those that are "necessary and appropriate." Gone is the preference renewal right system that effectively insulated concessioners from competition and enhanced their political clout, replaced by an open competitive bidding process. Gone, too, is the concessioners' possessory interest in their capital improvements, replaced by a new leasehold surrender interest provision that limits the Park Service's liability when terminating concession contracts, seemingly giving the agency more flexibility when dealing with its concessioners. A new "reasonable and appropriate" standard governs the rates that concessioners may charge, and franchise fees are retained by the parks rather than returned to the general treasury.[22]

Veteran major concessioners, who had long portrayed themselves as the "true champions of public access and the right of the people to use parks," objected vehemently to losing their favored position, but to no avail. A subsequent effort to overturn the Park Service's implementing regulations was soon rejected by the courts, but that did not stop Glacier's major concessioner from enlisting the Montana congressional delegation to request a legislative exemption from the new law. Just how this legislative overhaul will affect visitor services and the concessioner's role in national park policy remains to be seen, although the Glacier exemption incident suggests that local politics will continue to be a factor. Indeed, given the profit motive that drives private-sector decision making, a new commitment to resource protection is unlikely to displace recreational tourism as the market-oriented concessioner's principal concern.[23]

Although the relationship between the national parks and gateway communities has vacillated over the years, the inherent political nature of the relationship shows few signs of abating. Given the strong economic and other attachments between individual parks and adjacent communities, local business owners have not been shy about seeking political assistance from their congressional delegations whenever park policies might adversely affect them. The Wyoming delegation, as noted, has repeatedly responded to local complaints by aggressively attacking Yellowstone's snowmobile, wildfire, and wolf reintroduction policies. In doing so, these politicians have invoked the usual state sovereignty and property rights arguments, undeterred by counterarguments about the broader public interest, overcommercialization, ecological restoration, and fiscal prudence. The effect of such local congressional oversight can be significant, as journalist Michael Frome has observed: "Park superintendents walk on eggs. Each one knows that a congressman with clout, even a little, can bring him down." The challenge, in a political setting in which national parks are often regarded as "pure pork and plums," is to resist converting them into mere commodities, a fate that is entirely at odds with the Organic Act's nonimpairment stricture.[24]

MAKING PEACE WITH CONCESSIONERS

Concessions controversies have dogged the Park Service over the course of its long struggle to promote visitation while meeting its preservation responsibilities. In two high-profile cases, the agency's efforts to remove poorly located and decaying facilities to protect important park resources met quite different fates. At Sequoia National Park, after a frustrating fifty-year campaign to relocate concession facilities from the sensitive Giant Forest area, park officials finally succeeded in what has been hailed as a precedent-setting restoration effort. In Yellowstone, on the other hand, when park officials sought to relocate the Fishing Bridge facilities to restore important grizzly bear habitat, the proposal was blocked. In each instance, powerful private and local interests, invoking national park visitor concerns, mounted a vigorous campaign to downplay competing environmental concerns so as to protect their own commercial welfare.

Sequoia and the Giant Forest

Sequoia National Park, established in 1890 as the nation's third national park, takes its name from the giant sequoia trees it was designed to preserve. The jaw-dropping Giant Forest grove of sequoia trees, situated atop the park's main southern entry route, has long served as a major visitor attraction. Not surprisingly, the park's first major concessioner selected the Giant Forest as the site of its operations, putting forth a grandiose plan for a 76-room hotel, 125 cabins, and 775 tent cabins, in all, enough to accommodate three thousand visitors and eight hundred employees. The immodest goal was to make Sequoia the "greatest tourist attraction in the western United States." Before these plans were fully realized, however, the original operator was forced out of business, opening the door for Mather to entice Howard Hays, a longtime friend with Yellowstone concession experience, to take over the operations. Hays, along with his brother-in-law, George Mauger, proceeded to upgrade the existing Giant Forest facilities, creating "an imposing complex" that included Giant Forest Lodge and several hundred cabins, tent camps, and other structures. In their view, visitors savored the opportunity to sleep among the giant trees, just as John Muir had, making the Giant Forest complex vital to their business plan and profits.[25]

The legendary Colonel John White, who served more than twenty-five years as park superintendent, originally welcomed Hays and Mauger for the stability they brought to the park's concession operations, but as the concessioners pressed for more "pillows" and space, he began to question whether their operations among the giant trees fit with his preservation responsibilities. In 1931, faced with yet another expansion request, White said no, citing damage to the trees and the area's natural beauty, much to the concessioners' surprise. In fact, White told Hays and Mauger that they should prepare to move their facilities to another less sensitive location, referencing a study that showed that the buildings and human traffic were damaging the trees' shallow root system and thus endangering the park's namesake tree. Undaunted, Hays turned to his friend, Park Service director Horace Albright, who granted his expansion request over White's objections. Albright used the opportunity, however, to place specific overnight visitor limits on the lodge and the other nearby facilities, representing the first time the Park Service had ever "put a limit on tourism development in any of its parks."[26]

With the end of World War II and visitation to the parks accelerating again, White saw another opportunity to close down the Giant Forest lodging complex when the twenty-year concession contract came up for renewal. This time, he drew support from Park Service director Newton Drury, who also believed that the facilities needed to be removed from the grove. He also had more studies showing damage to the trees and recommending removal. By now, however, the concessioner had 180 structures in the Giant Forest and the Park Service owned 34 structures of its own, including its main visitor center and administrative office facilities. Moreover, still convinced that park visitors relished the opportunity to sleep among the giant trees, Hays and Mauger argued that they had incurred great expense rehabilitating their lodge and cabins in reliance on Albright's earlier decision that they could remain. As the contract negotiations dragged on, White became more intransigent. Hays once more used his connections with the superintendent's superiors, succeeding this time in forcing White's resignation, and he secured a twenty-year contract extension, with the proviso that removal of Giant Forest Village would be reconsidered in ten years.

Meanwhile, the Giant Forest situation continued to deteriorate. The postwar upsurge in park visitation prompted more traffic jams and overcrowding in this entry portal. When park officials discovered one of the iconic giant sequoias leaning dangerously toward some of the cabins, they had no choice but to remove the 2,222-year-old tree, representing the first time Sequoia's overseers had intentionally destroyed one of the primary natural wonders entrusted to their care. To many observers, the Park Service's commitment to visitation and its concessioners had triumphed over its preservationist responsibilities. Even as the Giant Forest concessioner situation persisted, however, the Park Service began to relocate some of its own facilities, including the visitor center and several campgrounds, to the Lodgepole area, a less sensitive location several miles farther into the park.

Curiously, when Sequoia revised its master plan in 1971, it did not contemplate removal of the Giant Forest concession facilities, even with the damage more apparent than ever. In fact, when the concession contract was renewed in 1972, it made no mention of removal. By then, Hays and Mauger had sold their interest to the Fred Harvey Company, which in turn sold to GSI in 1972, giving the Park Service a new corporate partner. In 1974, still plainly concerned with overcrowding and damage in the Giant Forest complex, park officials produced a draft Development Concept Plan that recommended relocating all overnight facilities to the Lodgepole site. Because the proposal would have significant environmental implications, the agency was now required to comply with the National Environmental Policy Act, which not only meant preparing an environmental impact statement but also public hearings on the matter. Unlike the earlier removal proposals, which were handled as closed-door negotiations between the Park Service and the concessioner, this proposal was both public and transparent. And this time, the new concessioner—GSI—did not object to removal except to flag the $11 million cost involved in moving its operations.

Over the course of the next decade, the Park Service's Development Concept Plan underwent a series of revisions and public hearings, with the agency eventually concluding that the Giant Forest facilities must be removed and the area restored. Despite recurrent funding concerns, the Park Service proceeded over the next ten years to eliminate most of the Giant Forest complex, demolishing 282 buildings, 24 acres of parking lots, a sewage treatment plant, and several miles of road. To address park visitor needs, the agency constructed new overnight accommodations six miles away at the less environmentally sensitive Lodgepole area. According to the park's historians, this removal effort succeeded where earlier ones had failed because the agency engaged the public in the process and ultimately secured its support for restoring the Giant Forest area. That scientific studies going back several decades documented ongoing environmental damage and that the new corporate concessioner did not oppose the relocation also helped bring the project to fruition.

The result is stunning. Today, the Giant Forest grove looks much like it did at the park's inception. Stately giant sequoia trees tower over a forest floor largely devoid of built structures, while park visitors—most of whom are unaware of the decades-long battle between the agency and its concessioners—find their accommodations several miles away outside this iconic setting. The now-resolved Giant Forest controversy stands as a powerful testament to the defining role that concessioners have played in shaping visitor facilities and experiences at the parks. It also suggests that a resolute Park Service, when steeled by the nonimpairment standard, supportive scientific studies, and an informed and engaged public, can successfully pursue large-scale ecological restoration projects, even in the face of strong economic forces and political opposition.

Yellowstone and Fishing Bridge

At Yellowstone, the Grant Village–Fishing Bridge saga had quite a different outcome from that at Sequoia. Originally conceived during the 1930s, the park's plan to develop new visitor accommodations on Yellowstone Lake languished until after World War II, when it was revived as part of the Mission 66 effort to upgrade and increase lodging facilities. As originally designed, the new Grant Village complex would contain seven hundred motel units, restaurants, stores, a gas station, a dormitory, and various other facilities for park visitors and employees. Although the project involved clearing virgin lodgepole pine forest adjacent to the lake, no one yet realized how important the location was to the park's grizzly bear population, which depended on the cutthroat trout spawning in the nearby streams as a seasonal food source. When the project began to fade again for lack of financing, Yellowstone superintendent John Townsley continued to champion it, convinced that more beds were needed to accommodate the park's growing visitation. The state of Wyoming also supported it, believing that new accommodations situated on the road running north from the Jackson

Hole area would attract more tourists to the park's southern entrance and thus boost local businesses.[27]

In 1974, the park completed its master plan, linking construction of Grant Village with removal of the aging Fishing Bridge facilities on Yellowstone Lake's northern shore, a short distance around the lake from the Grant Village site proposal. Park officials believed that removing the Fishing Bridge facilities would help the park's dwindling grizzly bear population, which had plummeted following closure of the park's garbage dumps. Citing the rising number of bear mortalities and incidents near the Fishing Bridge complex and the area's important habitat value for bears, the plan called for eliminating 100 dilapidated cabins, 310 campground sites, 360 recreational vehicle overnight sites, the park's first visitor center, the historic Hamilton Store building, and several other buildings. With the grizzly bear protected under the Endangered Species Act, the Park Service was required to consult with the U.S. Fish and Wildlife Service, which reluctantly issued a biological opinion approving the Grant Village project conditioned on removal of the Fishing Bridge facilities. In other words, any new visitor accommodations on Yellowstone Lake were contingent on cleansing the human presence from the Fishing Bridge site to safeguard the park's grizzly bears. Whether it was a good trade-off for the grizzly bear, given the lost spawning stream habitat values at the Grant Village site, is open to question because the arrangement never came to fruition.

Once the Fishing Bridge closure plan was announced, the town of Cody, Wyoming, erupted in opposition. According to town officials, without the Fishing Bridge accommodations, park visitors would bypass their eastern gateway community in favor of another park entry point, costing the local economy substantial revenues. With assistance from the Wyoming congressional delegation, Cody officials convinced the Park Service to modify its closure decision, sparing the 360-unit recreational vehicle campground, the Hamilton store, and the visitor center. An environmental lawsuit challenging this change in plans was dismissed by the local federal court,[28] meaning that Grant Village and the remaining Fishing Bridge facilities would both remain in place, notwithstanding their aggregate effect on grizzly bear habitat along the lake's northern shore. Meanwhile, the Park Service finally found a partner for the Grant Village project in TW Services, a corporate conglomerate with other park concessions. Although TW Services agreed to operate the motel and related businesses, it was not obligated to invest in the construction project itself, leaving the Park Service as the project's chief financier.

Few people have anything kind to say about the dreary pine forest location or the unattractive box-like buildings that are today Grant Village, a dismal reminder of the construction excesses spawned by the Mission 66 initiative.

The entire controversy yielded several discordant lessons in national park concession policy, community relations, and resource protection. First, even in the absence of a concession partner, the Park Service—when convinced that additional accommodations were needed—proved quite willing to ignore or discount contrary financial, ecological, and even legal concerns to push forward a visitor-focused development project that raised serious wildlife and aesthetic concerns. Second, when it comes to building or removing visitor accommodations in a national park, local communities and businesses often have the political ability to thwart even the best-intentioned plans. Third, although the available scientific evidence on grizzly bears supported the Fishing Bridge demolition proposal, subsequent developments—most notably a perceptible upswing in the park's bear population—have called that initial understanding into doubt, reinforcing the dynamic nature of the park's ecology. Further, despite the new Grant Village complex, Yellowstone historian Paul Schullery notes that the overall development footprint devoted to the park's visitors has actually diminished over the past hundred years. In fact, few people expect any major new in-park accommodations to be built in the foreseeable future.[29]

THE "GLITTER GULCH" SYNDROME

Uneasy relationships between individual national parks and adjacent gateway communities are more common than not across the park system. On the one hand, gateway communities historically have lived off the national parks while frequently showing little concern for park resources or for establishing a community presence compatible with the presence of a national park. The result is often glitzy and haphazard development, the loss or fragmentation of critical wildlife habitat, and a degraded scenic environment. On the other hand, the Park Service has often proved a high-handed and intrusive neighbor with little regard for local economic needs and even less engagement in community life. In short, the two entities have frequently found themselves at odds over local economic development efforts and other proposals, some of which could yield mutual benefits. These problems have been particularly evident outside Rocky Mountain, Great Smoky Mountains, and Grand Canyon National Parks.

Rocky Mountain and Estes Park

The town of Estes Park, Colorado, adjacent to Rocky Mountain National Park, established its tourism-oriented identity before the park was created. With its breathtaking mountain backdrop and the landmark Stanley Hotel, the town's early leaders focused on attracting tourists to their community, readily supporting creation of the park in 1915 as a means to promote visitation. They soon joined with park officials to develop local winter sports venues, including a ski area and skating rink in the park, hoping to extend the traditional summer tourism season for area lodges, but town and park officials did not always see eye-to-eye. During the 1930s, the town embraced the Bureau of Reclamation's Big Thompson Dam project, part of a larger project designed to transport water by pipeline through the mountains from the park's west side to the growing Front Range. With the town still recovering from the Great Depression, the project meant construction jobs and a new bureau headquarters building, so town officials discounted whatever effects it might have on the park. Park officials, however, strongly opposed the entire project, which they feared would set a bad precedent and scar the natural setting, and they eventually secured concessions that limited the project's effect on the park itself. Conversely, when park officials sought to expand the park's boundaries to address wildlife management concerns, the town routinely opposed these proposals, lamenting the loss of tax revenues and future development opportunities.[30]

In the aftermath of World War II, as park visitation mounted across the country, Estes Park businesses redoubled their efforts to attract new visitors and to prolong their stay. The business strategy—one routinely employed by national park gateway communities—was to encourage visitors to "eat, stay, and shop," bolstered by a phalanx of increasingly gaudy curio shops, amusement centers, and generally haphazard development. In Park Service vernacular, Estes Park (along with other gateway towns) was pejoratively dubbed "glitter gulch": an aesthetically unappealing community setting fundamentally at odds with the natural surroundings that attracted most people to the area. As commercial development expanded, so, too, did traffic congestion, often exacerbated by the park's cumbersome entry process. With the town and the park moving in opposite directions, the relationship between them soured even more. Rather than working together to address mutual concerns, they pointed fingers at each other, each blaming the other for their woes.[31]

By the late 1980s, however, it was even more evident that the park and its supporters could no longer ignore the town or its expansionist tendencies. Like most parks, Rocky Mountain National Park is not a complete ecosystem. Although elk and other park wildlife may spend their summers in the park's high mountains, they seasonally retreat outside the park for crucial lower elevation winter habitat. Estes Park real estate and other nearby private lands, however, were rapidly being subdivided and gobbled up for new homesites to accommodate an influx of retirees and other newcomers, who swelled the town's population by 35 percent during the 1980s. The opportunity to own property abutted by the park offered these new arrivals an "endless backyard," but their growing presence also fragmented critical habitat and severed migration corridors, putting the park's wildlife at ever-greater risk. Unaccustomed to living with migratory wildlife and the ever-present threat of summer wildfires, these newcomers routinely questioned park resource management policies, entreating park officials to protect them from the very natural environment that originally drew them to the area. Clearly, a more coordinated strategy to address both the town's and the park's needs was necessary.[32]

The relationship began to change during the early 1990s, prompted by more enlightened leadership on both sides. Local officials, tiring of the town's glitter gulch image that did not fit the surroundings, moved to upgrade the community and develop more compatible attractions, including a public golf course, hiking and biking trails, and a refurbished riverfront in the commercial district. They undertook a local visioning process that also involved the Park Service, while the park hired a land use specialist whose job was to work with its neighbors to reduce tensions and development impacts on the park. As superintendents and other managers have come and gone, the specialist has remained at the park, helping sustain trust between the park and the community. In 1995, Estes Park voters and other county residents passed a 0.25 percent property tax increase to support open-space acquisition, enabling town officials to purchase key wildlife habitat parcels. More recently, the Park Service has adopted an elk management plan designed to better control herd numbers, which has helped improve relations with town residents and neighboring landowners who were feeling overrun by these animals. Although the relationship is still a work in progress and serious transboundary air-quality and wildlife issues must still be addressed, Estes Park has begun to put park concerns into its development equation, and park officials are engaging more meaningfully with local concerns. With this improved coordination, the traditional gateway community image is receding, as are long-standing park-town tensions.[33]

Great Smoky Mountains and Its Neighbors

The glitter gulch problem is, however, still strikingly evident on the flanks of Great Smoky Mountains National Park, where gateway communities have long capitalized on the park's presence while pursuing commercial activities strikingly at odds with the area's natural surroundings. Indeed, the east Tennessee towns of Gatlinburg and Pigeon Forge are regularly hailed as extreme examples of tourism-driven gateway communities run amok. Gatlinburg traces its tourism heritage to the early twentieth century, when the area's once-thriving lumber industry folded and local residents began promoting the town's tranquil setting and scenic mountain backdrop. To help entice more visitors to the area, local entrepreneurs—with outside support from the Vanderbilt family and others—moved to revive a traditional mountain handicrafts industry, featuring weaving, quilting, and wood carving, even though these skills had long since faded from the local culture. The decision to recast the community as a quaint mountain haven wedded to a no-longer-existent past established a troubling pattern that still persists: money can be made by creating and marketing an image regardless of its connection to the surroundings or reality.[34]

When proposals for a new national park in the Great Smoky Mountains surfaced during the 1920s, Gatlinburg's community leaders readily supported the idea. Because hotels and other commercial establishments would not be allowed in the new park, they believed that the town's nascent tourism industry could capitalize on its prime location, situated not only adjacent to the proposed park but at the confluence of key roads running through it. Once the park was officially established in 1934, prominent Gatlinburg families, who owned much of the real estate in town, began building new hotels, restaurants, and other businesses catering to the growing number of visitors. The Park Service, with assistance from the Civilian Conservation Corps, built new roads and trails designed to better serve park visitors, disregarding complaints that it was opening too much of the fledgling park to automobiles. Park officials, evidently enthralled with the town's mountaineer heritage promotional message, incorporated these same themes into their vision for the new park, thus becoming a willing accomplice in perpetuating a false image of the area. After resettling the local inhabitants outside the park, the Park Service proceeded to restore the popular Cade's Cove area in this pioneer image, even though this small settlement had already progressed well beyond its early pioneer heritage.[35]

Like elsewhere, tourism soared at Great Smoky Mountains National Park in the aftermath of World War II, triggering a corresponding local economic boom and further transformations in Gatlinburg. As time passed, the resident families who had controlled the town and much of the tourism business were gradually bought out by outside entrepreneurs, including large national corporate chains in some cases. New construction aimed at attracting and accommodating even more visitors steadily reshaped the town, which eventually boasted a new convention center, a fifteen-story hotel, a space needle, an aerial tram, and even a ski area with a Bavarian motif. These development pressures pushed the town's boundaries ever outward, consuming more and more open space in the narrow valley and forcing development up the mountainsides against the park boundaries. In the course of the 1960s, an early effort to control development failed when the town created a regional planning commission but refused to give it any meaningful power, a reflection of the region's general conservatism and antipathy toward any land use regulation.

Throughout the postwar period, ever alert to new approaches for attracting visitors, the town abandoned any pretense of authenticity in its promotional efforts. During the 1950s, Gatlinburg's leaders sought to cash in on the public's interest in then-popular Hollywood images and television shows by seizing on a new hillbilly marketing theme. The town's mayor and other business leaders cheerfully assumed an unfamiliar "country bumpkin" identity and embarked on a series of auto caravans across the South to promote Gatlinburg as an authentic hillbilly haven. By the 1960s, with the nation engulfed in the civil rights struggle, another cultural craze—one related to Dixie and the Confederacy—swept across the region and was promptly incorporated into the town's promotional efforts. This new southern rebel thematic identity was no truer than the earlier ones, however: the mountain people of Gatlinburg and surrounding Sevier County identified with the Union, not the Confederacy, during the Civil War. Historian C. Brenden Martin sums up the town's evolution as "a classic case of interference in which outside interests restructured and re-interpreted regional culture for market appeal."[36]

Over the years, Gatlinburg's single-minded focus on capturing the local tourism market has permanently altered the appearance and character of this gateway community. What once was a quaint mountain village is now an overdeveloped commercial hub squeezed into a narrow mountain corridor that is dedicated to serving and entertaining visitors, with little apparent connection to the adjacent national park or the surrounding landscape. According to one perceptive assessment, "high-powered, high-volume tourism" has "transformed . . . [Gatlinburg] into an amusement park." Indeed, the town now features an astounding array of amusements and shopping opportunities, ranging from miniature golf, bungee jumping, and country music halls to factory outlet stores, wax museums, and western wear shops. The community's overwhelming commitment to tourism is reflected in its annual retail sales figures, which rose from $76 million in 1976 to $300 million in 1996. There are no longer any residential areas in Gatlinburg; rather, the town's housing consists almost entirely of rental properties and second homes. Most local jobs are in the tourism industry, offering low pay and only seasonal employment. And oversized commercial structures—including the space needle, high-rise hotels, and the aerial tram—obstruct scenic views from the town itself and mar park vistas. Simply put, the tourist industry dominates the town, relegating the park and its concerns to secondary importance.[37]

A few miles west of Gatlinburg, the town of Pigeon Forge has taken gaudy commercialism and rampant tourism to yet another level. Once a quiet agricultural community, Pigeon Forge embraced tourism during the 1960s after a major highway was routed through the town. With local residents controlling most of the land in nearby Gatlinburg, outside investors seeking a foothold in the area's booming tourist economy turned to Pigeon Forge, where land was still available and relatively cheap. Key major developments included the Rebel Railroad theme park built during the early 1960s, which was replaced by a large Wild West theme park and then by Dollywood, a country music theme park named after singer Dolly Parton, who was raised in the area. Dollywood quickly became Pigeon Forge's premier tourist attraction and soon spawned a bevy of look-alike competitors that transformed the town into a major entertainment center, abetted by an assortment of curio shops and other amusement establishments lining the main street. Dollywood and its competitors, with their pseudo trappings of local culture, allowed visitors to claim a mountain experience without actually visiting the nearby park or mountains.

In the early 1980s, the first factory outlet mall arrived, soon followed by others, and shopping now attracts more visitors to Pigeon Fork than does Great Smoky Mountains National Park. In fact, a visitor would be hard pressed to know that Pigeon Forge, festooned with entertainment centers, shopping malls, and fast-food joints, is a gateway community to the nation's most visited national park. Although the park and its mountain scenery once played at least some role in the community's economic vitality, that is no longer true. Hence, there is little local concern about the park or its welfare.[38]

That such rampant commercialism and unbridled development affects Great Smoky Mountains National Park and its visitors cannot be denied. Although park visitors can still enjoy nature once they enter the park, the park's viewshed is now dotted with visible built structures, and visitors cannot escape a carnival-like atmosphere once outside the park's boundaries. For many visitors who have been lured to the area by Dollywood and other commercial attractions, the park is just an afterthought; they may visit it briefly to enjoy a few scenic views, often without leaving their cars. What once was a unique local culture has long since given way to a mass commercial culture that bears no relationship to the national park experience. Moreover, subdivision and new construction outside the park has compromised traditional wildlife habitat and created new human-wildlife conflicts. The park's black bears, who depend on lower-elevation food sources to sustain themselves through the winter months, have found their migration routes severed, leading to more bears being killed by autos and frightened homeowners.

Even as park and town officials work to improve relations and curtail the worst development excesses, however, the park faces additional threats from beyond these gateway communities. Airborne pollutants—nitrogen oxide, sulfur dioxide, and ozone—that emanate from regional coal-fired power plants have severely impaired park vistas and vegetation, while also raising serious public health concerns. When automobile emissions linked to the area's booming tourism economy are added to the air pollution mix, the park finds itself awash in serious environmental challenges with roots that extend far outside its borders.[39]

Grand Canyon and Canyon Forest Village

The inherently parochial side of national park–gateway community relations was on full display at Grand Canyon National Park during the late 1990s. When the Park Service and others sought to reduce visitor pressures inside the park by supporting construction of new visitor facilities outside it, the proposal met sustained local resistance. In this instance, the gateway community of Tusayan, Arizona, derailed the so-called Canyon Forest Village project, which was designed to reduce serious overcrowding and traffic congestion problems. Park officials, allied with the Forest Service and several environmental groups, proposed constructing new overnight accommodations, additional visitor facilities, and a park transportation hub on nearby national forest lands outside the park's South Rim entrance. They were forced to abandon the project after opponents prevailed in a local rezoning ballot initiative, arguing that the environmental effects and local economic ramifications were too great. That the project would have substantially expanded and upgraded existing gateway facilities (widely regarded as subpar) and that the project was financed by a foreign investment firm virtually ensured that the local business community would oppose it. The decisive question had little to do with the park, its resources, or the need to address visitor pressures; instead, it was about local money and power. The incident stands as yet another lesson that shifts in park conservation policy, particularly those that affect gateway communities, must also take account of local economic concerns and not just park resource problems.[40]

RESOURCE PROTECTION AS GOOD BUSINESS

The Park Service, from its inception with businessman Stephen Mather at the helm, has been entwined with private enterprise in an ongoing effort to entice visitors to the national parks and to accommodate them once there. Today, that partnership involves not only large corporate conglomerates that run overnight facilities in the major national parks but also smaller concessioners and adjacent communities tied to the parks for their economic welfare. Despite their differences, they are all committed to building the tourist base, either in a quest for greater profits or for an enlarged budget and more funds. Whereas Mather actively enlisted the business community and adjacent towns in his promotional efforts, however, the Park Service now downplays its own promotional role, leaving the private sector to foster national park visitation in its own terms. In addition, agency policy no longer encourages more lodging or other visitor facilities inside the parks, deferring instead to gateway communities to provide these services. As a result, private enterprise now plays an even more prominent role in shaping park visitor expectations and experiences.

As much as one's idea of a national park may be defined by wilderness, beauty, or outdoor recreation opportunities, the stark reality is that the park visitor experience can rarely be divorced from commerce and the sense that the park itself is a marketplace commodity. Although most of the early resort-like excesses—tennis courts, golf courses, swimming pools, ski areas, and the like—have been eliminated, park concessioners are still driven by the same profit motives that prevailed during the early years, and their national corporate structure enables them to reach a broad domestic and international audience. Gateway communities and businesses, also deeply beholden to tourist dollars, have never limited their commercial or promotional activities only to those that are compatible with the national park setting, a point exemplified early on by White's City outside Carlsbad Caverns National Park and even today by Gatlinburg and Pigeon Forge. In short, the profit-driven pressures that motivate both park concessioners and gateway communities are quite real, intense, and universal.

The long-standing relationship between the Park Service and its concessioners remains uneasy, even in the wake of the successful Sequoia Giant Forest restoration project. The agency, committed by law to nature conservation as its first priority, must answer to the public interest; park concessioners, with their bottom-line concerns, are motivated by their own private interests and expansion opportunities. These interests collide over the Park Service's emerging policy of removing overnight accommodations from sensitive locations and relocating them to ameliorate ecological damage. Because the 1998 congressional reforms have reinforced concessioner property rights, the financial stakes are high whenever the Park Service contemplates closing or relocating facilities. And because the 1998 reforms did not fundamentally alter the bilateral nature of the concession contract negotiation process, the public has little opportunity to inject its interest into the process. Buoyed by these legal protections and their political allies, park concessioners continue to hold a strong hand in their dealings with the Park Service. As a counterweight to these forces, the Park Service must therefore be prepared to enlist the weight of public opinion whenever it contemplates limiting visitation or closing facilities, as Sequoia did so adroitly during the Giant Forest restoration effort.[41]

Of course, the Park Service–concessioner relationship can—and often does—operate as a true partnership. Consistent with Mather's original vision, the Park Service and its concessioners must continue to work together to provide visitors an array of accommodations, ensuring that the parks remain accessible and affordable to a broad cross section of the American public. The challenge going forward is to expand this relationship beyond its purely commercial bounds, creating a partnership that enhances the visitor's experience and connection to the park. One obvious way to do so is to incorporate an educational component into concession operations, one that is designed to educate the visiting public about the wonders they are exposed to during a national park visit. At Glacier Bay National Park, for example, Park Service interpreters accompany each cruise ship that enters the bay and explain to the passengers the significance of what they are seeing and the national park's role in protecting this unique place. Such a symbiotic relationship—one that injects an element of the public interest into the private-sector equation—gives additional credence to the important conservation obligations that take precedence in the national park setting. And it offers an opportunity to refocus the park visitor experience away from commerce and toward the world of nature.

The national park–gateway community relationship is also awash in complexity and controversy. As revealed in the disputes over Yellowstone's Fishing Bridge facilities and Canyon Forest village outside Grand Canyon National Park, gateway communities are acutely sensitive to park planning and resource management decisions and stand ready to challenge those that could affect their financial well being. Although gateway communities are firmly linked to nearby parks both economically and ecologically, the financial side of the connection generally predominates in this relationship. Any future effort to relocate or eliminate visitor facilities runs the very real risk of upsetting gateway businesses and their local political allies, regardless of the environmental benefits that may accrue. The same generally holds true, whenever park officials have sought to insert themselves into local planning or zoning matters in an effort to safeguard park resources.[42] Whether the national parks are perceived by gateway communities as a valuable asset or a meddlesome neighbor thus depends on establishing and maintaining a strong working relationship between park officials and community leaders.

There is, however, a certain irony in these clashes between the national parks and their gateway neighbors. Without the nearby national park, few gateway communities would enjoy the economic and other benefits that accrue from a relatively stable tourist industry. It is the presence of the park—not the community—that attracts most visitors, effectively making the national parks an "anchor tenant." Indeed, a 2011 study concludes that national park visitors spent $12.13 billion in local gateway regions in 2010 and that this visitor spending combined with Park Service payroll-related spending accounted for 189,000 local jobs in communities near parks.[43] Many community leaders recognize these realities and now acknowledge the critical role a park's presence plays in the local economy and quality of life. They may also acknowledge, sometimes begrudgingly, the ecological interdependencies between the national parks and surrounding lands. But too often they remain reluctant to curb local development projects that, although generating economic activity, can threaten these connections. The community of Estes Park, as we have seen, is gradually shifting from this single-minded focus on the economic side of the park-gateway community relationship, manifesting a broader understanding that the region's environmental well-being is vital to its longer-term welfare. Elsewhere, in places like Moab, Springdale, Jackson Hole, and West Glacier that abut a national park, a similar pattern is evident. As this more nuanced view of the gateway community's relationship with the national parks takes hold more widely, new opportunities to address environmental threats to park resources should arise.

At the end of the day, as reflected in the 1918 Lane letter instructions, the Park Service's relationship with its concessioners and neighbors must be guided by national, not local, concerns. The public interest in nature preservation must take priority over the private interest in profit. Whatever the economic consequences, the agency's principal obligation is to safeguard park resources from adverse effects, including those that originate in concessioner-run facilities and gateway communities. In short, the Park Service's relationships with these neighbors must be truly bilateral; the agency's diverse partners must acknowledge that the welfare of park resources merits the same attention as the economic bottom line, while agency officials cannot simply ignore legitimate local concerns. Without ecologically healthy national parks, concessioners and gateway communities would have little to sell, and the parks would lose much of their allure for visitors. The loss would be shared by the nation as a whole, confirming why the national interest must take precedence over local concerns whenever park resources are put at risk for purely economic reasons. A national park, after all, is a natural sanctuary, not a resort, playground, or mere commodity.

CHAPTER 6

"Ancestral Lands"

Nature, Culture, and Justice

Although the national parks are widely associated with the gateway communities adjoining them, the relationship between national parks and their American Indian neighbors is not as evident. With few exceptions, the early national parks were created without regard for competing Native American claims or concerns; entire tribes and families were routinely expelled from their ancestral lands, ironically, so as to protect these new nature enclaves from the taint of any permanent human presence. These original inhabitants did not stray far, however. By one account, nearly one-fourth of the national park units have a connection with Indian tribes, usually through a common border or established inholding rights. According to former Park Service director Russ Dickenson, there is not "a single major national park or monument today in the western part of the United States that doesn't have some sort of Indian sacred site."[1] Over time, once Indian tribes began to assert themselves in political and legal arenas, the Park Service has found itself confronting an increasing array of challenges linked to historic land claims, treaty rights, and sacred sites. Not only do these controversies raise important ownership, access, and social justice questions, but they also pose important questions about national park conservation policies that sharply separate people and nature.

A CHECKERED HISTORY

The evolution of the national park system is steeped in history, offering tangible evidence of the forces and attitudes that have prevailed during various eras in the nation's development. By any measure, the early national parks were a creation of the dominant Anglo-European culture that spread across the United States, imposing its will and values on the surrounding landscape and indigenous peoples. Not only were Native Americans routinely displaced to make way for new settlers, they also were dispossessed of their ancestral homelands in order to establish new national parks. The insult of this original banishment has not receded over time, as manifested in a steady assortment of tribal land claims, cultural site controversies, and treaty-based disputes involving the national parks. Other marginalized groups such as African Americans, Hispanics, and Appalachian mountain residents have also regularly found themselves standing outside the national park system. Although Park Service director Stephen Mather labored from the beginning to engage American citizens with the new parks, paradoxically, neither he nor his immediate successors extended this effort to all segments of the populace. As a result, the Park Service has found itself challenged to connect with Native Americans and other marginalized groups and to respond to their concerns.

In the Beginning

The Native American relationship with the parks has been difficult from the beginning. The concept of a national park was first articulated in 1832 by George Catlin, a frontier painter who foresaw a vanishing landscape as Anglo-American civilization advanced westward. After encountering the region's Indian inhabitants during a journey up the Missouri River, he envisioned a "nation's Park containing man and beast, in all the wild and freshness of their nature's beauty." Catlin's original vision never came to pass, however; federal Indian policy and national park policy took quite different paths. To make way for western settlement, the Indians were removed from their native lands and resettled on reservations, generally in out-of-the-way locations on lands that were deemed to have little economic value. The national park idea that took hold with the Yellowstone designation did not include people; rather, these new nature reserves were put off-limits to settlement or development activities, and any native inhabitants who occupied or used the area were removed and denied further access. This separation between people and nature has continued as a hallmark of national park policy and has served as a divisive matter in relations between the national parks and their Indian neighbors.[2]

The Native American experience with individual national parks shares several common features, although each setting presented its own unique circumstances. Without exception, the early legislation creating the nation's first national parks made no mention of existing Native American inhabitants or any provision for their continued presence in the new parks. In Yosemite, early park proponents not only ignored the valley's native inhabitants, who ironically were instrumental in creating the attractive open meadows with their light burning practices, but the Park Service also proceeded over several decades to evict the Indians once it took over management responsibility following passage of the Organic Act. In Yellowstone, despite clear evidence that local Indians regularly used the area for hunting and ceremonial purposes, park supporters propounded an erroneous "geyser taboo" myth—maintaining that the Indians feared the area's exploding geysers and boiling hot pots—that persisted across the decades. In Glacier, despite an 1895 treaty with the Blackfeet that expressly retained native hunting, fishing, and timber rights on ceded lands that were incorporated into the new park, Congress ignored these rights in the enabling legislation creating the park in 1910, and the Park Service has consistently opposed these tribal uses. Similar stories can be recounted at Mesa Verde, Grand Canyon, Mount Rainier, and elsewhere. Native Americans were simply written out of these early national parks, even as the parks were being created from Indian lands.[3]

To the extent that Indians were acknowledged at all during the early national park era, it was only at a few park sites such as Mesa Verde and, in retrospect, at various national monuments. The Antiquities Act of 1906 was specifically designed to help safeguard examples of Native American culture, including the cliff dwellings and other structures, rock art, and various artifacts that littered the Southwest but were rapidly being looted by private collectors.[4] In quick succession, presidents designated such places as Devil's Tower, Chaco Canyon, Rainbow Bridge, and Hovenweep as new national monuments and entrusted the Park Service with their safekeeping. As the designated guardian of these sites, the agency focused solely on the past and on preserving native artifacts and ruins, while only occasionally recognizing the historic Native American presence in the large natural parks. Agency officials were simply not concerned with protecting contemporary native cultures, whether that involved accessing native sacred sites, maintaining traditional ceremonies, or negotiating fishing, hunting, or other treaty rights. Nor were they inclined to seek out traditional native knowledge about the lands and resources now under their care. The Park Service's view of the Indian role in the national parks was simply as a matter of history; neither culture nor justice concerns were part of their relationship.[5]

Federal Indian Policy over Time

Over the years, while national park policy has stayed relatively constant, federal Indian policy has fluctuated radically with profound implications for the national parks. Historians divide federal Indian policy into several eras, each reflecting a quite different approach to the nation's native inhabitants. The period of conquest, which extended through the 1870s, was marked by heavy-handed treaty negotiations that created the Indian reservation system; most tribes were removed from their traditional homelands and resettled in remote areas on low-value lands. In 1887, hoping to promote Native American assimilation into the dominant Anglo culture, Congress adopted the General Allotment Act, which transferred some tribal lands to individual members who could then sell the parcels and which also allowed the federal government to dispose of the remaining—or "excess"—reservation land to non-Indian settlers. This new allotment policy hastened the process of land dispossession, effectively breaking apart many reservations and undermining tribal cultures, but doing little to ease the chronic poverty that plagued most Indian reservations. Coincidentally, these reservation and allotment policies mirrored the national park dispossession efforts that removed Native Americans from park lands and denied them any ongoing access rights.

During the 1930s, chagrined by the dire poverty that prevailed across Indian country, the federal government reversed course. In 1934, Congress passed the Indian Reorganization Act, ending the sale of reservation lands and calling for new tribal constitutions designed to promote self-determination and enhance self-respect. Although the administration of Franklin D. Roosevelt sought to improve the plight of Native Americans, the Great Depression and World War II dominated the nation's attention, undermining the administration's Indian policy initiatives. Once the war ended, Harry S. Truman's administration joined Congress to chart a new policy course to hasten, once again, the assimilation of Indians into the mainstream culture. This new "termination" policy was harsh; it explicitly sought to sell off tribal lands, withdraw federal support for Indians, and eliminate any semblance of a separate Indian culture. To implement the policy, Congress promptly passed an array of tribal termination acts that eliminated more than a hundred tribes with more than eleven thousand members and released more than 1.3 million acres from the reservation system.[6]

Most tribes opposed this new federal termination policy, fearing the loss of their reservation lands, treaty rights, and self-identity. They responded by adopting aggressive new legislative and legal strategies designed to maintain their cultural identity, combat age-old discriminatory policies, and assert long-dormant treaty rights. Native American leaders, drawing on the widening African American civil rights struggle as a model, turned to the courts and a newly sympathetic Congress to gain recognition of tribes as sovereign entities entitled to chart their own destinies. A series of U.S. Supreme Court and lower court decisions not only immunized tribal activities and members from state taxation and jurisdiction, but also protected tribal treaty rights, including the high-profile Boldt decisions (named for the federal judge who presided over the case) that recognized tribal fishing rights for salmon in the Columbia River basin. Congress, once sensitized to Indian concerns, responded with a plethora of new laws: the Indian Civil Rights Act of 1968, the Indian Self Determination and Education Assistance Act of 1975, the American Indian Religious Freedom Act of 1978, the Indian Child Welfare Act of 1978, the Indian Gaming Regulatory Act of 1988, the Native American Graves Protection and Repatriation Act of 1990, and the Tribal Self Governance Act of 1994. Taken together, these judicial and legislative milestones marked a new era of tribal sovereignty and self-determination, catapulting Indians into a lead role in matters affecting reservation lands, treaty rights, and historic cultural sites.[7]

A New Era Aborning

As this new era of self-determination has unfolded across Indian country, the relationship between tribes and nearby national parks has undergone a remarkable transformation, one that is manifest across the system given the tribes' historic association with the national parks. Flush with a reinvigorated sense of identity and a bundle of new sovereign rights, the tribes and their members have seized the initiative and begun asserting powerful legal and moral claims to access the national parks, to reclaim park lands, and to play a role in management decisions. Indeed, the strength of some of their claims begs the question whether the national park idea should also embrace the notion of parks as native homelands.

Originally silent on Native American concerns, national park policy now reflects this evolving relationship, acknowledging new tribal roles and individual rights. Of course, these policies are set against the backdrop of the extensive federal laws, court decisions, and executive orders that provide the framework for understanding Indian rights and tribal authority. Since the 1970s, every U.S. president has supported tribal self-governance and has issued formal Indian policies promoting it. President Clinton's Executive Order 13,175, for example, requires federal agencies to "respect Indian tribal self-government and sovereignty, honor tribal treaty and other rights, and strive to meet the responsibilities that arise from the unique legal relationship between the Federal Government and Indian tribal governments." The Park Service has incorporated these admonitions into its management policies. For example, agency officials must consult with tribes on a government-to-government basis whenever proposed actions may affect tribal interests; they must "strive to allow American Indian . . . access to and use of ethnographic resources;" they must "accommodate access to and ceremonial use of Indian sacred sites by religious practitioners;" and they must consult with tribal governments on "planning, management, and operational decisions that affect subsistence activities, sacred materials or places, or other resources." Moreover, the Park Service has reoriented its original limited archaeological focus by establishing an ethnography division, which includes trained anthropologists and ethnographers concerned with living cultures, and by creating an American Indian Liaison Office. Although designed to improve its sensitivity to contemporary tribal concerns, these new policies, however well-meaning, must still be reconciled with the Park Service's fundamental resource conservation obligations, an ongoing process that has provoked difficult questions concerning tribal land claims, treaty rights, and access to park resources.[8]

REVERSING HISTORY AT GRAND CANYON NATIONAL PARK

The presence of Native Americans and their concerns may have been willfully ignored when early parks like Yosemite and Glacier were established, but more recent park creation and expansion efforts have been forced to confront the new reality of tribal political power. This change was apparent as early as the mid-1970s when the Park Service's campaign to extend Grand Canyon National Park's boundaries to include the picturesque Havasu Falls area ran smack into a resilient Havasupai tribe intent on righting historic wrongs and regaining its lost lands. Having seen its original reservation reduced from 38,400 acres to a mere 518 acres in two short years during the 1880s, the Havasupai tribe had endured a contentious relationship with the Park Service ever since Grand Canyon National Park was created in 1919. By 1939, the Park Service was invested in expanding the park's boundaries to include the tribe's lands, having purchased an old mining claim near the waterfall area and constructed its own campground on the site. During the late 1960s, with broad support from the environmental community, the Park Service released a master plan that called for park expansion onto the reservation, arguing the need for adequate tourist facilities in the increasingly popular falls area. The tribe, smarting from past injustices and recurrent conflicts with the Park Service, which included agency employees razing Indian cabins in an effort to evict tribal members from Supai Camp on the South Rim, responded by proposing to expand the reservation onto national park lands. It enlisted Arizona senator Barry Goldwater and congressman Morris Udall to support its expansion proposal, which tribal members believed would improve their dismal economic circumstances.[9]

After considerable political wrangling, the Havasupai prevailed when Congress passed the Grand Canyon Enlargement Act of 1975. Although expanding the park by 400,000 acres, the legislation granted the tribe 185,000 acres of national park and national forest land as well as exclusive use rights to another 95,000 acres of park land, giving tribal members access to the mesa lands above the canyon. By one account, these additions to the Havasupai's postage stamp–sized reservation represented "the largest Indian restoration act in U.S. history."[10] The legislation also contained several key restrictions, however: the additional 185,000 acres must remain "forever wild," which precluded logging, mining, and other development activities; and the tribe was required to develop a land use plan, subject to Interior Department approval, that protected "scenic and natural values" on the additional 95,000 acres. In short, although it regained substantial acreage from the park, the tribe was denied complete control over how the land was used.[11]

To no one's surprise, the reservation expansion did not resolve all the contentious local resource management issues, nor did it presage an economic bonanza for the tribe. The Havasupai tribe and the Park Service are still sorting out their relationship on the exclusive-use national park lands, and they still spar over the Indian presence and living conditions at Supai Camp. Nonetheless, invoking persuasive moral arguments, the tribe triumphed over the powerful alliance that sought park expansion onto reservation lands and recovered lost ancestral lands from the park. The episode marked a significant shift in local park-tribal relations, with potential implications for other national parks situated on dispossessed lands.

TREATY RIGHTS AT GLACIER NATIONAL PARK

The question of Native American access to the national parks to hunt, fish, or otherwise use park resources has provoked several long-standing controversies. In most instances, the Indian claims are derived from early treaties that contained language seemingly reserving these rights to tribal members. Such is the case at Glacier National Park in northwestern Montana, where the Blackfeet reservation abuts the park's eastern boundary. The tribe has long asserted that even though its 1895 treaty with the United States relinquished the "ceded strip" that runs south and east along the spine of the Rocky Mountain Front, tribal members retained hunting, fishing, and timber-cutting rights on these lands. When Glacier National Park was established in 1910, however, the enabling legislation contained no reference to the treaty or to Blackfeet rights, nor was the tribe consulted by Congress about the legislation. Rather, following the pattern of earlier national park legislation, the Glacier bill outlawed hunting and timber harvesting in the new park and imposed fishing restrictions. According to federal lawyers at the time, Congress had the power to do so under the terms of the 1895 treaty: the Blackfeet retained hunting, fishing, and timber rights only as long as the ceded land "shall remain public lands of the United States," which ceased to be the case when the lands were removed from the public domain to create the national park.[12]

The Blackfeet tribe and its members, never believing they relinquished such vital rights, have found themselves at odds with the park ever since. Glacier officials have long complained about illegal Indian hunting and cattle trespass on east-side park lands, and tribal members have periodically been arrested for these and other offenses. Efforts to test the treaty in court have resulted in a series of decisions that basically sustain the park's position. Early on, the Park Service recognized the reservation boundary as a potential trouble spot for elk and other park wildlife that seasonally migrated onto the lower-elevation tribal lands, where they were hunted. In addition, Glacier officials, worried that the nearby presence of the impoverished reservation might deter visitation, were concerned about the proliferating tacky developments, inadequate tourist facilities, and general lack of economic development. In an effort to seize the initiative, Directors Mather and Albright separately endorsed early legislative proposals designed to expand the park eastward onto the reservation, a move they contended would better align the park's boundaries with biological realities. These proposals, however well-intentioned from a wildlife management perspective, were routinely opposed by the tribe, who viewed them as yet another land grab. They ultimately went nowhere in Congress.[13]

The Blackfeet tribe, with a historic unemployment rate hovering at the 80 percent level, has long seen Glacier in economic development terms, but to little avail. Once the park was established, the Great Northern Railway captured much of the early tourist traffic at its rustic lodges, regularly employing Blackfeet tribal members to appear in their native ceremonial garb to greet arriving passengers and to entertain them with traditional dances. The Indians also sold souvenirs and trinkets, and a few were employed in menial positions at the park. Although confronted daily with this lucrative tourist market on its doorstep, the tribe reaped little benefit from it, and that has not changed much over the years. An assortment of tribal economic development proposals, including one during the early 1960s that involved constructing a resort complex on the eastern shore of St. Mary Lake near the park's eastern entrance, have failed repeatedly. More recent efforts by the Park Service to hire tribal members and to grant concession contracts to Native American entrepreneurs—an arrangement often resisted by the existing businesses—have achieved some progress. But the treaty rights conflicts still cloud park-tribe relations, and the Park Service, bound by its conservation obligations, shows no sign of opening park lands to Indian hunters or for other tribal resource use purposes.[14]

Given the park's location adjacent to the reservation, however, park resources are quite vulnerable to the Blackfeet tribe's land use and related development decisions. It turns out that the Blackfeet reservation sits astride the Bakken Shale Formation, which has yielded significant natural gas deposits farther east. In its quest for more jobs and revenues, the tribe has leased nearly the entire reservation for energy development, and oil companies have already drilled exploratory wells next to the park. A major find could lead to full field development, effectively industrializing the park's ecologically important eastern border and imperiling wildlife, water quality, scenic values, and other resources. Undeterred by these environmental concerns, the tribe appears intent on pursuing its new energy development agenda, even in the face of stiff intratribal opposition. So far, an emergent conservation vision for this "Crown of the Continent" region has not provided any relief for the park or otherwise.[15] Given the tribe's history and dire economic circumstances, few opponents of this pending development are willing to speak out forcefully. Plainly, the park-tribe relationship as neighbors is a two-way association that can either benefit or damage park resources and interests.

THE DEVIL'S TOWER SACRED SITE CONTROVERSY

It is no surprise that Native Americans, having long regarded national park landscapes as their ancestral lands, should view some park lands as sacred sites of profound cultural importance. It is also no surprise that Native American tribes, having survived hard-hearted federal assimilation policies designed to exterminate indigenous cultural values, should now seek to access and safeguard these sacred places. That is exactly what has transpired at such diverse locations as Devil's Tower, Rainbow Bridge, and several other national park locations. The key questions, in most instances, are whether tribal members are entitled to use specific sites for their own religious or cultural purposes and whether their interest in the site entitles them to exclude others from it. Beyond the U.S. Constitution's opaque First Amendment religion clauses, Congress provided some guidance to these questions in the American Indian Religious Freedom Act of 1978, which provides that "it shall be the policy of the United States to protect and preserve for American Indians their inherent right to freedom to believe, express, and exercise the traditional religions . . . , including but not limited to access to sites . . . and the freedom to worship through ceremonials and traditional rites." And the courts have further defined the scope of sacred site rights.[16]

At Devil's Tower National Monument in northeastern Wyoming—proclaimed the nation's first national monument in 1906 by President Theodore Roosevelt—the Park Service found itself in the middle of a heated controversy between Native American spiritualists and rock climbers. During the 1980s, rock climbing grew exponentially as a sport, and Devil's Tower gained a reputation as a world-class climbing destination. Climbers began flocking to the tower to test their physical prowess against its sheer granite walls, and the tower's acclaim spawned an important local guiding industry that provided all-important jobs in this rural area. At the same time, Park Service officials noticed a visible increase in Native American prayer offerings left at the monument. Devil's Tower is an important cultural location for several tribes, some of whom linked their creation myths to this scenic promontory that they called Bear's Lodge. For the Indians, the rock climbers' mere presence recreating at the tower, sometimes punctuated by vulgar exclamations that reverberated off the rock face, defiled this spiritually sacred site and disrupted their solitary prayer vigils. For both sides, the park's management policies assumed monumental importance.[17]

Once the conflict was apparent, the Park Service set about identifying some common ground through a four-party collaboration process. The initial negotiating positions adopted by the climbing, environmental, tribal, and local government representatives were quite strident, showing little appreciation by anyone for opposing perspectives. Over time, though, the collaborative effort paid off, serving both to educate and sensitize the parties to each other's real concerns. The process concluded with an originally unimaginable agreement: to address Native American concerns, the climbers agreed to a voluntary climbing closure at the monument during the month of June, when the spiritually important solstice occurs. They also agreed to educate their fellow climbers about the monument's sacred standing among Native American tribes, and the Park Service agreed to stop issuing commercial climbing permits during June. For their part, the tribes did not press for a mandatory closure, explaining that a climber's decision not to disturb their prayer vigils should be made voluntarily out of respect for their religious beliefs, not as a result of a compulsory government order.

Although a seemingly reasonable compromise, the agreement was unacceptable to local guide businesses, who faced the prospect of lost revenues during the prime climbing season. They turned to the conservative Mountain States Legal Foundation and sued the Park Service, alleging that the month-long closure constituted an illegal governmental religious accommodation that violated the First Amendment's establishment clause. In the end, after the litigation had traversed two levels of the federal judiciary, the courts endorsed the Park Service's revised position, which made the commercial closure voluntary after the district court enjoined the agency's mandatory commercial closure decision, finding that it violated constitutional religious neutrality principles.[18] Since then, park officials regularly distribute educational literature to visitors explaining the site's spiritual importance to Native Americans as well as the voluntary climbing closure, which data indicate is 85 percent effective. Viewed as a carefully crafted Park Service effort to accommodate Native American religious and cultural values within constitutional limits, the outcome at Devil's Tower demonstrates that compromise solutions sensitive to legitimate native spiritual concerns and park values can be found and made to work. It also suggests, however, that when private economic interests are at stake, the agency's most well-intentioned efforts to address legitimate Native American concerns may not always prevail, reconfirming the influence that powerful local and commercial interests can assert over park policy.[19]

RECLAIMING ANCESTRAL LANDS AT DEATH VALLEY NATIONAL PARK

At Death Valley National Park, the Park Service has come full circle in its relationship with the Native Americans who originally inhabited this stark desert environment that straddles California and Nevada. The Timbisha Shoshone presence in Death Valley dates back nearly a thousand years, when the modern band's hunter-gatherer ancestors took up part-time residence in this heat-blasted landscape, wintering in the valley and then retreating to the nearby mountains during the summer. Once miners and ranchers arrived during the mid-1800s and started to stake ownership claims to the land, however, the Indians found themselves being evicted from their ancestral lands, most notably at Furnace Creek, one of the few reliable water sources around. By the late 1800s, the early mining activities had given way to large-scale borax mining, which yielded more than $30 million worth of ore before the industry went bust early in the twentieth century. With the end in sight, Pacific Coast Borax converted its mining operations to a luxury inn and ranch at Furnace Creek, solidifying its claim to the precious water and disregarding the Indian presence. In fact, the Timbisha Shoshone were ignored throughout the entire nineteenth-century Indian treaty and reservation era, leaving the small tribe without any legal recognition or land of its own.[20]

In 1933, President Herbert Hoover declared the Death Valley National Monument and turned 1.6 million acres over to the Park Service for management. Although the monument proclamation acknowledged the area's "unusual features of scenic, scientific, and educational interest," it failed to even mention the Shoshone presence, implying that the land was historically uninhabited. Once the Park Service took over, it incorporated the area's mining history into its interpretive program, but ignored the Native American presence or history. Instead, the agency viewed the disheveled presence of the poverty-stricken Indians at Furnace Creek, already the prime tourist attraction, as a public health menace and a potential public relations disaster.

To address the situation, the Park Service and the Bureau of Indian Affairs arranged to move the one hundred or so Indians who remained in the monument to a forty-acre plot removed from the main tourist hub, where the Civilian Conservation Corps constructed nine adobe houses. Although this arrangement provided the remaining band members with improved housing, it did not acknowledge any native rights, still leaving the tribe legally unrecognized and practically landless. By the 1950s, with the federal termination policy in full swing, the Park Service began demolishing vacated village houses, anticipating that the remaining band members would eventually just leave the area.

They did not leave, though. In fact, during the 1960s, the Western Shoshone tribes began aggressively pursuing their land claims, an effort that included the Timbisha band members. The Park Service, increasingly sensitive to its "slumlord" relationship with the Timbisha, finally extended electric power to the small village, completing the project in 1978. Then, as the 1980s were winding down, the Bureau of Indian Affairs granted tribal recognition to the Timbisha, providing the roughly fifty band members remaining in Death Valley with the legal standing to assert their own land claim inside the monument. The next breakthrough came in 1994, when Congress passed the California Desert Protection Act, which not only converted the monument to a park and added 1.3 million acres to it, but also directed the secretary of the Interior to "conduct a study . . . to identify lands suitable for a reservation for the Timbisha Shoshone Tribe that are located within the Tribe's aboriginal homeland area within and outside the boundaries of the Death Valley National Monument and the Death Valley National Park."[21]

The ensuing study and related negotiations occupied another five years, in part because the Park Service initially resisted the precedent-setting notion of ceding national park lands to a Native American tribe. Finally, after the Timbisha band members joined the Alliance to Protect Native Rights in National Parks to publicize their cause, the two sides struck a bargain that granted the tribe three hundred acres of trust lands at Furnace Creek, where they could construct homes, a modest hotel, a cultural museum, a gift shop, and a government center. The tribe also received another three hundred thousand acres—designated the Timbisha Shoshone Natural and Cultural Preservation Area—that was to be comanaged with the Park Service. Several conditions were attached, however: gambling casinos and hunting were outlawed in the park, but juniper berry and piñon nut gathering was permitted, as were sacred site closures, although limited in size and duration. As the negotiations proceeded, the Park Service conceded that the Timbisha were serious about protecting the lands they sought, and the Indians eventually trimmed their acreage demands, still acquiring enough well-located lands to derive some badly needed economic benefits. In 2000, Congress gave its blessing to the deal.[22]

The significance of the Park Service–Timbisha arrangement cannot be overstated. An impoverished band of forgotten Indians numbering fewer than fifty members on site succeeded in reclaiming a portion of their ancestral homelands inside the Death Valley National Park and legitimizing their permanent presence in the park. Never before has a tribe, without a legally recognized presence on the land, reclaimed national park land as its own native homeland. Establishment of the comanaged cultural preservation area provides an important opportunity to experiment with the joint management of park resources, subject to explicit conservation limitations that still enable the Timbisha to pursue many of their ancient uses and rituals. With the Timbisha now permanent park residents, the opportunity exists to integrate the Native American story into the park's interpretive programs and cultural history, with the tribe taking responsibility for telling its own story and relationship to the land. The arrangement, by acknowledging an early and ongoing Native American presence on these national park lands, not only helps redress this disturbing past injustice but also significantly alters this park-tribal relationship.

A BADLANDS TRIBAL NATIONAL PARK PROPOSAL

In a further sign of the changing times, Native Americans are seeking management responsibility at some park locations. This development is not new, however, dating at least to the unique relationship between the Park Service and the Navajo at Canyon de Chelly in northeastern Arizona. Situated within the expansive Navajo Reservation, Canyon de Chelly is rich in Native American ruins and scenic splendor, encompassing three large canyons, dramatic rock formations, and a reliable water source. Home to Navajo farm families for several centuries, the canyon drew increased tourist interest during the early twentieth century, alerting tribal leaders to the area's economic potential. Cognizant of its own limited management experience, the Navajo tribal council responded with an extraordinary decision: it endorsed creation of a national monument on these reservation lands. In 1933, following congressional authorization, President Hoover proclaimed a new national monument, one where the tribe retained ownership rights to the lands and minerals, while the Park Service was directed to oversee "the care, maintenance, preservation, and restoration of the prehistoric ruins or other features of scientific or historical interest." Although not without problems, this joint management relationship has persisted over the years, setting a precedent for other such arrangements acknowledging the historical Native American presence and cultural connections at other national park sites.[23]

This unique comanagement model took another step forward when in April 2012 the National Park Service announced that it was supporting creation of the first Tribal National Park in the Badlands. After nearly a half century of contentious relations and unfulfilled promises, the Park Service and the Oglala Sioux tribe jointly endorsed converting the so-called south unit of the 242,000 acre Badlands National Park from exclusive federal oversight to a tribally administered entity subject to conventional national park management standards. Incorporated into a recent general management plan, which also maps out an ambitious ecological restoration and visitor access agenda for the unit, the Tribal National Park proposal would require congressional approval, which remains uncertain at this time. The south unit, long regarded as the "bastard child" of the Badlands, has a dark history shrouded in social and environmental injustice that the proposal begins to address.[24]

The Badlands are located in western South Dakota and extend into the Oglala Sioux's Pine Ridge reservation. The northern portion of the Badlands landed in federal ownership in 1939 when Congress—at the behest of the Park Service, which coveted this barren, desiccated, and yet stunning landscape—authorized creation of Badlands National Monument, which President Franklin D. Roosevelt then proclaimed a few months later. Soon thereafter, park officials sought to expand the new monument onto the adjacent Pine Ridge reservation, only to be rebuffed by the tribe. Once the United States entered World War II, however, the Air Force went looking for a practice bombing range, and it found one just south of the new national monument on the Pine Ridge reservation. Citing the wartime emergency, the military forcibly leased the tribe's lands and then condemned the interspersed private lands, forcing 125 Indian families to relocate on short notice. Because it was the Depression era, land prices were at rock bottom, and the displaced Indians received a relative pittance, not enough to either replace their lost lands or for many to even retain their cattle herds.[25]

By the mid-1960s, no longer needing the bombing range, the military decided to dispose of it as surplus federal property. The Park Service, perceiving a second opportunity to expand Badlands National Monument, expressed keen interest. After lengthy negotiations that rejected returning any of the lands to the original individual owners, Congress agreed to a land exchange between the Interior and Defense Departments that retained the lands in tribal ownership subject to Park Service management. In a controversial 1976 Memorandum of Agreement, the tribe granted the agency an easement over what is now the 133,000-acre south unit in return for promised improvements and economic development assistance that included half of the park's annual entrance fees.[26] Even though Congress converted the area to national park status in 1978, the Park Service and the tribe have been locked in an increasingly unhappy relationship that has witnessed few economic returns for the tribe and a federal management regime described as "benign neglect" by one knowledgeable observer.[27]

The north and south units of Badlands National Park are simply two quite different places, yet they have both been administered under the same general management plan since 1982. The north unit lies just off Interstate 90 where, thanks in part to widespread promotional efforts by Wall Drug and other nearby merchants, it attracts nearly one million visitors annually. Few of these visitors ever make it to the more distant south unit, however, owing to the road configuration and the lack of evident tourist attractions. Besides, unexploded ordnance is still scattered across the south unit, and visitors are chillingly advised to stay on the established roads and trails, even though a federally funded cleanup was initiated in 1995. A promised south unit visitor center at the White River junction has never been built, ostensibly due to Congress's failure to appropriate the necessary funds. And rather than encountering bison and other wildlife on the south unit, those visitors venturesome enough to get there are greeted by cattle. In the tribe's view, grazing lease revenues are more reliable than tourist dollars.

The Pine Ridge reservation—the second largest land-based Indian reservation in the United States—is not only one of the poorest places in the country, but it has also been a hotbed of controversy. The reservation's unemployment rate regularly exceeds 80 percent, and the principal county in which the reservation is located has consistently ranked at or near the bottom for per capita income in recent census reports.[28] During 1973, federal law enforcement officials squared off against radicalized American Indian Movement members in a high profile standoff at Wounded Knee, site of the notorious 1890 massacre by the Seventh Cavalry of 150 Indians for performing the outlawed Ghost Dance.[29] Since then, Native American traditionalists within the reservation community have frequently been at odds with tribal leaders and others more interested in capitalizing on tourism and other economic opportunities. These tensions were on display in 2002, when several tribal members occupied park lands to protest recent agency decisions that would allow off-road vehicles to traverse across sacred burial sites and would permit a fossil excavation on other sacred lands. Asserting that the entire south unit of Badlands was a sacred site, the protesters presented a loud and strong case for tribal control over the area.[30]

In response, the Park Service and the tribe entered into formal government-to-government negotiations over the future of the south unit, beginning with an agreement to prepare separate management plans for each unit. In 2006, as the relationship warmed, they set about jointly drafting a new south unit general management plan that would establish ambitious resource conservation goals. Released in April 2012, the final plan pointedly observes that "the park is managed holistically as part of a greater ecological, social, economic, and cultural system."[31] It describes a management strategy focused on ecological restoration, expanded visitor access, and cultural preservation, including educational and interpretive programs featuring the Oglala history and culture. The plan designates nearly 90 percent of the land as a Natural Area/Recreation Zone that blankets the park's backcountry, and 10 percent is treated as a Development Zone in the frontcountry. Hunting by tribal members will be allowed, subject to park regulations. The plan also supports building a new Lakota Heritage and Education Center as the principal visitor center, where cultural artifacts that are now housed off-site would be displayed and Native American interpreters would explain Sioux traditions, beliefs, and culture. An important but unspoken goal of the plan is to enhance economic opportunity for tribal members.

Most important, the plan calls for creation of the first Tribal National Park, which would ultimately give the Oglala Sioux tribe management responsibility for the entire south unit. The proposal, a dramatic departure from traditional federal management of national park units, has been endorsed both by the Park Service and tribal officials and has met with little outright opposition. Contingent on congressional approval, the plan calls for a gradual transition from Park Service to tribal administration, with Park Service personnel providing job training to tribal members who would eventually manage the new park. The plan anticipates that the new park, sporting both the National Park Service's arrowhead symbol and the Oglala Sioux tribe's logo, would be funded by congressional appropriations and entrance fees. Federal laws and policies, including the Organic Act, the National Environmental Policy Act, and the Endangered Species Act, would continue to apply across the park, trumping any inconsistent tribal laws or policies. Under this legal regime, the tribe would also be obligated to provide meaningful public involvement opportunities in its decision-making processes and allow for legal recourse, which could require the tribe to waive its sovereign immunity. Whether Congress is prepared to approve this new national park concept with legislation remains to be seen, as does the precise form such a law might take.[32]

Regardless, the Badlands tribal national park proposal represents a bold step toward reconnecting Native Americans with their ancestral lands while retaining traditional national park conservation imperatives. That the Park Service has taken this unprecedented step into the unknown reflects further evolution in the larger relationship between the national parks and their tribal neighbors, and it serves as a concrete illustration of the inherent ecological, economic, and social connections between the national parks and the surrounding landscape. Just as the Park Service cannot ignore the off-site migratory needs of the wildlife it is charged with safeguarding, it also cannot ignore the concerns and developments occurring on adjacent reservation lands. Of course, that the Badlands National Park's south unit was still part of the Pine Ridge reservation gave added weight to the Oglala Sioux tribe's management interests, as did the deadly presence of unexploded ordnance within the park, a grim reminder of the federal government's earlier land grab. As a matter of both social and environmental justice, this first tribal national park proposal thus heralds another crucial chapter in the unfolding relationship between the parks and their Native American neighbors.

Aligning Nature, Culture, and Justice

These examples of how Native Americans are renegotiating their relationship with the national parks represent but a few of the many controversies that are helping redefine this relationship. Across the system, the Park Service and the tribes are engaging at new levels: at Apostle Islands, where the Chippewa rebuffed a Park Service–led effort to incorporate tribal lands into the new park; at Olympic, where the Park Service has joined with the Lower

Elwha S'Klallam tribe to begin removing two dams blocking salmon migration routes; at Little Big Horn, which has been renamed from the Custer Battlefield National Monument to acknowledge the Native American role in this historic battle; and the list goes on. In part, these efforts respond to historical misdeeds that eliminated any active Native American presence from the parks; and, in part, these initiatives are forward looking, intended to reconnect parks with an original human presence, promote better coordination between both parties, and help improve tribal economic circumstances. Although it is premature to recast the national parks as Native American homelands, it is clear that American Indians and their culture are being written back into the national parks. By any standard, the Native American role in the national parks is evolving, and in the process, it is challenging the conventional national park narrative.

As a matter of geography and history, the national parks and Native Americans are linked together. In part, that connection is as neighbors who inhabit a common landscape and share a common responsibility for wildlife, water, and other ecological resources. What happens on tribal lands can often affect park ecosystems, just as what occurs on park lands can affect tribal interests. When the Blackfeet decide to permit oil wells on their reservation lands within Glacier's shadow, the ramifications for migratory elk, mutual watersheds, and air quality redound within the park. Likewise, when Glacier officials deny tribal hunting rights on park lands, it has an effect on the tribe's wildlife management options. In part, the connections are based on history, in the form of treaty rights, sacred site access, or land claims. Whether the Native American claims are framed in social justice or in cultural heritage terms, the Park Service ignores these claims at its peril. The experience at Grand Canyon is instructive. After initially seeking to expand the park at the expense of the neighboring Havasupai tribe, the Park Service was ultimately compelled to relinquish more than 280,000 acres of park land to the Havasupai on the strength of their historic claim to these ancestral lands. Native American tribes are not only park neighbors; they can be powerful forces with sovereign rights and their own agendas, sometimes putting them in a more pivotal position than gateway communities or other park neighbors.

The Park Service's evolving relationship with Native American tribes is having an evident effect on nature conservation policy. From the beginning, the national parks have been conceived and managed largely as wilderness-like reserves devoid of any permanent human presence (except, of course, the roads, hotels, stores, and other accommodations that have been provided for park visitors).[33] It is now widely agreed, however, that Indians historically not only occupied and used park lands but also altered these landscapes, thus undermining the myth of pristine nature that has prevailed at least since the 1963 Leopold report. Indian treaty rights and sacred site access claims virtually ensure an ongoing Native American interest in the parks that cannot be ignored. For the most part, though, these Native American claims have not proven to be inconsistent with national park resource protection obligations. Although accommodating a solitary prayer vigil inside a park may inconvenience some visitors, it is not likely to harm park resources. In fact, accommodation of such cultural or religious practices actually extends the Park Service's traditional preservationist role into the realm of cultural preservation, which is not only consistent with its statutory obligations but can also complement its nature conservation efforts. By pursuing a thoughtfully integrated approach to Native American cultural concerns, the Park Service can begin to break down its historic nature-culture divide, thus acknowledging that humans are part of nature, but without jeopardizing the very resources that the parks were created to protect.

Given the history of dispossession that pervades most park-tribe relationships, Native American national park neighbors can often assert particularly strong social justice claims to park lands, resources, and related economic benefits. Indeed, Indian justice claims have occasionally proven strong enough to regain lost lands for individual tribes, as has occurred at Grand Canyon and Death Valley. Similar powerful arguments have put the Oglala Sioux in the position of securing a management role over half of Badlands National Park, which would provide an opportunity to begin explaining the Sioux history and culture to visitors and to realize some much needed economic benefits. For the most part, these relatively new arrangements—generally the result of government-to-government negotiations—are being structured in a manner that keeps park resource conservation goals paramount in future management plans. An important challenge going forward, however, is to ensure that tribal economic interests are also being met, particularly given the potential tourism revenues at stake or the prospect of environmentally damaging development on reservation lands adjacent to the parks. Although national parks play a vital commercial role in the lives of their non-Indian gateway communities, as we have seen, the same is not generally true for park tribal neighbors. To address this problem, the evolving relationship between national parks and their tribal neighbors must be conceived in a way that ensures meaningful mutual benefits—both economic and ecological—for each party.

Although few people expect Native Americans to regain ownership of those ancestral lands that are now national parks, there is nonetheless an evident need for better coordination between the parks and their tribal neighbors. Clearly, the parks do not exist as islands but must instead be understood as part of a larger ecological and human landscape. Besides sharing common watershed, wildlife, and other valuable resources, both entities also have a common interest in the visiting public, whether for economic, educational, or other purposes. The emerging comanagement arrangements for the proposed first tribal national park at Badlands exemplifies how such a relationship might be structured to achieve both nature and cultural conservation goals as well as new cultural educational opportunities. Despite long-standing concerns about tribal capacity, the reality is that many tribes have gained significant experience in self-governance, resource management, and cultural preservation over recent years. Many tribes also have significant economic, educational, and other resources at their disposal through gaming, energy development, and various commercial ventures, further strengthening their capacity for meaningful engagement with their national park neighbors. That much of Native American culture revolves around natural and cultural resource preservation will often lend a common perspective to both parties' management interests, helping promote conservation as a shared goal.

To be sure, Native American and national park interests will not consistently align with each other, but that does not obviate the need to seek common ground on shared ecological and other matters. Workable compromises that address multiple interests can be achieved, as reflected in the Devil's Tower sacred site arrangement. Perhaps nowhere is this alignment of interests more evident than at Olympic National Park, where the Park Service joined with the Lower Elwha S'Kallam tribe in a coordinated campaign to remove two old dams on the Elwha River and restore native salmon runs. Originally constructed to provide inexpensive local power, the dams inundated park lands with impounded water and destroyed the tribe's treaty fishing rights. Working together, the Park Service, the tribe, and environmental groups convinced Congress to decommission the dams and return the river to its natural state. Dam removal will both restore the tribe's historic fishery and the park's ecological integrity, while also improving the tribe's economic position and meeting local economic concerns. In the evolution of national park–Native American relations, the Elwha restoration project stands as a positive example of how legitimate Native American claims and national park conservation interests can be served when two powerful entities join forces and engage with parties beyond their respective boundaries.[34] Even when the national park idea is conceived in "ancestral lands" terms, the parks are still quite able to fulfill their basic conservation purposes and even restore lost ecosystem components.

CHAPTER 7

"Nature's Laboratory"

Experimentation and Education

Long regarded as an ideal outdoor laboratory, the national parks have not been consistently administered with science or education in mind. Even from the earliest days, scientists and others recognized that these protected settings provided a rare opportunity to study the natural world and to learn from it, but for most of its first fifty years, the National Park Service showed little interest in scientific inquiry except as it might enhance the agency's modest visitor education programs. Early resource management policies rested more on conventional wisdom than rigorous data-based research and experimentation. As long as the parks offered spectacular scenic vistas and diverse recreational opportunities, park managers were satisfied that they had fulfilled their conservation responsibilities. After all, these splendid settings were not set aside as research or educational facilities, nor with much regard for on-the-ground ecological realities. These views, however, were gradually supplanted by a persistent cacophony of voices calling for more scientifically rigorous management and an explicit scientific mission for the national parks, a position finally enshrined in law at the end of the twentieth century. But the national parks' educational role and potential have yet to receive the same attention. Indeed, the Park Service's interest in incorporating either science or education into its mission has wavered over the years, as will become evident.

SCIENCE, RESEARCH, AND EDUCATION IN THE PARKS

It is quite ironic, given the National Park Service's well-documented and historic indifference toward science, that the national park idea and the world's first national park can be traced to early scientific expeditions. The 1870 Washburn-Langford-Doane expedition—whose reports on the little-known Yellowstone country gave credibility to early accounts of its extraordinary thermal features and bizarre geology—observed the region's tremendous scientific potential: "As a field for scientific research, it promises great results; in the branches of geology, mineralogy, botany, zoology, and ornithology, it is probably the greatest laboratory that nature furnishes on the surface of the globe." Intrigued by such reports, Dr. Ferdinand Hayden, an intrepid government surveyor with scientific training, promptly embarked on his own expedition into the Yellowstone country, where he assembled "extensive collections in geology, mineralogy, botany, and all departments of natural history." In his final report, printed shortly after Congress designated Yellowstone a "national park," Hayden wrote: "This noble deed may be regarded as a tribute from our legislators to science, and the gratitude of the nation and of men of science in all parts of the world is due them for this munificent donation."[1]

Science in the Early Years

These early Yellowstone reports linking science to nature preservation were not the first time this connection was drawn. In his 1865 report extolling the Yosemite country's scenic attractions, Frederick Law Olmstead noted "the value of the district in its present condition as a museum of natural science and the danger . . . that without care many of the species of plants now flourishing upon it will be lost and many interesting objects be defaced or obscured if not destroyed." Although moved by such reports to protect these two special places as national parks, Congress made no mention in its enabling legislation of their scientific and educational potential or of the need to employ science in managing them.[2]

In 1906, however, Congress did connect preservation with science when it adopted the Antiquities Act. The scientific community, led by Professor Edgar Lee Hewett, convinced Congress of the need to preserve endangered Native American relics and dinosaur bones that were disappearing at an alarming rate across the Southwest and elsewhere. The Antiquities Act vested the president with nearly unbridled authority to protect "historic landmarks, historic and prehistoric structures, and other objects of historic or scientific interest that are situated upon the [public] lands" as national monuments. Explicitly acknowledging the scientific importance of these objects, Congress established a permit system allowing only professionally qualified researchers to excavate these sites, but only if the research was "undertaken for the benefit of reputable museums, universities, colleges, or other recognized scientific or educational institutions, with a view to increasing the knowledge of such objects." Although several cabinet departments initially shared responsibility for these new national monuments, Congress transferred them wholesale to the National Park Service once it was established ten years later.[3]

Passage of the National Parks Organic Act in 1916 provided Congress with an opportunity to vest the new National Park Service with a scientific component to its mission, but it did not do so. In fact, the debates surrounding passage of the Organic Act contained few references connecting the national parks to science; rather, the act's proponents focused almost exclusively on the need to protect scenery, promote tourism, and ensure efficient management practices. Supporters of the organic legislation—perhaps influenced by the rival Forest Service's explicit embrace of scientific principles to manage timber and other resources—eschewed science as a basis for creating new parks or administering them. In fact, the 1918 Lane letter directed the fledgling Park Service to turn "for assistance in the solution of administrative problems in the parks relating both to their protection and use [to] the scientific bureaus of the Government." Without an explicit scientific component to its mission, the new agency embarked on a course that emphasized scenic preservation and tourism, a decision that would eventually open it to strident criticism for endangering vital natural resources. Thus, although recognized as natural outdoor laboratories, the early national parks were administered with little regard for science or their research potential.[4]

To be sure, science did surface in the initial national park interpretive programs, which represented an important early agency foray into the realm of education. Although the Lane letter did not expressly enumerate science as part of the Park Service's mission, it did state that "the educational . . . use of the national parks should be encouraged in every practicable way," contemplating scientific study visits by university classes and museum displays of park animals, trees, and the like. Director Stephen Mather hired Robert Sterling Yard, a New York newspaper editor and longtime friend who had served as best man at Mather's wedding, to head the fledgling agency's public education section. To promote the new park system, Yard promptly produced the *National Parks Portfolio*, with an introduction by Secretary Lane asserting, "It is the destiny of the national parks, if wisely controlled, to become public laboratories of nature study for the nation." A 1922 superintendents' communiqué observed that the parks were "opening the doors of Nature's laboratory" and "offer[ing] to the American public . . . constantly increasing opportunities for acquiring information on many phases of natural history and science."[5]

To instill this educational component into the national park mission, Mather endorsed the idea of building park museums and establishing visitor education programs, including guided nature walks, lectures, and the much-revered ranger campfire talk. Congress was skeptical, however, of even this limited scientific and educational role; it initially refused to appropriate any funds for park museums, so the early ones were funded privately. By 1930, however, following an Interior-initiated comprehensive study, the Park Service had created an outside advisory board on the role of education in the parks as well as a new Branch of Research and Education to coordinate the agency's educational efforts across the growing park system. Significantly, the advisory board, perhaps sensitive to intruding into the domain of universities and other educational institutions, recommended that "the Federal Government should handle only such educational matters as may not be cared for adequately by other means." Moreover, the new branch concluded that educational programming should be the responsibility of individual parks, a decision that gave the agency's education and interpretation efforts an ad hoc quality and effectively relegated them to a second-level concern at many parks.[6]

George Wright Makes the Case

During the early 1930s, under Horace Albright's leadership as the Park Service's second director, the role of national parks as experimental laboratories surfaced from the shadows, although not as an enduring mission. In 1930, facing increased criticism from academic scientists like the University of California's Joseph Grinnell and from within his own ranks, Albright established the agency's first Branch of Research and Education, which set about "gather[ing] the scientific information necessary to the development of the museum, educational, and wildlife administration programs of the national parks." At the same time, ranger-naturalist George Melendez Wright emerged from the ranks and, using his own financial resources, convinced Albright to support a wildlife survey to help park managers better understand and manage the animals, birds, and fish they were charged with protecting. To do so, Albright created a new Wildlife Division in the Branch of Research and Education and secured funding support from Congress to continue the survey. Wright died tragically in 1936, however, and his Wildlife Division was dissolved following his untimely death. At the end of the decade, the Park Service counted 34 permanent naturalists on staff, most of whom were engaged in presenting natural history programs to the visiting public, and it boasted 115 museums, confirming the increasingly important role that nature education—but not research—was coming to play in the parks.[7]

Any real opportunity to expand the Park Service's mission into the worlds of science and education languished during the war years. Following World War II, however, the agency and its allies sought to upgrade the role of education. A 1945 report by the National Parks Association recommended that "scientific, educational and inspirational values dictate the major uses of primeval parks" and that "no visitor . . . should leave without having been informed about the special significance of that particular area, as well as of the system as a whole."[8] Director Conrad Wirth's Mission 66 construction program included numerous projects with educational dimensions, including 114 new visitor centers and several new park museums. In addition, Wirth engaged Freeman Tilden, an accomplished senior newspaper editor deeply attracted to the national park idea, to enhance the Park Service's nascent public education program. Tilden responded by writing *Interpreting Our Heritage*, which is still regarded as a classic among Park Service interpreters. Tilden's book not only set forth key principles for effective engagement with the visiting public, but also pushed the agency to include environmental education as part of its interpretive programs. In Tilden's view, interpretation should be seen as "an attempt to reveal the truths that lie behind the appearances," a challenge that in the national park setting could only be met with some degree of scientific knowledge.[9]

The Park Service itself nevertheless continued to disregard science as a part of its mission, triggering discontent. By the 1950s, the U.S. victory in World War II, hastened by development of the atomic bomb and other technological advances, had validated science as the pathway to future national progress. Other federal resource management agencies, including the Forest Service, Bureau of Reclamation, and Biological Survey, anchored their management policies in science, and Aldo Leopold, with publication of his groundbreaking text on wildlife management, established a very clear role for the biological sciences in wildlife conservation. Within the Park Service, a few of the agency's naturalists were engaged in field research, perhaps most notably Adolf Murie with his Mount McKinley wolf and Yellowstone coyote studies. In addition, in an unusual move, the Park Service approved a university-driven study of wolf-moose interactions at Isle Royale National Park designed to examine predator-prey relationships, establishing an early basic research initiative that has evolved into an uninterrupted decades-long project that demonstrates just how valuable the national park laboratory setting can be in promoting real understanding of the natural world.[10]

These research efforts received only modest support, however, and most were directed toward addressing immediate management problems rather than developing a formal research agenda. More often, agency scientists were assigned to work on visitor education programs, a situation fraught with professional frustration and one that prompted several to resign. Well aware of the problem, Professor Stanley Cain publicly chastised the Park Service at the 1959 Sixth Biennial Wilderness Conference for "missing a bet in the lack of an adequate natural history research program that would regularly feed into [its] interpretation programs the basic information which they now do without or get only by happy chance." If science was not valuable in its own right, then the Park Service should at least recognize its value in the agency's growing and popular visitor education efforts.[11]

The Leopold Report and Its Aftermath

As Mission 66 was running its course, the Park Service faced new controversies involving its traditional wildlife management policies, further calling into question the role of science in park management. Although the agency no longer eliminated predators from the parks, it still employed traditional range carrying-capacity principles to control wildlife numbers, as reflected in Yellowstone's regular practice of shooting its excess elk and bison. During the early 1960s, faced with public outrage over the annual slaughter, Interior secretary Stewart Udall appointed two outside scientific commissions to review the matter. He charged one, chaired by A. Starker Leopold, with reviewing the Park Service's wildlife management policies and asked the other, convened by the National Academy of Sciences, to address the agency's "natural history and research needs." The ensuing reports not only fundamentally altered national park resource management policy, but also rekindled interest in elevating the role of science within the agency.[12]

The Leopold report, with its lyrical and oft-quoted language, has become a classic in Park Service lore. Upon examining national park wildlife management policies, the Leopold commission proposed a fundamental overhaul to the agency's resource management policies: "As a primary goal, we would recommend that the biotic associations within each park be maintained, or where necessary recreated, as nearly as possible in the condition that prevailed when the area was first visited by white man. A national park should represent a vignette of primitive America." A primary element in the proposal was the notion of ecological restoration, which envisioned the Park Service employing adaptive management strategies under the supervision of "biologically trained personnel." Noting serious shortcomings in the agency's existing science program, the report urged "the expansion of the research activity in the Service to prepare for future management and restoration programs." It also elaborated on the restoration activities that it envisioned, including recovery efforts for extirpated predators to re-create historic predator-prey conditions, use of controlled burning to mimic nature's dynamic processes, and elimination of nonnative species. By linking ecological restoration in the parks with a new scientific research agenda, the report laid the groundwork for a new direction in national park policy, one that would incorporate science into the agency's resource management agenda.[13]

The National Academy report—the first hard look by an outside body at the role of science in the national parks—proved a stunning indictment of Park Service practices. Echoing the Leopold report, the Robbins report—named for committee chair William J. Robbins—concluded that the agency's science program was inadequate and lamented that science was not being appropriately employed in making park resource management decisions:

> An examination of natural history research in the National Park Service shows that it has been only incipient, consisting of many reports, numerous recommendations, vacillations in policy, and little action. Research by the National Park Service has lacked continuity, coordination, and depth. It has been marked by expediency rather than by long-term considerations.. . . In fact, the Committee is not convinced that the policies of the National Park Service have been such that the potential contribution of research and a research staff to the solution of the problems of the national parks is recognized and appreciated.

Concerned that "several . . . [of] the national parks will be degraded to a state totally different from that for which they were preserved," the report observed that less than 1 percent of the Park Service's budget was devoted to research, whereas other comparable agencies routinely expended 10 percent of their budgets on research.[14]

The Robbins report then offered a series of recommendations designed to integrate science and research into the Park Service's routine management practices. They included creating a "permanent, independent, and identifiable research unit" within the agency, conducting an inventory of natural history resources in each park unit, supporting mission-oriented research to address resource management issues, establishing research centers in individual parks, promoting research by university and other outside scientists, and consulting routinely with researchers over management decisions. The message was clear: the Park Service was guilty of ignoring science, the national parks were suffering as a result, and significant changes were essential.

What ensued were a series of unsuccessful, short-lived agency efforts to incorporate science and research into national park administration. Director George Hartzog, with Secretary Udall's blessings, sought congressional support for an expanded research program, but he encountered resistance and was forced to fund it from another agency account. To establish an independent research unit, Hartzog created an office of chief scientist, but the position only lasted for three years before it fell victim to another reorganization initiative. As a result, most agency scientists found themselves working for park superintendents, where their professional efforts were directed toward solving daily resource management problems, not the type of basic research that might detect potential systemic problems or provide much useful time-tested data. And, because they were dispersed throughout the agency, the scientists lacked any real clout and felt marginalized.[15]

Meanwhile, a nasty confrontation between the Park Service and two prominent independent researchers over grizzly bear management in Yellowstone cast serious public doubt on the agency's commitment to science-based decision making. Brothers John and Frank Craighead, both well-known independent wildlife researchers who had spent years studying the park's grizzly bears, objected to the agency's abrupt closure of the garbage dumps in the aftermath of the Leopold report, arguing that the now-habituated bears would inevitably come into conflict with visitors and end up being shot. Yellowstone officials rejected the brothers' publicly expressed doubts, closed the dumps, and then witnessed a large spike in bear deaths. The ensuing media attention only intensified the controversy, prompting the Park Service, in a heavily criticized move, to expel the Craigheads from the park. One point of contention in the controversy vividly illustrated the ongoing tension over scientific research and aesthetic preservation in the parks:

Yellowstone's superintendent objected to the Craigheads' practice of radio collaring bears for tracking purposes, believing that it detracted from the visitor's ability to observe wildlife in a natural setting. The Craighead bear study termination decision not only left park officials with limited scientific data to use in managing the park's diminishing bear population, but it suggested that the agency was not open to truly independent scientific research that might conflict with its management practices.[16]

As the years passed, the reports critical of the Park Service's commitment to science and research mounted, confirming that the agency still was not taking science seriously. In 1972, the Conservation Foundation released *National Parks for the Future*, expressing deep concern over the Park Service's inadequate knowledge about basic environmental conditions in the parks and calling for fully funded ecological research programs in each park to ensure that critical resources were adequately protected. In 1980, the Park Service acknowledged its own science deficit in its seminal *State of the Parks* report, identifying wholesale environmental threats to individual parks from adjacent development activities while conceding that the agency had little data to support its conclusions or frame a response. Seven years later, the General Accounting Office reported that little had changed: only half of the parks had completed resource management plans, and few parks had any long-term inventory or monitoring programs in place to assess resource conditions.[17]

Some progress on the scientific front was evident, however. During the 1970s, the Park Service had joined with several regional universities to establish cooperative park studies units designed to leverage the agency's meager research funds by enlisting university scientists in park research projects. In 1988, the agency revised its management policies and added new inventorying and monitoring requirements. The revisions also acknowledged the ecologically dynamic nature of park environments and collectively represented a further nod to science as an important resource management tool.[18]

Moreover, during the decades following World War II, the Park Service was attentive to its educational responsibilities. Building on its early visitor programs, the agency developed a well-regarded visitor interpretation program focused primarily on the natural history or historic events associated with each park unit. These programs—including campfire talks, guided hikes, and museum exhibits—introduced millions of Americans to the wonders of the natural world and helped instill in many a lifelong appreciation for nature. Only rarely, however, did park interpreters address controversial issues, a fact highlighted by the absence of discussion about adjacent development threats, wildlife habitat loss and migration route blockages, or distant pollution sources. This shortcoming, although partly attributable to political sensitivities, also reflected the agency's late entry into the field of scientific research and management that limited its ability to incorporate the latest scientific information into its public programs.

A Science Mandate, Finally

In 1992, when the National Academy of Sciences revisited the role of science in the national parks, it still found "crucial problems in the NPS [National Park Service] research program . . . rooted in the culture of the NPS and in the structure and support it gives to research." These problems, according the academy's *Science and the National Parks* report, called for a "fundamental metamorphosis . . . a new structure [and] . . . new culture." To achieve such a radical transformation, the academy recommended a new legislative mandate giving the Park Service an explicit research mission, separate funding and reporting autonomy for the science program, and designation of a high-level chief scientist to ensure credibility and sufficient independence. In its own report, also released in 1992 as the "Vail Agenda," the Park Service generally agreed with these recommendations. Citing an inadequate "information and resource management/research capability," the "Vail Agenda" endorsed the idea of a legislative research mandate and use of the "best available scientific research" for management decisions and educational programs, including new research initiatives when necessary.[19]

These renewed calls to elevate science in the Park Service soon bore fruit, but in an unexpected way. The Academy's report coincided with Bill Clinton's election to the presidency, and he made clear that science—not politics—would govern his approach to public land policy, including the national parks. Clinton's Interior secretary, Bruce Babbitt, promptly embraced the scientific mantle and created a new agency named the National Biological Survey. Babbitt's goal was to elevate ecological research within the Department of the Interior to a level comparable to that attained by the U.S. Geological Survey over the past century. When confronted with strenuous opposition from a hostile Congress, however, Babbitt implemented his plan administratively by transferring the Department of the Interior's scientists to this new agency, thus depleting the Park Service of its already-meager research capability. Although the relocated scientists were still expected to engage in research for the Park Service and other Interior agencies, the effect of Babbitt's order was devastating. By separating scientists from the agencies they served, the Biological Survey initiative represented a major setback in the Park Service's efforts to establish its own credible and effective science program.[20]

The stage was set, however, for yet another dramatic turn of events. In 1997, Park Service historian Richard West Sellars published *Preserving Nature in the National Parks*, a thoroughly documented chronicle of the agency's long and neglectful history of science in the parks. The book explains how scenic preservation—"façade management" in Sellars's terminology—and visitor concerns had consistently trumped the ecological sciences within the agency's engrained culture. Sellars's conclusion was both clear and troubling: "In both philosophy and management, the National Park Service remains a house divided—pressured from within and without to become a more scientifically informed and ecologically aware manager of public lands, yet remaining profoundly loyal to its traditions." Not surprisingly, the book created quite a stir and is widely credited with convincing Congress to finally bestow an explicit science research mission on the Park Service.[21]

In 1998, Congress passed the National Parks Omnibus Management Act, which "authorized and directed [the Secretary of the Interior] to assure that management of the National Park System is enhanced by the availability and utilization of a broad program of the highest quality science and information." To achieve this goal, the legislation required that science be employed in park management decisions (adding this requirement to annual superintendent performance reviews), mandated a systemic inventory and monitoring program to establish baseline resource conditions within the parks, and provided for a network of multidisciplinary cooperative study units that would enlist universities in this new national park research mission. Under these new congressional instructions, the Park Service was finally authorized to bring science to the fore, particularly in its resource management decisions.[22]

The Park Service responded by issuing the Natural Resources Challenge designed to rebuild and strengthen its science capacity. Billed as an effort to better integrate science and resource management, the challenge views scientific research, inventorying, and monitoring as essential to preserving and restoring national park ecosystems. It is built around several key issues, including endangered species, exotic species, air quality, water quality, stream flows, collaboration, and public education. Because the agency was devoting less than 7.5 percent of its budget to natural resources management, the challenge sought additional congressional funding to upgrade the role of science and ensure that park managers had the necessary information to tackle difficult resource issues, many with ramifications beyond park boundaries. Congress responded by appropriating $14 million in new funds, some of which has gone to support seventeen cooperative ecosystem studies units at regional universities and to establish new resource learning centers designed to further promote scientific research in the parks, disseminate scientific research findings, and integrate science into resource management decisions. With tight budgets and continued hostility to science in some political circles, it remains to be seen whether the Natural Resources Challenge will finally succeed in fulfilling the vision of national parks as natural laboratories devoted to independent scientific inquiry that has eluded the agency and its science advocates for so long.[23]

Education to the Fore

As the twenty-first century dawned, the drumbeat was intensifying to formally acknowledge in law that education is also part of the national park mission. The National Park Service Advisory Board, under the leadership of Professor John Hope Franklin, a widely respected historian and scholar, issued a visionary report entitled "Rethinking National Parks for the 21st Century" that was quite clear on this point:

> The Park Service should be viewed as . . . an [educational] institution. Parks are places to demonstrate the principles of biology, to illustrate the national experience as history, to engage formal and informal learners throughout their lifetime . . . to stimulate an understanding of history in its larger context . . . as the sum of the interconnection of all living things and forces that shape the earth.

In fact, as its first recommendation, the report asserted that "education should become a primary mission of the National Park Service." Noting that the parks "help us understand humanity's relationship to the natural world," the report urged the agency to "present human and environmental history as seamlessly connected. How one shaped the other is the story of America; they are indivisible." In 2009, the Second Century Commission embraced these recommendations, observing that "the national park system encompasses an unparalleled range of educational assets." It then went one step further and called on Congress to "affirm in legislation that education is central to the success of the National Park Service mission, and that the Service has a fundamental role to play in American education over the next century."[24]

For its part, the Park Service has significantly enhanced its education and interpretation policies, more closely aligning them with resource conservation. The agency's revised 2006 *Management Policies* admonish that national park educational programs must "be based on current scholarship and research about the history, science, and condition of park resources." Rather than viewing education and interpretation programs as a local park matter, they are seen as a shared national and local responsibility, one designed to "make national parks even more meaningful in the life of the nation" by fostering dialogue and understanding among visitors. Noting that some resource management and historical issues are inherently controversial and have implications beyond park boundaries, the policies instruct that the "parks should, in balanced and appropriate ways, thoroughly integrate resource issues and initiatives into their interpretation and education programs," giving weight to the historical and scientific evidence. The policies, in short, view the laboratory-like national parks as a natural classroom for improving public understanding about the natural world and the ecological connections linking parks with the surrounding landscape.[25]

Moreover, the Park Service has taken concrete steps to expand its educational efforts, primarily through creative programming in individual parks. At Santa Monica Mountains National Recreation Area, park officials have reached out to the Los Angeles public schools and regularly bring inner-city school children to the park, where they are introduced to its natural history and afforded hands-on experiences planting trees and doing other restoration work. These types of initiatives are quite timely, given the dwindling connection children have with the natural world, a phenomena that has been labeled "nature deficit disorder."[26] At Grand Canyon, Great Smoky Mountains, and other parks, rangers are now explaining to visitors the park's air pollution problems, including potential sources and ongoing efforts to alleviate the problem. Further, in its 2011 *Call to Action* document released in anticipation of the agency's centennial, the Park Service devotes an entire section to explaining how to "strengthen the Service as an education institution," noting various new opportunities to teach about nature and history to diverse audiences.[27] Agency officials are plainly beginning to recognize that the parks offer an obvious teaching forum to connect the general public with these landscapes and to garner support for the management policies necessary to preserve them.

Such programs and initiatives are none too soon given the mounting resource threats confronting the national parks, ranging from growing development pressures on adjacent lands to global climate changes that will plainly affect future generations and the world they inherit. Whether similar education programs can be established and maintained across the system is a matter very much tied to our evolving view of the national parks, one that links their place in the larger natural and human landscape with their place in the world of science, knowledge, and ideas. In any event, the foundation is in place for the national parks to embrace an explicit educational mission that ensures that they remain a relevant and vital institution capable of addressing the myriad conservation challenges that lay ahead, including those emanating from beyond park boundaries. [28]

PUTTING SCIENCE TO WORK

Slow to embrace science, either for basic or applied research purposes, the Park Service has found itself and its policies regularly ensnarled in an array of science-driven controversies. In the aftermath of the Leopold and Robbins reports, these controversies have frequently involved the agency's ecological restoration policies. Even when the scientific data and predicted outcomes offer strong support for a particular restoration initiative—be it wildfire, wolves, or watersheds—the opposition can be intense. Whether similar controversy will ensnarl climate change research in the parks remains to be seen, but the climate issue is already enmeshed in politics, which does not bode well. In short, the marriage of science and management in the national park setting is fraught with complexity, controversy, and challenge.

Wildfire Science and Policy

Wildfires were commonplace in the ecosystems now found in most western national parks. Long before Euro-Americans appeared on the scene or contemplated the national park idea, lightning-ignited fires regularly scorched California's mixed-conifer Sierra Nevadas and periodically raged through the Yellowstone plateau's lodgepole pine forests. Once these lands were set aside as national parks and forests, the federal government instituted an aggressive fire-suppression policy aimed at extinguishing all fires to safeguard the valuable scenery and timber from destruction. Early federal suppression efforts proved less than comprehensive, however; lightning-ignited blazes were routinely allowed to burn in remote and inaccessible backcountry locations where fire posed little threat to human life or private property. During the 1930s, however, the road and trail construction projects of the Civilian Conservation Corps helped open the backcountry, enabling the agencies to extend fire-control efforts into more remote areas. The post World War II period saw the arrival of surplus aircraft, bulldozers, chain saws, and other hardware that further extended suppression efforts across the landscape.[29]

Meanwhile, scientists began to question the wisdom of all-out fire suppression. Noting that fire was a natural force that had long shaped the West's forested ecosystems, the scientists cautioned that suppression policies had created an accumulating fuel load that portended even more intense wildfires. The seminal 1963 Leopold report, observing that "overprotection from natural ground fires" had altered Sierra Nevada ecosystems, called on the Park Service to use fire in a controlled way that was sensitive to the potential fuel hazard. According to these respected scientists, fire restoration in park ecosystems was essential to re-create a more natural setting that would also help improve wildlife habitat. Park ecosystems, having evolved in the presence of fire and other dynamic natural forces, could only be truly preserved if these forces were allowed to operate as they had historically. Other scientists, including the University of California's Harold Biswell, a highly regarded forestry professor and fire researcher, concurred with these conclusions. A professional consensus believed that these ecosystems were sufficiently altered through fire suppression to endanger the very giant sequoias and other majestic trees the Park Service was committed to protecting.[30]

During the late 1960s, science and policy converged in Sequoia and Kings Canyon National Parks. In 1964, drawing on the Leopold report and new scientific studies, park officials ignited the national park system's first controlled burns in Kings Canyon's Redwood Mountain Grove. Follow-up studies confirmed the ecological benefits derived from the fires, which had reduced the growing fuel load and returned the forest to a more natural trajectory. These findings encouraged the Park Service to initiate other controlled burns in Sequoia and Yosemite. After helping plan these burns, park scientists monitored the aftermath to better understand the role fire played in the ecosystem and the effects related to human-ignited burns. In 1972, Yosemite officials took the next step by designating a new natural fire zone where wildfires would be allowed to burn unimpeded, although subject to monitoring to guard against a runaway blaze that could endanger park neighbors. Paradoxically, with these science-driven shifts in its fire policy, the Park Service stood in the vanguard of the federal land management agencies, even as it otherwise remained reluctant to embrace science.[31]

Emboldened by the Sierra Nevada results, the Park Service began instituting fire-tolerant management policies elsewhere. Not only were some parks setting controlled burns to reduce fuel accumulation, but they were also allowing lightning-ignited fires to burn in the backcountry. These new fire policies, however, were subject to defined prescriptions designed to keep wildfires under control and to limit when controlled burning might be used. In 1985, an independent panel of scientists validated the Sierra Nevada parks' fire policies. The review panel found that fire restoration was critical to these ecosystems, but recommended adjustments to reduce unattractive scorching on the giant sequoia trees and further scientific studies to clarify fire history, dynamics, and related matters. With this vote of confidence, Park Service officials began allowing more fires to burn and began setting controlled burns across the system. Despite some complaints about smoke-obscured vistas and smoke-related breathing problems, the new fire policies otherwise proved unremarkable.

That changed, however, during the summer of 1988, when a series of fires raged across Yellowstone National Park and the surrounding national forests. With national headlines blaring that the park was being destroyed, the Park Service's fire management policies were exposed to public scrutiny and political attack. Although scientists assured that periodic conflagrations were a regular historic occurrence in Yellowstone's high-elevation lodgepole pine ecosystem and although some of the fires were initially fought as illegal human-caused ignitions, the common perception was one of a charred and devastated park, the result of a failed policy. Once the political posturing subsided, however, the agencies reaffirmed the general policy of reintroducing fire to the landscape, although faulting how it had been implemented and recommending tighter limits on natural fires and controlled burns. Seeing the fire-scarred park as fertile research ground, scientists predicted a natural recovery that would help restore the ecosystem to its historic condition. By most accounts, these predictions have proven accurate: the park's forests are returning, fire-created open spaces have provided wildlife with new meadow-like habitat, and some native vegetative species are expanding their range. The Yellowstone fire experience provided a remarkable opportunity to educate the public about the dynamic role of fire and ecological processes in shaping the park environments that they enjoy visiting. In addition, it vividly illustrated that the park was part of a much larger fire-sculpted ecosystem.[32]

These lessons were sorely tested a few years later when a Park Service–ignited controlled burn escaped containment at Bandelier National Monument. Carried along by unexpectedly high winds, the Cerro Grande fire roared across the landscape, ultimately destroying several hundred homes in Los Alamos, New Mexico, and endangering the Los Alamos National Laboratory. Although the recriminations were swift and the follow-up investigation revealed several errors in implementing the controlled burn policy, the final conclusion was to reconfirm once again the standard national park fire policy, subject to additional constraints to protect against runaway blazes. Even more than the Yellowstone fires, the Cerro Grande blaze illustrated how national park ecological restoration policies could transcend park boundaries with ramifications for adjacent public and private lands. Although science has thus far carried the day for fire policy, human values clearly cannot be ignored in establishing resource management policies, especially those with significant ramifications outside the parks.[33]

Wolf Ecology and Restoration

Wolf restoration in Yellowstone, although well-grounded in the biological sciences, provoked a firestorm of controversy that still persists. Not long after the Park Service eliminated wolves from Yellowstone in the late 1920s, the Murie brothers and other biologists began questioning the wisdom of this predator extermination policy. Most notably, during the early 1940s, Adolph Murie completed a year-long study of Mount McKinley National Park's wolves, concluding that they were not mere "beasts of destruction" but rather preyed mainly on weakened Dall sheep and other ungulates, and they did so without threatening the sheep population's survival.[34] Such studies helped convince Professor Aldo Leopold, by then one of the nation's most respected authorities on wildlife management, to reevaluate his own early predator extermination views, prompting him in 1944 to propose restoring wolves to Yellowstone National Park.[35]

Although Leopold's wolf restoration idea gradually gathered respectability among biologists, it made little headway in policy circles until the mid-1970s, when endangered species recovery became official federal policy. Under the Endangered Species Act of 1973, once the wolf was listed as a federally protected animal, the U.S. Fish and Wildlife Service set about developing a legally mandated recovery plan. The plan, based on the latest scientific information about wolves and their habitat needs, called for reintroducing the species to Yellowstone and other remote wilderness locations in the northern Rocky Mountains. After some initial uncertainty, the Park Service endorsed the plan, viewing wolf recovery as consistent with its post–Leopold report commitment to ecologically based management of the parks, which included restoring extirpated species. The wolf reintroduction proposal, however, was met with immediate and stiff resistance from the ranching community and the affected states.[36]

After Congress added a new experimental reintroduction provision to the Endangered Species Act in 1980, wolf recovery in Yellowstone was merely a matter of time. The new provision—known colloquially as Section 10(j)—provided additional flexibility for managing wolves (or other controversial reintroductions), helping reduce local resistance by giving ranchers the ability to defend their livestock against depredating wolves. Moreover, a series of scientific reviews, including a much-scrutinized environmental impact statement, confirmed that the Yellowstone ecosystem, with its abundant prey base, was ideal wolf habitat. Despite last-second litigation designed to stop the reintroduction, Yellowstone inaugurated a new era in mid-January 1995 when eight Canadian wolves were trucked into the park and placed in a holding pen to be gradually prepared for reintroduction into their new home. Once the courts cleared the way, the wolves were finally released into the park on March 21, 1995, where they set about reestablishing long-severed predator-prey relationships, particularly with the park's abundant and unsuspecting elk population.[37]

Release of these initial wolves, however, was just the first step in a much larger and well-documented ecological renaissance in the park. Over the intervening years, the park's wolf population has flourished, growing to fifteen packs and at least ninety-six wolves by 2009, and several hundred wolves roam the surrounding national forests and other nearby lands. Wolf recovery has provided scientists an extraordinary opportunity to study this keystone predator's role in the ecosystem and better understand vital ecological relationships. Within the park, the growing wolf population has already dramatically altered the prevailing ecology. Aspens, cottonwoods, and willows are making a comeback in the park's northern riparian areas, most likely because elk, fearful of the wolves in these open locations, no longer linger when feeding on the young tree shoots. As this riparian vegetation recovers, long-absent beavers have reappeared and are reshaping park stream systems with new dams, and songbirds are again frequenting these areas; wolves have significantly reduced the park's coyote populations, which has reduced pressure on antelope and other coyote prey species; and wolf-killed prey has benefitted an array of other species that feed on carrion. The lessons gleaned from this trophic cascade are proving invaluable from a scientific perspective. Moreover, as originally predicted, the park's wolves have become a magnet for visitors, who by some estimates are adding $7 million to $10 million annually to the local economy and extending the tourist season by several months.[38]

Controversy over the wolves nevertheless persists, as illuminated in the ongoing struggle over returning management responsibility to the surrounding states. According to the Endangered Species Act, once a protected species is recovered, the federal government will relinquish management to state wildlife officials, who ordinarily would oversee the animals. A "delisting" decision does not alter the Park Service's role managing wolves inside the park, but it does allow the surrounding states to establish wolf-hunting seasons and quotas outside park boundaries, which the three states bordering Yellowstone have done. Although fully protected inside the park, any wolves that cross the boundary are thus potential targets.[39]

The states of Montana, Idaho, and Wyoming have been eager to resume control of the wolves. State management would give area ranchers more flexibility in protecting their livestock from wayward wolves, and it would create a new hunting opportunity that will further control wolf numbers, reducing the effect wolves are having on local elk herds. State-sanctioned hunting outside the park can reverberate inside it, however, as occurred in late 2009 when hunters killed the alpha male and female from the park's most studied wolf pack, effectively eliminating that pack and truncating those scientific studies. In August 2010, a Montana federal court blocked the federal government's state-by-state delisting effort, ruling that the Endangered Species Act requires species recovery at an ecosystem level, a decision that would require greater management coordination among the park's neighboring states and with the park. Congress, though, responding to intense local political pressure in the aftermath of this court ruling, legislatively delisted the wolves in Montana and Idaho, returning full management authority to these states. After several false starts, Wyoming's wolf management plan has finally received the U.S. Fish and Wildlife Service's blessing, so it, too, has assumed responsibility for the wolf. Fearing the effect that increased hunting pressures might have on the wolf population, some observers have called for a buffer zone outside the park where wolf hunting would be curtailed or not allowed.[40]

By almost any measure, the Yellowstone wolf reintroduction effort has been a resounding success. The Yellowstone wolf population is officially classified as "recovered," highlighting the important ecological restoration role of the national parks in conserving wildlife. The restored wolves have begun to reshape the park ecosystem, providing rich opportunities for scientific study and wildlife viewing. As was evident from the beginning, however, biology is only part of wolf restoration; politics and litigation are also elements of the equation, and these forces are still at work. The affected states and other local interests—notwithstanding opposition from most environmental groups—have now gained much greater control over the expanding wolf population. As new state management policies take effect, the park's wolves will be increasingly vulnerable when they stray outside the boundary line, confirming yet again that the park itself is an imperfect wildlife reserve, one that is quite sensitive to external influences and events.

The successful Yellowstone wolf reintroduction has spurred calls to reintroduce wolves elsewhere in the national park system. At Rocky Mountain National Park, where wolves were exterminated at the beginning of the twentieth century, wolf advocates contend that a restored wolf population would help control the park's burgeoning elk numbers, thus avoiding the need to cull elk from the herd. Although park officials considered wolf reintroduction when revising their elk management plan, they decided against it, concluding that the wolves would create too many conflicts with nearby communities and ranchers, who did not support this option. In the Park Service's view, both the park and its elk herd were too small, when compared with Yellowstone, to support a robust wolf population, a view shared by other federal, state, and local agencies that also objected to the restoration option. Of course, wolf restoration at the park might have served as a laboratory for other restorations in the southern Rocky Mountains, such as southern Colorado's San Juan Mountains, but Park Service officials did not see this role as appropriate for the park, especially when their primary concern was elk management, not wolf restoration.[41] The wolf restoration idea has also surfaced at Olympic and Grand Canyon National Parks. Despite its own management policies, which support wildlife restoration efforts to reestablish natural processes under defined conditions, the Park Service has yet to take any meaningful action on these proposals. Were wolves to disperse naturally into these or other any national parks, however, they would be protected under the Endangered Species Act as well as the Organic Act. And their presence, besides helping restore impoverished park ecosystems to a more vital state, would afford additional research opportunities in these natural laboratory settings.[42]

Restoring River Ecosystems

The national watershed restoration effort extends to the national parks, where the notion of restoring natural flows squares with current scientific thinking. As integral parts of larger watersheds, several national parks have seen their hydrologic character altered by upstream dams and diversions over the years, upsetting the ecological order. The seminal incident occurred at the turn of the twentieth century in Yosemite National Park, where construction of the O'Shaughnessy Dam inundated the entire Hetch Hetchy Valley to provide the city of San Francisco a secure water supply. Although the dam project cost Yosemite a valley that John Muir described as "a wonderfully exact counterpart of the great Yosemite [Valley]," the incident spawned passage of the National Park Service Organic Act with its stringent nonimpairment mandate. Given the hard realities of California water politics, it may be too late to restore Hetch Hetchy to its former glory, but at least one former Interior secretary floated the idea in 1987, suggesting that the offending dam be removed and the valley restored. In any event, the Park Service is already deeply engaged in several major watershed restoration efforts, including the massive Everglades restoration project that will be examined later as an example of the emerging role national parks are assuming as the core of larger ecosystems.[43]

Among the western parks, Grand Canyon and Olympic are each engaged in precedent-setting initiatives to restore vital river corridors where upstream dams have disrupted water flows and significantly altered the ecological setting. Shaped over the millennia by annual spring floods on the Colorado River, the Grand Canyon is no longer subject to these formative events, thanks to the upstream Glen Canyon Dam. The dam, completed in 1963, has tamed the river below it. Because the dam effectively blocks spring floods and traps sediment, the downstream beaches and backwater habitat that were typically formed from the silt-laden flood waters have disappeared. In addition, because the dam releases much colder water than historically flowed downstream, which in turn has enabled the introduction of nonnative fish species, the river's native warm-water fish species—namely the razorback sucker and humpback chub—that depended on this aquatic habitat have been devastated and are now listed as endangered species. In response, scientists proposed revising the Bureau of Reclamation's dam operations by periodically releasing extra water in an effort to mimic historic high-flow spring runoff conditions, a proposal that Congress endorsed in the Grand Canyon Protection Act of 1992.

This restoration experiment has since yielded mixed results. High-flow dam releases in 1996, 2004, and 2008 provided valuable scientific information and helped temporarily rebuild some beach sites while also improving habitat for the warm-water fish species. But scientists have concluded that periodic experimental releases alone will not restore the river ecosystem; such restoration will require an integrated sustained habitat improvement effort. As a result, the Bureau of Reclamation, Park Service, and U.S. Fish and Wildlife Service are implementing a more aggressive and comprehensive adaptive management strategy. First, they are establishing long-term protocols for testing high-flow dam releases to determine whether multiple high-flow events can help rebuild and conserve sandbars, beaches, and associated backwater habitats. Second, they are assessing how to improve habitat for the native fish and how to protect them from predation or competition by the nonnative fish. Meanwhile, convinced that the Glen Canyon Dam was a monumental mistake, dam opponents have called for its dismantling, an unlikely outcome but one that would return the downstream park's river ecosystem closer to its historic condition.[44]

In Olympic National Park, as we have seen, the Park Service is in the process of removing two dams on the Elwha River that block historic salmon migration to spawning habitat within the park's interior. Built in 1913 and 1927, before the Olympic National Monument was redesignated a national park, the Elwha and Glines Canyon Dams served as a power source for the nearby town of Port Angeles and a local lumber mill. Once constructed, the dams plugged the river and prevented four hundred thousand salmon from annually moving upstream, eliminating an important species from the park's ecosystem and destroying the nearby tribe's treaty fishing rights. The dams have outlived their usefulness, however, enabling Congress to conclude that the environmental and other benefits of removal outweighed the costs involved in demolishing the dams, including local financial assistance and alternative energy arrangements. According to park scientists, once the dams are removed, salmon will be able to swim seventy miles upstream into the park, where they can spawn and provide a food source for other wildlife. The Park Service began demolishing the dams in late 2011, a process anticipated to take three years to complete and to cost in excess of $300 million.[45]

Unlike the Grand Canyon restoration effort, the dams on the Elwha are located inside or proximate to the park and provided modest local economic benefits. Moreover, most of the Elwha River watershed lies within Olympic National Park, where it is managed as wilderness and provides rare pristine wildlife habitat, all of which has enabled dam removal proponents to align science with politics to move the project ahead. These two restoration initiatives illustrate an important larger point, however: both Grand Canyon and Olympic are inherently connected, ecologically and economically, to the surrounding landscape and thus cannot rely solely on science to drive their restoration agendas.

The Climate Change Challenge

The impending reality of global climate change presents the national parks with unprecedented scientific and conservation challenges. With the world's temperatures creeping upward and precipitation patterns shifting in response to greenhouse gases trapped in Earth's atmosphere, scientists predict profound changes in the natural systems that support life on Earth. Without significant carbon reductions in our energy-driven economies, these changes will be felt across the spectrum of ecosystems, although more acutely in some regions than others. The national park system, extending across an incredible array of largely undisturbed ecosystems, is not only being affected by these changes, but it serves as a barometer for how natural systems respond to climate disruption. The parks are thus truly nature's laboratory for improving our understanding of climate-related changes and for devising responsive conservation strategies.[46]

Several national parks already manifest the effects of climate change, and other changes are widely predicted across the system. Glacier National Park's namesake glaciers have been disappearing at an unprecedented rate and may be gone by 2030, and other national parks, including North Cascades, Mount Rainier, Olympic, and Yosemite, are also experiencing glacial loss. Predicted changes in the timing, amount, and duration of precipitation events will affect much of the Southwest and drastically alter the Colorado River ecosystem through the Grand Canyon. Having documented warming temperatures at Joshua Tree National Park, scientists predict that the park's namesake Joshua trees will soon disappear entirely. In Yellowstone, a drought-driven pine bark beetle infestation is killing the park's high-elevation whitebark pine trees, eliminating a key seasonal food source for grizzly bears and putting them at increased risk of deadly human encounters as they venture farther afield in search of alternative food sources. Similar heat- and drought-related stress has destroyed piñon pine trees in Bandelier and Mesa Verde and has also affected bird life and other small mammals in these parks.

Even other effects are forecast. As the polar ice caps melt, scientists predict a significant rise in sea levels, which could imperil coastal parks stretching from Biscayne Bay and the Virgin Islands to Channel Islands, Olympic, Acadia, and Alaska. Species dependent on specific climatic conditions—such as pikas, which require high-elevation mountain habitat, and desert bighorn sheep—could be lost or displaced as temperatures rise. Increased drought will trigger more frequent and intense wildfires that will change vegetative patterns and hence wildlife habitat and food sources. These changes will enable more nonnative species to find their way into the parks, creating additional competition for already-stressed native species. Moreover, changes in precipitation patterns as well as prolonged drought conditions will mean less water in park rivers and streams, altering the critical in-stream and riparian habitats that harbor sensitive aquatic, amphibian, and terrestrial species. In short, climate-related changes in the national parks will not only fundamentally change these ecosystems but will eliminate some of their most prominent and popular features.[47]

Understanding these climate-induced changes is a crucial first step to addressing them effectively. The national park system, with its strikingly diverse assortment of intact ecosystems, provides scientists with an ideal laboratory environment for gathering critical baseline information and assessing how a changing climate affects natural systems, including vegetative conditions, native species, water levels, and wildfire frequency. As warmer temperatures take hold across the national parks, scientists have an unparalleled opportunity to monitor and assess how these relatively undisturbed natural systems respond to rising temperatures, including their role as potential carbon sinks, and to then share this basic research information. The lessons derived from these baseline studies will be germane not only to the national park system but beyond it as well. Once we more fully comprehend and can demonstrate the dire implications of global warming, we should be better able to design the adaptation and mitigation strategies necessary to preserve at-risk species and other resources while securing public support for these measures. The Park Service cannot effectively preserve our natural heritage if it does not understand how increased temperatures are affecting it.[48]

Beyond basic research, the national parks have another important role in adapting conservation policy to meet the climate change challenge. And just as the Park Service has historically geared its scientific research agenda toward addressing resource management problems, future climate change research should be geared toward meeting the agency's conservation obligations. Just as the Park Service used the Leopold report to embark on a new natural regulation management experiment, so, too, the climate change crisis presents an opportunity to view the parks as an experimental laboratory for testing new climate-based resource management policies. As temperatures warm and changes occur in the parks, conservation policy must be redesigned to ensure resilient ecosystems, secure wildlife corridors that enable displaced species to move to more suitable habitat, and provide other adaptations necessary to protect park resources. To meet these conservation challenges, according to most observers, the sheer scale of climate change will require coordinated landscape-scale planning efforts involving the national parks and their neighbors. None of it will be easy, but the parks offer an ideal laboratory setting for designing and pursuing new conservation policies in the forthcoming heat-stressed environment.[49]

GIVING SCIENCE AND EDUCATION THEIR DUE

In today's world, the case for science and scientific management in the national park laboratory setting is compelling. Indeed, from George Wright's *Faunal Survey* to the Leopold and National Academy reports, the case for science-based conservation policies was made convincingly enough so that Congress finally enacted legislation in 1998 to incorporate science into the Park Service's organic mandate. The effect of science on the agency's culture and policies is evident: resource management goals that are now focused on conserving dynamic ecosystems rather than preserving static scenery, diverse basic and applied research initiatives that are redefining conservation and the role of national parks in these efforts, new ecological data linking the national parks to the surrounding landscapes with attendant management implications, and the opportunity to enhance national park education programs and the public's understanding of contemporary conservation challenges. The rise of science has also provoked criticism of the Park Service's approach to scientific research, however, and even the best science cannot dictate resource management policy without taking account of competing human concerns and values. Nonetheless, the national park as an outdoor laboratory has spawned a more objective and ecologically focused approach to conservation in the national parks and elsewhere.

The national parks are an ideal laboratory to examine and better understand the natural world and its dynamic processes. As the Park Service strengthens its commitment to science, basic and applied research opportunities abound, many with management implications. Extending from the Rocky Mountains and the southwestern deserts to eastern hardwood forests and tropical Caribbean beaches, the national park system encompasses a diverse array of ecosystems, each shaped by different natural processes and reflecting its own distinct biological relationships. These largely undisturbed park settings offer scientists a unique baseline for studying and understanding how these systems behave and how human activities can affect them. As noted, Park Service scientists and their academic counterparts are increasingly availing themselves of this opportunity, assembling important data on wildfire behavior, predator-prey interactions, watershed conditions, climate change ramifications, and the like. Once the data are subsumed, better informed resource management decisions and the ability to anticipate new conservation issues should result.

The ecological sciences are critical to forging a deeper understanding of the national parks and their myriad connections to the surrounding landscape. The scientific research accompanying the Park Service's fire management, wolf reintroduction, and river restoration initiatives has not only identified linkages between the parks and the larger landscape, but has also aided resource managers in molding their policies accordingly. In addition, as scientists gain greater understanding of how climate change will affect park wildlife populations, this knowledge will aid in designing future landscape-level conservation efforts to reduce long-term survival risks. Without a solid science-based foundation for these ecosystem-level restoration efforts, park officials might find it difficult to withstand the inevitable political pressures that such initiatives generate.

Laboratories also serve teaching functions, and the Park Service has long been in the business of educating the American public about the natural world. In fact, the national parks provide an ideal outdoor classroom where the agency can introduce visitors to unbridled nature and educate them about conservation issues. Drawing on its historical commitment to interpretation as well as its new scientific research responsibilities, the Park Service's challenge is to distill and impart cutting-edge knowledge to the public and to link that knowledge to its own conservation concerns. Nowhere is this challenge more compelling or more controversial than in the area of climate change, the science and policy of which has become enmeshed in partisan politics. Through its long-standing nature interpretation program, the Park Service stands alone among the federal land management agencies, uniquely situated to promote greater public understanding of what a changing climate means for diverse ecosystems and the available options for sustaining them. The agency is thus in a unique position to not only educate the public about nature conservation in today's world, but also to marshal the public's support for science-based policies designed to meet tomorrow's challenges.

Laboratories promote science, and science inevitably generates criticism, even in the national park setting. As the Park Service has adopted a more science-based mission, critics have accused the agency of embracing "bad" science and rejecting contrary independent research. (Some of the most contentious criticisms have been directed toward the science underlying Yellowstone's wildlife management policies, a subject that is explored in the next chapter.) Critics also have chastised park officials for not supporting independent research, noting unduly close connections between the park's science, management, and education programs. Citing the Craighead incident and other cases, they maintain that park officials have ousted critical independent researchers from the parks and have denied others the permits necessary to conduct park-related research. One academic researcher believes that "views contrary to park positions have been looked on as threats and avoided."[50]

The national parks can only fulfill their natural laboratory role if park officials foster an atmosphere of independent scientific inquiry and are willing to countenance the opposing views that will inevitably emerge. The robust debate over fact and policy that ensues can only redound to the benefit of the parks and the public that treasures them. Whether the research is undertaken by park or nonpark scientists, researchers must enjoy sufficient autonomy to draw their own independent conclusions, regardless of potential management implications. As long as the parks are open to independent researchers, the opportunity to test accepted fact and to investigate alternative explanations should protect against "bad" science and forestall unwarranted conclusions. When science forms the basis for management decisions, that science can—and should—be tested through the mediums of peer review and National Environmental Policy Act–based environmental analyses that are subject to public scrutiny and further testing. Such accountability is essential to ensure that the Park Service's scientific conclusions and conservation efforts are well-grounded in fact and rest on the best available data. Not only is this approach consistent with standard scientific protocol, but it should also promote sound resource management policies and decisions.

Although science is playing a more salient role in the national parks, it does not mean that Park Service resource management policies can be based solely on science. As laboratories, the parks provide an important experimental setting where the prevailing conservation policy is to allow nature to take its course with minimal human intervention. Although this resource management approach may yield important scientific knowledge otherwise unobtainable, park managers cannot discount potentially competing human values and concerns. In fact, the proximity of the national parks to other federal and private lands that are managed quite differently poses inevitable conflicts, ones that often pit scientific truths against economic and other concerns. Wildfires may represent an important regenerative ecological force in Yellowstone's lodgepole pine forests, but when flames cross the park's boundaries and endanger nearby communities, fire becomes a destructive force with potentially disastrous human consequences. In short, although scientific knowledge can provide managers with a factual foundation for evaluating their options, it is not a substitute for the human values—economic welfare, social stability, aesthetic considerations, and more—that must also be factored into any resource management decision. At least now, with the ascendance of science in park conservation policy, the true benefits and costs—including long-term ecological sustainability concerns—are part of the resource management calculus.

The idea of the national park as nature's laboratory is reshaping our image of conservation policy and the national park itself. Giving science a more conspicuous role in policy making has refocused the Park Service's conservation efforts, both allowing nature freer rein within the parks and introducing new ecological restoration efforts that are fundamentally remaking the park environment. Although tourism, recreation, and economic concerns are still very much on the agency's mind,[51] these concerns must now be counterbalanced against a science-driven conservation agenda that puts value in nature and its processes. The point is not that one idea—nature's laboratory—has triumphed over others, but that science is now acknowledged as a vital dimension of nature conservation, forcing us to view the national parks in a new light. With this fresh look, informed by the evolving sciences of ecology and climate change, we can better understand how best to conserve park resources and just how connected parks are to the world around them.

CHAPTER 8

"Fountains of Life"

An (Imperfect) Wildlife Reserve

From the beginning, the national parks have served as a sanctuary for wildlife, a primary feature that has long drawn visitors to the parks. Whether it is a roadside bison jam in Yellowstone or a chance encounter with a solitary bear in the Great Smoky Mountains backcountry, the experience of viewing native wildlife in its own habitat has thrilled generations of park visitors. Wildlife conservation has a checkered history in the national parks, however. Not only have Park Service wildlife management policies flip-flopped over the years, but the agency has regularly faced intense political pressures in its quest to safeguard park wildlife. Animals may generally be protected in the national parks, but that did not stop the Park Service from eliminating wolves and other predators during its early years, nor has it stopped the agency from culling excess elk and deer to restore park ecosystems. Nor has the Park Service's preservation mandate protected park wildlife when they wander beyond park boundaries, something that occurs regularly in the seasonal quest for food. Whether viewed from a historical, scientific, or political perspective, the national parks are imperfect wildlife reserves that nonetheless play a crucial conservation role.

WILDLIFE POLICY IN THE NATIONAL PARKS

During the late nineteenth century, when the national park movement was gathering momentum, the general philosophy toward wildlife was utilitarian. Like other natural resources, wild animals were valued in terms of what they contributed to human welfare: hides in the case of bison, food in the case of elk, and predation in the case of wolves and other predators. The result, during the nation's westward expansion, was an all-out assault on the animals that occupied the landscape. Several species, including the bison, grizzly bear, and wolf, were driven to near extinction, only finding refuge in remote terrain far removed from the market hunters, bounty seekers, and advancing settlers. One such refuge was Yellowstone National Park, where a handful of bison, elk, and grizzly bears survived the onslaught, but even then they faced an uncertain future. Although protected in theory, Yellowstone's wildlife were still at the mercy of local residents and other fortune seekers, who regularly killed the new park's animals for sustenance, profit, and sport. What conservation ethic that existed rarely extended to wildlife, just to the geysers, rugged mountains, and other scenic wonders that dotted this unusual landscape.

Wildlife Conservation in the Early Parks

Indeed, wildlife conservation was not on Congress's mind much in 1872 when it established Yellowstone National Park. Not only did the Yellowstone enabling legislation not prohibit the killing of park wildlife (it only outlawed the "wanton destruction of fish and game within said park"), the new park's straight-line boundaries took no account of wildlife habitat needs. Once it became apparent how vulnerable the park was to poachers and other interlopers, the U.S. Cavalry was enlisted to oversee the area and to protect park wildlife and other resources. But with few roads and a vast landscape to patrol, the troopers found it difficult to effectively police the area. Nor could they look to the law for assistance, because the newly written park management regulations set forth no penalty for killing park animals.[1]

That soon changed, however, after a cavalry patrol encountered a notorious local poacher who had just finished slaughtering several park bison. Although the commanding officer lacked the legal authority to arrest the poacher, the reporter and photographer accompanying the patrol recorded the incident for George Bird Grinnell's popular *Forest and Stream* magazine. The resulting story and photos triggered an intense public outcry, prompting Congress to respond with the Lacey Act in 1894, which outlawed hunting in Yellowstone and thus set a no-hunting standard for future national parks.[2] Even then, though, it took years to control poaching in the fledgling national parks; as late as 1906, Yosemite's acting military superintendent reported that Yosemite Valley was a "death trap" for all wild game.[3]

There were also signature examples of enlightened park caretakers undertaking early wildlife conservation initiatives. In fact, some of the nation's earliest and most successful wildlife restoration initiatives are linked to the national parks. At the end of the nineteenth century, as the bison bloodbath wound down, Yellowstone offered sanctuary for the nation's nearly extinct bison population; only a handful of wild bison remained alive, having found refuge in the new park. Recognizing their perilous state, the cavalry officer in charge took the remnant bison herd under his wing. He established the Buffalo Jones Ranch at Mammoth and then supplemented the remaining animals with domestic bison imported from Texas and Montana. Over time, these efforts succeeded in nurturing the bison population back to health, and the park is today home to more than five thousand of these emblematic beasts, which have long graced the Interior Department's official seal.[4]

With its protective wildlife policies, Yellowstone has also served as a big-game reservoir or incubator. During the early twentieth century, several western states, including Colorado, Montana, and Washington, rebuilt their depleted elk populations with transplanted park animals. The replenished herds were an unequivocal boon to local hunters, who were the obvious direct beneficiaries of Yellowstone's protective wildlife management policies. In fact, Grinnell regularly urged his hunter-readers to support the national parks movement, arguing that the wildlife protected in the parks would eventually leave these sanctuaries and become quarry available for hunters.[5]

When the Park Service was created in 1916, Congress incorporated wildlife conservation into the Organic Act as one of the new agency's primary responsibilities. One year later, when Congress established Mount McKinley National Park in Alaska, it specifically stated that "the park shall be . . . established as a game refuge" in an effort to safeguard the area's Dall sheep and caribou that were rapidly falling prey to trophy hunters.[6] Curiously, though, the seminal 1918 Lane letter provided little guidance for how the Park Service would meet its wildlife-related responsibilities; it merely outlawed hunting in the parks and instructed agency officials to consult with other government bureaus about "the care of wild animals, and the propagation and distribution of fish."[7] With its attention directed toward promoting visitation, the Park Service was keen to establish wildlife management policies that would enable park visitors to see and experience the animals.

The result was an amalgam of visitor-oriented—or anthropocentric—wildlife policies, few of which withstood the test of time. Park Service director Stephen Mather permitted construction of a zoo at Yosemite that allowed visitors to come face-to-face with the park's elk and other animals. He also approved importing nonnative tule elk into the park, where they were confined in a fenced enclosure. The Buffalo Jones Ranch in Yellowstone, where rangers herded the park's bison much like domestic livestock, rapidly degenerated into more of a tourist attraction than a working ranch. In Yellowstone, Yosemite, and elsewhere, the Park Service put bears on display by feeding them at garbage dumps where bleachers were built to accommodate hordes of curious visitors. These same park visitors were not discouraged from feeding the bears, which only further habituated them to humans and prompted inevitable bear-human conflicts that usually ended badly for the bear. To make sure that plenty of elk, bison, and other "good" animals were on hand, the agency also pursued a relentless predator eradication policy to rid the parks of wolves, cougars, and other "bad" animals. To promote sportfishing, the Park Service—following the lead of the early military caretakers—transplanted rainbow trout and other exotic fish species into park waters so that anglers would have game fish to catch during their visits.[8]

It was soon apparent, however, that many of these early wildlife management policies lacked any biological rationale. A vivid illustration of the problem occurred at Grand Canyon, where the federal predator eradication campaign proved so successful that the area's deer population exploded in the absence of any cougars or wolves to keep herd numbers in check. Faced with an ecological disaster on Arizona's Kaibab Plateau as the deer ravaged the available forage, the Forest Service began shooting the excess deer, a move that prompted the state of Arizona to sue, asserting that it—not the federal government—was responsible for managing wild game, even on federal public lands. The federal position eventually prevailed in court, and the culling proceeded; the legal precedent was later invoked by the Park Service to support its own deer-culling campaign at Carlsbad Caverns over the strenuous objection of New Mexico's wildlife officials.[9] Not all the early policies were such evident failures, however. In Yosemite, at John Muir's urging, the park's initial military caretakers closed the park to the domestic sheep that swarmed across its mountain meadows, thus removing them as competitors with the park's foraging wildlife. Nonetheless, the Park Service's principal focus was plainly on protecting scenery and "good" animals, with little regard for the biological consequences that might ensue.

Proposing a New Direction

The absence of scientific rigor from the agency's early wildlife management efforts was not surprising because the Organic Act contained no reference to science. The resulting problems were initially brought into the open during the early 1930s, although without conclusive resolution. Professor Joseph Grinnell (a distant cousin of George Bird Grinnell, editor of *Forest and Stream*) from the University of California at Berkeley was an early critic, arguing that the national parks "should not be artificial" but should be managed according to scientific principles with minimal human intervention. To promote his agenda, Grinnell organized the agency's first naturalist conference in 1929, bringing together scientists and Park Service field personnel to address natural resource management matters. At Grinnell's urging, Yosemite officials closed the park's zoo in 1932 and relocated its exotic tule elk population a year later. Several of Grinnell's students also went to work for the Park Service, where they sought to reverse the agency's heavy-handed wildlife policies. None was more important than the independently wealthy George Melendez Wright, who used his own funds and personal charm to spearhead the reform effort.[10]

A biologist by training, Wright was convinced that the Park Service needed to elevate the stature of scientists within the agency and incorporate scientific principles into park management policy. With Horace Albright's support, Wright along with his colleagues, several of whom were also former Grinnell students, prepared a groundbreaking *Faunal Survey* report that not only reviewed the current status of park wildlife populations but also critically examined the prevailing policies. Wright's report amounted to a near complete repudiation of long-standing national park policy as captured in his argument that the ultimate management objective should be "to restore and perpetuate the fauna in its pristine state by combating the harmful effects of human influence." The opportunity for visitors to view wildlife in the parks, he contended, was no more important than preserving animals and plants in their primitive state.[11]

Wright's specific recommendations went even further. In his view, each park should contain the required habitat to maintain minimum viable species populations on a year-round basis, park boundaries should be designed to follow natural features, wildlife should not be artificially fed unless necessary to save a species, native predators should not be eradicated and should be restored where previously extirpated, exotic species should be eliminated from the parks, wildlife management strategies should entail minimal human interference, and management policies and adjustments must be based on scientific research. In a subsequent report, Wright went on to advocate for the creation of buffer areas around national parks to safeguard park wildlife that depended on surrounding lands for their seasonal habitat needs.[12]

The visionary Wright report launched the agency on a new path, but one that would not be fully realized for more than thirty years. One of Wright's key proposals—the formation of a new wildlife division within the National Park Service—was implemented almost immediately with Wright at the helm, providing the agency with technical expertise it had not previously enjoyed. Unfortunately, Wright himself died in an automobile accident in 1936, just as several of his recommendations were being incorporated into official policy. By 1938, the new division had more than thirty biologists, and wildlife rangers were stationed in each of the national parks, representing a notable departure from the Park Service's long-standing practice of hiring primarily landscape architects, who focused on protecting park vistas not wildlife. Even with this infusion of new scientific talent, however, the Park Service workforce stood at four hundred landscape architects compared to a mere twenty-seven biologists.[13] In the policy realm, the Wright report helped convince the agency to alter its predator control policies, close its bear-viewing sites, acknowledge a role for fire on the landscape, stop introducing exotic fish species into park waters, and establish a scientific research program.[14] With the Great Depression in full swing and then with the advent of World War II, however, agency officials faced severe budget constraints and soon retreated from any major reform effort.[15]

Conjoining Science and Policy

Once the war ended, as we have seen, the Park Service turned its attention elsewhere and embarked upon the ambitious Mission 66 program, directing its energy and resources toward visitor facilities and other infrastructure repairs. Meeting the needs of the tourists who were flooding into the national parks took priority over wildlife and resource management concerns, which continued to languish until the early 1960s. Wildlife controversies still dogged the national parks, though. None was more prominent than charges that Yellowstone officials were mishandling the park's northern range elk herd, which was annually culled by Park Service sharpshooters to keep population numbers in check. In response to growing criticism over this policy, Interior secretary Stewart Udall commissioned two studies, both to be overseen by outside scientific experts. As previously noted, one was led by prominent biologist A. Starker Leopold and was charged with reviewing the Park Service's wildlife management policies, and the other was directed by the National Academy of Sciences to review the agency's "natural history and research needs."[16]

Released on the eve of what was soon to become a national environmental movement, the Leopold and Robbins reports proved watersheds in national park wildlife management policy. The concise yet eloquent Leopold report made a powerful case for revising the Park Service's resource management policies, echoing many of the views found in Wright's early reports. After declaring that "a national park should represent a vignette of primitive America," the report called for "an overall scheme to preserve or restore a natural biotic scene." According to the scientists, this would entail restoring missing species, eradicating exotic species, stopping artificial feeding programs, reducing road construction, eliminating unnecessary tourism facilities, and enhancing the Park Service's scientific research capabilities. To accomplish these policy objectives, the report acknowledged that intensive management, based on the best ecological data available, would be necessary in some instances, including the controlled use of fire and the shooting of excess ungulates. Moreover, the report noted that most parks were too small to meet the essential habitat requirements of resident species and that past human manipulations or intrusions had so altered ecological processes that active intervention would be necessary to restore anything approaching a natural ecological order.[17] Secretary Udall responded by instructing the Park Service to take appropriate steps "to incorporate the philosophy and the basic findings into the administration of the National Park System."[18]

Released a few months later, the Robbins report largely reinforced the Leopold report's recommendations, both as to appropriate resource management goals and the role of science in achieving them. The report called for the "maintenance of natural conditions" in the parks, explaining that "each park should be regarded as a system of interrelated plants, animals, and habitat (an ecosystem) in which evolutionary processes will occur under such control and guidance as seems necessary to preserve its unique features." In addition, the report recommended that Park Service research efforts pay "specific attention to significant changes in land use, in other natural resource use, or in other economic activities on areas adjacent to national parks, and likely to affect the parks." It also promoted the concept of "cooperative planning" among the federal, state, and private entities responsible for conservation and recreation.[19]

The Leopold and Robbins reports had an immediate, far-reaching, and lasting effect on Park Service resource management policies. In 1964, citing the reports, Udall directed the agency to manage national parks "toward maintaining and where necessary reestablishing, indigenous species" while "preserving the total environment."[20] In 1968, the Park Service issued a policy document providing that national parks should be managed as ecological entities. The document stated that "the concept of preservation of a total environment, as compared with the protection of an individual feature or species, is a distinguishing feature of national park management." Noting that the national parks were becoming "islands of primitive America" affected by development activities on surrounding lands and by escalating visitor numbers, the document called for "active management" of the natural environment. It then concluded that such an approach will entail "application of ecological management techniques to neutralize the unnatural influence of man, thus permitting the natural environment to be maintained essentially by nature."[21] Since then, the agency has sought to adhere to these basic principles in its wildlife management policies, although not without continuing controversy over appropriate goals and managerial strategies.

Just as the Park Service began to implement the Leopold report's science-based recommendations, the agency was also confronted with a full-fledged national environmental movement that gave additional impetus to these policy changes. During the 1960s and early 1970s, Congress passed an array of new environmental laws that included the Wilderness Act, National Environmental Policy Act, and Endangered Species Act, which have had significant effects on the agency's resource management policies and practices. With these laws on the books, the Park Service was now forced to develop and explain its resource management decisions in the full glare of public scrutiny. Not only were agency officials required to follow the National Environmental Policy Act public participation procedures in their decision-making process, but they also had to be prepared to defend their decisions in court, as these laws opened the courthouse doors for almost anyone dissatisfied with an agency decision. Moreover, the Endangered Species Act cemented into federal law the principle that all species are important and must be protected against extinction, thus further eroding the tendency toward single-species management and prompting a much broader commitment to biodiversity conservation. In addition, the Endangered Species Act, following a 1982 experimental population amendment, provided the legal foundation for reintroducing extirpated predators to the national parks, most notably the long-absent wolf to Yellowstone, where its presence is reshaping the park's ecology.[22]

The Hunting Issue

After the Leopold report, the question of hunting in the national parks finally began to recede as an issue, but it can still generate controversy for park managers. Although the Organic Act did not directly address hunting in the parks, the 1918 Lane letter expressly rejected hunting as an acceptable activity, a position the Park Service has adhered to ever since. During the 1950s and early 1960s, however, with the Park Service engaged in large-scale ungulate reductions in Yellowstone and other western parks, state game commissions and hunter groups campaigned to open these parks to sport hunting, arguing that state-licensed hunters should be allowed to do the shooting as a recreational activity. State-supervised sport hunting is allowed on all other federally owned lands, including wilderness areas and national wildlife reserves, but opponents of hunting in the national parks feared the slippery slope; if sport hunting was introduced into the parks, then even more consumptive activities, including mining, logging, and livestock grazing, might claim a right to similar access. Further exacerbating the matter is that Congress, in 1950, approved hunting in Grand Teton National Park.[23]

Sensitive to the mounting pressure, Secretary Udall asked Leopold and his commission to address the hunting issue. Their answer was unambiguous: sport hunting had no place in national parks.[24] The issue did not die there, though. The National Rifle Association and sportsmen groups filed several lawsuits challenging the Park Service's prohibition on hunting and trapping in the national parks, and the issue is regularly raised when new parks are proposed. However, the courts have consistently upheld the agency's general ban: "Notwithstanding that the goals of user enjoyment and natural preservation may sometimes conflict, the [Park Service] may rationally conclude, in light of the Organic Act and its amendments, that its primary management function with respect to wildlife is preservation unless Congress has declared otherwise."[25] As a result, outside the Alaska parks where subsistence hunting is permitted, the national parks have remained closed to hunting and trapping unless Congress has specifically authorized such activity. Congress has, in fact, allowed hunting in the enabling legislation for several national recreation areas and national preserves, creating an important distinction between them and the traditional national park.[26] Otherwise, the national parks are true wildlife sanctuaries, at least while the animals remain within park boundaries and do not overpopulate the area and at least until Congress changes its mind and decides to authorize hunting in the parks, a matter that has been the subject of several thus far unsuccessful bills in recent years.[27]

Absorbing Ecology's Lessons

Neither the Leopold report nor the immediate changes it prompted has resolved all the Park Service's myriad wildlife management challenges. Despite the agency's current commitment to science in managing park resources, it has been faced with ongoing controversy over how much or how little managerial intervention is appropriate in the national park setting. These controversies have focused on the agency's approach to managing burgeoning native elk, bison, and deer populations in individual parks, specifically whether a hands-off policy can be scientifically and historically justified. At the same time, when the agency has intervened to control nonnative wildlife populations, it has also faced criticism from those who enjoy these species and question the underlying scientific justification. Moreover, as agency officials have come to understand the ecological consequences that buildings, roads, and other developments can have on wildlife, park managers have sought to reduce the human footprint in several parks, but not without provoking opposition. Giving wildlife priority over visitor preferences and local economic concerns is not always an easy or popular policy choice.

By the time the twentieth century drew to a close, the Park Service had consciously shifted its wildlife management policies away from individual charismatic species to focus on park ecosystems with their dynamic processes. Simply put, agency policy was now aimed toward preserving all native species in an ecosystem context, a position driven by evolving scientific knowledge about ecological integrity and health as well as related endangered species concerns. To reinforce the point, in 2001, the blue-ribbon National Park System Advisory Board recommended that the Park Service "adopt the conservation of biodiversity as a core principle in carrying out its preservation mandate."[28] This policy shift is captured in the agency's 2006

Management Policies, which instruct park managers to "maintain as parts of the natural ecosystems of parks all plants and animals native to park ecosystems." To do so, the Park Service will "not attempt to solely preserve individual species (except threatened or endangered species) or individual natural processes; rather, it will try to maintain all the components and processes of natural evolving park ecosystems, including natural abundance, diversity, and genetic and ecological integrity of the plant and animal species native to those ecosystems." This commitment to preserving natural processes also extends to restoring them, including reintroducing extirpated native plant and animal species, as long as the effort is environmentally sustainable and does not endanger human safety or property.[29] Within the system, Great Smoky Mountains has taken the lead in developing an all-taxa biodiversity index designed to inventory the full array of animals, plants, insects, and other organisms found in the park and to assess their condition so as to better understand and protect them.[30] To bolster support for this ecologically comprehensive approach to natural resource management, the Park Service will also have to educate the visiting public about the value of noncharismatic wildlife species as well as wildfires and other natural processes.

As the twenty-first century dawned, however, it was obvious that the Park Service could not accomplish its new ecological management goals without extending its vision outside the conventional park boundary line. Well before the groundbreaking Leopold report, it was increasingly evident that the national parks were no longer isolated enclaves that could exist without regard to what was happening on adjacent lands. The problem was starkly highlighted in the 1980 "State of the Parks" report, which chronicled an array of external activities that were seriously degrading park resources, including "industrial and commercial development projects on adjacent lands; air pollutant emissions . . . ; urban encroachment; and roads and railroads."[31] The 1992 "Vail Agenda" report reinforced these conclusions, recommending that "the prevention of external and transboundary impairment of park resources and their attendant values should be a central objective of Park System policy."[32] In 2001, the National Park Service Advisory Board admonished the agency to engage with its neighbors to help protect park biodiversity resources.[33] In addition, the 2009 Second Century Commission report flatly stated that "we can no longer draw a line on a map and declare a place protected."[34] The net result of escalating adjacent development activities, nearly everyone agrees, is the loss of critical wildlife habitat, migration corridor blockages, and further landscape fragmentation, all problems that will only intensify as climate change takes hold.[35] Before the national parks can be viewed as secure wildlife sanctuaries, the Park Service will therefore have to ensure that its resource management plans pass scientific muster and that it is prepared to address habitat loss, off-site pollution, and related strains on lands and waters that extend outside the existing boundary line.

THE YELLOWSTONE WILDLIFE CONTROVERSIES

Just as Yellowstone set the stage during the late nineteenth century for national park wildlife conservation policy, it continues to occupy center stage in the evolution of these policies. As a historical matter, Yellowstone's status as the nation's first national park and its central role in numerous high-profile wildlife conflicts with system-wide implications give the park's wildlife conservation policies an exalted status in agency lore and policy circles. That the seminal Leopold report had its origins in the Yellowstone elk management controversy adds further luster to the park's role and reputation. As a biological matter, Yellowstone is one of the largest and most intact parks in the contiguous states, and it now boasts the full complement of native wildlife that inhabited the area before Euro-American explorers arrived on the scene. Quite simply, with its prodigious elk and bison herds, endangered grizzly bear population, and exotic-fish problems, Yellowstone has provided a fertile testing ground for defining the role of national parks as wildlife reserves.[36]

The Northern Range Elk Herd

Yellowstone's abundant and highly visible elk herds—a regular highlight for the park's visitors—are the focus of a protracted controversy over appropriate wildlife management practices. The primary issues are whether the park has too many elk for the available habitat and whether the agency should actively control elk population numbers. The answers to these questions turn on hotly disputed scientific conclusions and equally contentious policy judgments. What is clear is that the park's elk numbers, particularly on the northern range, have grown dramatically since the 1960s. Equally clear, the elk population growth coincides with the Park Service's decision to adopt a noninterventionist wildlife management policy, which stopped further culling of the herd and has allowed it to evolve naturally. This decision, by one strident account, has the Park Service "playing god" in the ecosystem and, in fact, doing a terrible job of it.[37]

At the beginning of the twentieth century, the predominant wildlife management concern was to rebuild big-game herds that had been decimated by the recently ended market hunting frenzy. By then, with the U.S. Cavalry in charge of Yellowstone, the park's elk and other animals were mostly secure from illegal poaching and commercial hunting. To nurture these still-fragile wildlife populations, military officials initiated a supplemental feeding program, providing hay to the park's elk to help them survive the harsh winter conditions. As predator control efforts ramped up and the wolf population declined, the park's elk herds began to grow even more in size, confirming a predator-prey relationship that many biologists believe helped check herd numbers. Before long, with the park's elk population on the rebound, Yellowstone officials began shipping excess elk out of the park to help state game officials rebuild their depleted herds, ultimately providing park elk to forty states and three Canadian provinces.[38] In short, the park's elk were being managed in the same way that ranchers managed their livestock.

As the elk numbers continued to grow, however, concern mounted that the northern range was being overgrazed and could not sustain this population level. Having already witnessed the Kaibab Plateau deer explosion and ensuing die-off just outside the Grand Canyon, Park Service officials believed that the park's elk population was habitat limited, a view that squared with conventional range management wisdom. As settlement increased north of the park, the elk found their traditional migration routes to lower-elevation winter habitat blocked, while local ranchers objected to the elk competing with overwintering livestock for limited hay and other forage. When Park Service biologist George Wright and his team issued their second report in 1934, concluding that the northern range was in "deplorable condition" and steadily deteriorating,[39] the stage was set for a new policy based on actively limiting elk numbers by culling the herd. By then, the elk transplantation program had successfully run its course; few states wanted any more park animals. So, in the mid-1930s, the Park Service terminated Yellowstone's winter feeding program and dispatched park rangers to shoot excess animals, while the surrounding states extended their hunting seasons to enable hunters to take more elk outside park boundaries. Drawing on earlier elk population estimates, park officials settled on an optimal population of four thousand to five thousand animals for the existing range conditions and proceeded to cull the herd accordingly.

During the late 1950s, as elk numbers continued to swell, the Park Service's culling efforts intensified, prompting hunters in the adjacent states to begin agitating to open the park to sport hunting. As the political pressures mounted, Secretary Udall turned to Starker Leopold to reexamine the Park Service's wildlife management policies. The ensuing Leopold report, as noted, proposed that national parks be managed according to "an overall scheme to preserve or restore a natural biotic scene" and adamantly rejected the idea of sport hunting in the parks, but it also acknowledged that intensive management, based on the best ecological data available, would be necessary to accomplish its proposed naturalness goal. Observing that most parks were too small to contain the diverse habitat types required by resident species and that past human intrusions had significantly altered ecological processes, the report concluded that active intervention—including the shooting of excess ungulates—would be necessary to restore anything resembling a natural ecological order.[40] This endorsement of culling did not end the matter, though: the sight of park rangers killing the animals they were charged with protecting soon proved highly unpopular.

During the mid-1960s, following a series of graphic television news reports and critical magazine articles, the Senate Appropriations Committee convened a widely publicized field hearing to investigate the Yellowstone elk situation. In response to the hearing, the Park Service announced that it was terminating active elk management and that its ungulate herds would be left to exist on their own across the national park system. Although a notable departure from the Leopold report recommendations, the agency rationalized its decision as consistent with returning park landscapes to their natural or original pre-European condition, when the human hand was largely absent. At Yellowstone, park biologists came to regard this decision as a "grand experiment," labeling it "natural regulation" and asserting that ungulate numbers would be controlled by habitat and winter weather conditions. They also concluded, over strenuous objections from some scientists, that the park's northern range was not being overgrazed and that it could support a larger elk population.[41]

Those conclusions were soon put to the test as elk numbers steadily mounted. From around five thousand animals in 1967, the park's northern herd grew to nearly twenty-two thousand elk in 1981, far exceeding the agency's original projections. Critics charged that the park's elk population was now much too large and that overbrowsing was permanently altering the northern range's ecological character. They asserted that the park's aspens, willows, and other herbaceous vegetation were being eliminated by the hungry elk, as were the beaver, white-tailed deer, and bird populations that also relied on these trees and shrubs. In short, the critics believed that the proliferating elk had taken over the northern range and were eating it into ecological collapse. They blamed the deteriorating conditions on the park's natural regulation theory and challenged the basic premises underlying it, noting that the park was no longer in a pristine or natural state. Over the years, the park environment had been significantly altered, not only by Native American hunting and burning practices during the prepark era, but also by the absence of wolves and other predators that had historically helped control elk numbers and by increased development on adjacent lands that hindered elk migration patterns. The solution, they asserted, was a return to active management of the elk herd.[42]

As the controversy heated up during the 1980s, two questions predominated: Was the Park Service's scientific analysis correct? Did natural regulation represent a viable natural resource management policy? A series of studies addressing elk population dynamics, the historical condition of the northern range, and the underlying causes for noticeable ecological changes sought to answer these questions. Not surprisingly, with the scientific community already at odds over these issues, the answers varied depending on the researcher's predisposition. The critics, including Professor Frederic Wagner and his student Charles Kay, continued to find that the park's northern range was in ecological decline from an overabundant elk population,[43] while park supporters remained convinced that the range was simply undergoing regular periodic fluctuations in response to precipitation patterns, fire events, and the like. In 2002, the National Research Council reviewed the evidence and concluded that "Yellowstone is not in ecological trouble . . . no[r] on the verge of crossing some ecological threshold beyond which conditions might be irreversible."[44] As a result, the Park Service has reaffirmed its commitment to the natural regulation policy, which continues to govern management of Yellowstone's elk. Plainly frustrated, the critics have responded that the agency is not committed to rigorous scientific inquiry, pointing to the inbred nature of Yellowstone's independent researchers, who must rely on park officials for permission to conduct in-park research projects.[45]

In recent years, however, the terms of the debate have begun to shift as the behavior of Yellowstone's northern range elk herd appears to be changing along with ecological conditions. According to several studies, aspen and willow vegetation seems to rebounding on the park's northern range because the elk's browsing patterns are changing. The reintroduced wolves are the likely cause of this change in elk behavior because elk seem to now avoid lingering in heavily vegetated riparian areas where they are easier targets for preying wolf packs. In addition, ever since the 1988 fires and the ensuing harsh winter that killed nearly 6000 elk,[46] elk on the northern range have begun migrating out of the park again in significant numbers, thus reducing winter feeding pressure inside the park. The looming specter of climate change could further affect elk behavior; warming temperatures would likely expand the winter range within the park as forested zones move farther upslope, and shifts in spring runoff patterns could affect riparian and other vegetation.[47]

Meanwhile, although the debate over the ecological health of Yellowstone's northern range and its hands-off management policies persists, the Park Service continues to treat wildlife conservation as part of its broader ecological management policy agenda. This policy holds that dynamic ecological change, including population fluctuations, constitutes a routine dimension of modern wildlife conservation efforts and thus does not impair park ecosystems.

Ungulate Management Elsewhere

Elk abundance concerns are evident elsewhere in the Yellowstone region. In Grand Teton National Park, which is situated south of Yellowstone, the Park Service also does not intervene directly to control the elk population. The park does, however, host an annual congressionally mandated elk-hunting season intended to keep numbers in check. The elk regularly migrate out of the park, usually heading for the nearby National Elk Refuge just outside the town of Jackson, Wyoming, where they are fed artificially during winter. The elk refuge was originally established in 1912 when local leaders realized that the growing town and adjacent ranches had effectively blocked the traditional elk migration route southward, leaving the animals few feeding options during the area's harsh winter months. The net effect is that Grand Teton's elk exist in a heavily manipulated environment, even if the Park Service is not actively involved itself in shooting the animals. This arrangement, paradoxically, recalls the agency's earliest wildlife management policies: winter feeding at the National Elk Refuge has become a spectacle that draws thousands of visitors who take sleigh rides among the animals to view them up close, not unlike the early national park zoos and bear-feeding grounds.

Outside the Yellowstone region, Park Service elk and deer management policies vary widely, a reflection of the difficulties inherent in pursuing a natural regulation policy where national park boundaries diverge from on the ground ecological realities. At Rocky Mountain National Park, a proliferating elk population has created conflicts with the neighboring community of Estes Park and adjacent landowners, with critics maintaining that excessive elk numbers have seriously degraded the park's native trees and grasses. The Park Service, however, long resisted culling the herd, relying instead on a natural regulation management strategy. In 2008, however, park officials changed this policy to allow limited culling inside the park and to enable specially trained hunters to participate in the cull. At the same time, they rejected a proposal to reintroduce wolves to the park as a means of controlling the elk. In a subsequent lawsuit alleging that these decisions violated the Organic Act and the National Environmental Policy Act, a Colorado federal court ruled that the agency was not required to consider the wolf alternative and had reasonably justified culling as a conservation measure.[48] At Theodore Roosevelt National Park, the Park Service has approved an extensive culling effort to reduce the burgeoning nine-hundred-strong elk herd by more than half, while at Wind Cave, park officials are relying on hunting outside the park to control its elk overpopulation problem.[49] In the eastern national parks and historical sites, where abundant deer populations are intruding into nearby residential areas and becoming a nuisance, the Park Service is actively culling these herds with public and judicial support.[50] Each of these approaches appears consistent with the Park Service's 2006 *Management Policies*, which provide that intervention may be necessary when "a population occurs in unnaturally high or low concentration as a result of human influences . . . and it is not possible to mitigate the effects of the human influences."[51] The different ungulate management approaches also reinforce that few national parks represent complete ecosystems in themselves, a reality that requires some adjustment in applying the nonimpairment principle.

The Bison Wars

Yellowstone's restored bison herds—one of the nation's earliest wildlife restoration successes—have proliferated over recent years and are pushing out of the park, where they come under intense scrutiny as potential disease vectors. The park's bison brucellosis controversy reaches across the greater Yellowstone region, pitting federal and state agencies, ranchers, and wildlife advocates against one another. At the heart of the conflict is the concern that the park's free-ranging bison could transmit the brucellosis bacteria to local cattle herds. Brucellosis is a bacterial disease that can cause spontaneous abortions in domestic livestock, with substantial economic consequences for ranchers who could lose their cattle herds and face an export embargo. The primary question is whether the Park Service, having ceased intensively managing the park's bison herds in the aftermath of the Leopold report, must now aggressively control these iconic creatures that symbolize our frontier heritage.

In the late 1800s, as the once-multitudinous bison herds vanished from the plains, a few remnant bison took refuge in the Yellowstone backcountry. But even in the park, they were not entirely secure from poachers who still stalked them for their valuable hides. In 1901, to avert an imminent extinction, the park's military caretakers took charge of the remaining bison, as we have seen, establishing the Buffalo Jones Ranch as a bison sanctuary at Mammoth and also importing other bison for resettlement in the Lamar Valley. Once the Park Service assumed responsibility for the park, it continued to manage these bison herds much as domestic livestock and to keep them separate. In 1917, when brucellosis was detected in the bison herd, park officials started occasionally testing and slaughtering those animals that registered positive for the disease. (Paradoxically, the park's bison originally contracted brucellosis from nearby cattle.) After World War II, the Buffalo Jones Ranch operation was discontinued and the bison were allowed to intermingle. Park managers, however, still kept the population at a low level by culling the herds, a practice that reflected the prevailing view that browsing ungulates had to be controlled to meet the park's range carrying capacity.[52]

During the 1960s, in response to the Leopold report, the Park Service stopped trimming the bison herd. Adhering to the new nonintervention wildlife management policy, park officials left the bison to survive on their own, assuming that the park's harsh winters and limited food supplies would keep population numbers in check. Bison numbers climbed sharply, however, growing from 397 animals in 1967 to more than 2,500 in 1988 and then approaching 5,000 in 2009. Although they originally wintered in the park, the bison gradually started to migrate outside it in greater numbers, seeking forage at lower elevations on the adjacent national forests and nearby private ranch lands. Once outside the park, the wandering bison were at risk of contacting cattle and possibly contaminating them with brucellosis.

Fearful that their livestock industries could be embargoed from exporting cattle, the surrounding states of Montana, Idaho, and Wyoming sought to halt the annual bison exodus from the park. Joined by the federal Department of Agriculture, which oversees livestock diseases, the states challenged the Park Service's nonintervention bison management policy, arguing that park officials would have to resume actively managing the herds, both to control the population size and to curtail bison from leaving the park. Although the park's elk herds also carry the brucellosis disease and have occasionally transmitted it to local cattle, no one seemed concerned about them, nor did anyone seem fazed that there has never been a documented case of transmission from bison to livestock in the wild. The Park Service nonetheless agreed to reestablish control over its bison. What ensued were futile efforts by the agency to haze the wandering bison back into the park, a state-sanctioned bison hunt that turned into a public relations nightmare for Montana, and gridlock among the involved entities, each of whom pressed for its own solution to the dilemma. Under a series of makeshift plans, bison deaths continued to mount; in the winter of 1988–1989, more than 550 bison were slaughtered, and the death toll continued in subsequent years amidst mounting public outrage over the carnage.[53]

After protracted litigation and negotiations, the Park Service and its adversaries finally settled on an interagency bison management plan in 2000 that would eventually allow one hundred bison to migrate out of the park during the winter months. In the interim, however, any bison leaving the park remained subject to being shot, quarantined, sent to slaughter, or hazed back behind the boundary line. As a practical matter, the Park Service's short-lived experiment in natural regulation of the Yellowstone bison herd was over, and agency officials were once again intensively managing the herd, but for political, not biological, reasons. For the nation's largest and most visible free-ranging bison herd, there simply was not sufficient room in the sanctuary and no tolerance outside it. The plan did not solve the problem, however. During the winter of 2007–2008, more than fifteen hundred bison were killed after exiting the park, again triggering public protests and calls for a more humane and durable solution. Park visitors can nonetheless see the bison meandering through the park, unaware that any animals straying across the invisible boundary line are no longer protected and may be dispatched to safeguard the region's waning cattle industry. Despite its role as a wildlife reserve, Yellowstone's restored bison could rightly view the park more as a zoo than as an ecologically dynamic landscape.[54]

Recent developments, however, signal a profound shift in Yellowstone bison management policy, one that takes account of the animal's behavioral patterns and ecological realities. Bison advocates, working with the diverse federal and state agencies overseeing Yellowstone bison management, have secured access to additional winter habitat outside the park, and the state of Montana, for the first time, has allowed disease-free bison to be transported to a new home on a nearby Indian reservation. Through a combination of livestock grazing buyouts and lease retirements, which have removed cattle from lower-elevation winter range outside the park, the bison will have access to more than one hundred thousand acres of critical habitat in the nearby Gardiner Basin and Horse Butte areas outside the park.

With the state's decision that seemingly allows disease-free park bison to be relocated outside the area, the park is poised to resume its historic role as an incubator for distant wildlife restoration efforts, this time also helping reestablish an important tribal cultural connection with the bison. Along with Montana's move to initiate a statewide bison conservation dialogue, these developments suggest that the responsible agencies are prepared to not only treat the park's bison as the wild animals that they are, but also to take a more realistic view toward the steps necessary to control the risk of brucellosis transmission. Local ranchers have nonetheless sued to block these reforms, and that litigation is pending. Assuming that these new policies prevail, Yellowstone's bison may begin roaming much more freely across park boundaries, reestablishing their unique ecological niche in this expansive landscape.[55]

Recovering the Grizzly Bear

The Yellowstone grizzly bear, the epitome of a charismatic megafauna and palpable symbol of the park's wilderness character, has played a prominent role in bringing an ecosystem perspective to wildlife management. The history of Yellowstone's grizzly bears vividly captures the post–Leopold report evolution in national park resource management policies, including the elevation of science in setting conservation goals and policy. It also demonstrates the important role that the federal Endangered Species Act plays in wildlife recovery efforts, including in the national park setting. A wide-ranging omnivore, the grizzly bear does not confine its wandering to the park, where it is largely secure, but instead has used the greater Yellowstone ecosystem as its habitat. Because the bear poses a risk to human safety and property, its mere presence can produce fear, making the grizzly bear a true wildlife management challenge.

Although nearly fifty thousand grizzly bears once roamed across the western United States, their numbers declined rapidly once settlers arrived on the scene. Pushed to the brink of extinction, the bear found refuge in Yellowstone and a few other remote locations, where it could make a living with little human contact. A popular tourist attraction during Yellowstone's early years, the park's bear population dropped precipitously following closure of the garbage dumps, owing largely to an increased number of human-bear encounters that were time and again fatal for the bears involved. Although the dump closures were controversial, the Park Service held fast, and the bears were forced to subsist on their own. Not surprisingly, once the Endangered Species Act became law, the grizzly bear was added to the list of threatened animals in 1975, giving it substantial legal protection and obligating federal officials to develop a recovery plan to rebuild the population and avert the risk of extinction. The ensuing recovery effort—overseen by the U.S. Fish and Wildlife Service—brought the various federal and state agencies responsible for the bear and its habitat together to plan a coordinated recovery strategy.[56]

As mandated by the Endangered Species Act, science has figured prominently in the recovery effort, which is directed toward reestablishing a viable population of bears that can sustain itself over time. As a matter of biology, grizzly bears have a low reproductive rate, which makes safeguarding individual bears, particularly sows with cubs, an important dimension of the recovery effort. The Yellowstone bears are also isolated genetically from other grizzly bear populations that have persisted in Canada, Glacier, and farther west along the U.S.-Canada border. Many biologists, concerned by these facts, have long maintained the need to link these grizzly bear population segments together to ensure the long-term genetic viability of the isolated Yellowstone bear population. Secure habitat is also critical to the recovery effort. As omnivores, grizzly bears forage seasonally across the landscape for food. The bear's primary food sources include winter-killed carrion that they consume upon leaving their winter dens, spawning cutthroat trout that they eat during early summer, and whitebark pine seeds that are available in the fall just before they hibernate for the winter. The grizzly bear's quest for food can take it across the landscape without regard to existing boundary lines, potentially putting it in conflict with people, especially when it leaves the park. When that occurs, the bear often ends up being killed, making bear mortality numbers another concern. Thus, the recovery effort not only needed to ensure that population targets were met, but it also needed to ensure that adequate secure habitat was available to sustain the bear population.[57]

As the Yellowstone grizzly bear recovery effort unfolded, however, it initially faltered, and bear numbers continued to spiral downward. The causes were several. The park's people-habituated bears kept getting into trouble with park visitors and local residents, encounters that frequently resulted in a dead bear. Because the bears roamed widely, they needed habitat that was not always available outside the park. During the 1970s, the surrounding national forests were abuzz with logging, mining, and livestock grazing activities, which reduced the available habitat and caused mounting bear fatalities. If the bear's limited food sources failed in any given year, then the bears sought alternative food sources, usually at lower elevations where most people lived. Local communities and residents did not regularly safeguard their garbage, pet food, and the like from bears, offering the roaming bears an attractive nuisance that would get them into trouble. Prior to 1982, there was also little interagency coordination in the recovery effort, which meant different and sometimes incompatible management standards prevailed on adjacent lands, depending on which agency was the responsible manager. Moreover, despite having a good bit of scientific knowledge about the bear, scientists did not yet fully understand the bear's habitat and other needs or the management strategies necessary to meet these needs.[58]

In the mid-1980s, with the bear population still dwindling, the Congressional Research Service issued a scathing report lamenting the apparent imminent demise of the Yellowstone grizzly bear population and blaming it on the agencies for failing to coordinate their recovery efforts on an ecosystem scale. At the same time, the Park Service, notwithstanding notable progress keeping bears and visitors separate and improving backcountry camping practices, drew blistering criticism for reneging on its initial commitment to close the Fishing Bridge campground that was located in prime grizzly habitat. Park officials, as we have seen, reversed their closure decision in response to political pressure from local merchants and politicians who were plainly motivated by commercial rather than biological concerns. In the nearby national forests, logging, mining, and other activities continued apace, alongside the extensive road network needed to provide access, all of which was further fragmenting important bear habitat outside the park. By now, however, biologists understood that grizzlies were quite wary of roads, which brought more people into backcountry areas and meant more poaching, traffic fatalities, and lethal bear-human encounters. In addition, the area's extraordinary scenic, wildlife, and other amenities had been discovered, prompting a noticeable uptick in subdivision activity that was chopping up previously open spaces where bears might roam without problems. Although the critical Congressional Research Service report impelled a short-lived interagency effort to promote a new ecosystem-level management initiative that could have scaled back the level of development, that effort failed due to local political intervention. But the agencies also responded by revising their bear recovery efforts, finally putting the region's grizzly population on an upward trajectory.[59]

After an initial effort to remove—or "delist" in legal parlance—the Yellowstone grizzly bear from the endangered species list failed during the mid-1990s,[60] the agencies redoubled their recovery efforts with a focus on securing additional bear habitat. In the national forests, market forces and new policies contributed to a drop in timber sales and numerous road closures, while energy exploration interest evaporated and sheep grazing allotments were retired. Local communities outside the park started to clean up garbage dumps and other bear attractants, thus further reducing the likelihood of lethal bear-human encounters. The bears responded by expanding throughout the greater Yellowstone ecosystem and soon occupied most of the identified bear management units. Bear population numbers shot upward, rising to more than five hundred bears by 2006 and surpassing the various recovery targets, while bear mortality figures fell to within the acceptable range to sustain the population. In 2007, the U.S. Fish and Wildlife Service responded by once again proposing to delist the Yellowstone grizzly bear population, convinced that it had recovered, that its habitat was now secure, and that the surrounding states of Idaho, Montana, and Wyoming each had in place workable grizzly bear management plans that would protect the bear population going forward.[61]

All was not well in Yellowstone grizzly country, though. The area's whitebark pine trees were dying, both inside and outside the park, the result of an extensive bark beetle infestation, a fungal infection, and warming temperatures. The park's cutthroat trout population was also in decline, outcompeted in Yellowstone Lake by illegally transplanted nonnative lake trout. If these important grizzly bear food sources were unavailable, then the bears could be expected to seek food elsewhere, most likely venturing out of the park and into conflict with nearby communities and landowners. Moreover, although acknowledging the genetic value of a corridor connecting the Yellowstone bear population with its more northern cousins, the interagency conservation strategy made no commitment to establish such a linkage and only provided for transplantation after 2020 if there was no evidence by then of genetic interchange with other grizzly bear populations. Although each of the surrounding states had adopted its own grizzly bear management plan, skepticism persisted about the plans' effectiveness or enforceability, particularly in the national forests, where federal law took precedence over state law. Ultimately, a federal court ruled that the U.S. Fish and Wildlife Service had failed to adequately explain why potential loss of the whitebark pine as a critical food source would not jeopardize the bear's recovery and blocked the delisting effort, at least for a while.[62]

Clearly, Yellowstone National Park occupies a central role in the grizzly bear recovery effort. From the beginning, park lands have served as core bear habitat, and park bear management practices have helped set the standard across the ecosystem. Even if the bear is delisted and subjected to hunting under the state management plans, the park will remain off-limits to hunters, providing bears with a safe refuge. The science is clear, however, that the park does not provide sufficient habitat to maintain a viable grizzly bear population over time, a fact underscored by the decline of whitebark pine trees and spawning cutthroat trout within the park. Although the Interagency Grizzly Bear Committee has promoted coordinated management responses to the recovery effort during the past several decades, it is unclear whether that same level of coordination will actually continue once the states assume responsibility for the grizzly outside the park. Difficult as it has sometimes been for park officials to coordinate with their federal counterparts in the Forest Service, U.S. Fish and Wildlife Service, and Bureau of Land Management, those difficulties will only increase when they must work with three different states, each with its own management goals and strategies.[63] Nonetheless, the Yellowstone grizzly bear's long-term fate plainly rests on some form of coordinated landscape-scale planning and management that also includes a linkage with other bear populations. Absent such a commitment, the effort expended thus far to recover the grizzly bear may prove to have been for naught.

The Nonnative Species Problem

A central tenet in the concept of the national park as a wildlife reserve is the Park Service's unambiguous policy of protecting and restoring native—but not exotic or nonnative—species. According to the agency's 2006 *Management Policies,* exotic species "will not be allowed to displace native species if displacement can be prevented." This policy provides for controlling and even eradicating exotic plant and animal species when feasible to do so and when the exotic species "interferes with natural processes and the perpetuation of natural features, native species or natural habitats." During the 1920s, the Park Service first sought to eliminate exotics when it began removing wild burros from Grand Canyon National Park. Since then, it has moved against other exotic species—wild boars in Great Smoky Mountains and feral pigs in Hawaii Volcanoes, for example—in efforts to maintain native species and natural ecological processes and to curtail ecological damage. These exotic species control efforts have not been without controversy, however, and critics have challenged the scientific, legal, and humane basis for several of them. But in each instance, the courts have sustained the Park Service's exotic species control policy and practices against attack, although the agency has frequently opted for live capture and removal strategies over just killing the intruders.[64]

In Yellowstone, the committal to native species conservation is most evident in how the park has approached fisheries management. During the late 1800s, the park's military caretakers introduced nonnative trout to improve sport angling opportunities. Committed early on to sportfishing as a recreational activity, the Park Service continued to stock park waters across the system and even maintained fish hatcheries in some parks. In 1936, however, the agency reversed course and began to prohibit such exotic species introductions and to close down the hatcheries. Since then, it has been gradually addressing this nonnative fisheries legacy, seeking to remove all exotic species from park waters.

In 1994, Yellowstone officials discovered exotic lake trout in Yellowstone Lake, presenting a major threat to the lake's keystone cutthroat trout population and its entire fragile ecosystem. An extremely large and aggressive fish, lake trout regularly prey on smaller fish, such as the cutthroat trout, which is a critical food source for bald eagles, white pelicans, and other birds as well as grizzly bears, which rely heavily on spawning cutthroat. When implanted elsewhere, lake trout have often taken over a fishery, literally driving other fish species into steep decline. Upon determining that the lake trout were introduced illegally into park waters and placed the native cutthroats at real risk, Yellowstone officials—after consulting a panel of scientists—decided to physically remove them from the lake. Since then, the Park Service has progressively expanded and improved its gill netting efforts; in 2007 and 2008, more than seventy thousand lake trout were removed, many caught in strategically located nets before they could complete their spawning cycle.

Scientists nonetheless believe that it will be impossible to remove all the lake trout from the system, so the Park Service's efforts are directed toward controlling them through intensified suppression efforts that include professional fishers and additional monitoring of both lake trout and cutthroat trout. At least for now, the Yellowstone Lake ecosystem is not yet irreparably damaged, giving agency officials additional time to control these exotic intruders. Judging from the park's native fish conservation plan process, the public not only supports this aggressive lake trout control effort, but park officials have concluded that it does not violate the agency's nonimpairment mandate.[65]

At Olympic National Park, however, public controversy has dogged the Park Service's efforts to eliminate mountain goats as an exotic species, albeit one that provides a popular wildlife viewing opportunity for many visitors. Believed to have been first introduced on the Olympic peninsula during the 1920s to provide local sportsmen with a new hunting opportunity, mountain goats have proliferated over the decades both inside and outside the park. During the 1970s, when scientists detected that the growing goat population was beginning to damage native plants, park managers initiated several studies to determine whether the goats were native to the area, the extent of damage to native plant life, and potential management options. The conclusions were clear: the goats were exotic to the park; serious ecological damage was imminent, including the potential loss of several rare plant species; and the only effective and economically viable management option for removing the goats was by shooting them from helicopters, live trapping them with nets from helicopters having failed.

Critics responded, however, that other historical data showed that the goats were actually native to the area, that the ecological damage was negligible, and that alternatives other than shooting were available to control the population. In fact, the goats also occupy neighboring national forest lands, where they are hunted, and they still disperse into the park, drawn by its attractive habitat, where they remain popular with park visitors. For their part, the Forest Service and state wildlife officials decided not to cooperate with the Park Service so as to maintain goat-hunting opportunities in the area. Consequently, Olympic officials have put their plans on hold, pending further scientific review. Any boundary-based approach to eliminating Olympic's goats is not likely to succeed.[66]

A CONTENTIOUS AND IMPERFECT RESERVE

Over the years, the national parks have afforded wildlife a vital sanctuary, despite fluctuating—and sometimes controversial—management policies. Vested with exclusive jurisdiction over its lands, the National Park Service has been able to implement and maintain a strict no-hunting policy, thus sparing park wildlife the threat of being shot when they expose themselves. This policy makes the national parks unique because hunting is allowed on other federal lands, including nonpark wilderness areas and national wildlife refuges. As a protected reserve, the national parks have also served as important wildlife restoration sites, providing an ideal venue for nursing declining species back to health or for restoring missing species. Besides the successful efforts to restore bison, grizzly bear, and wolf populations in Yellowstone, other important wildlife restoration efforts are under way, including the California condor in Grand Canyon and salmon habitat in Olympic. With their expansive and largely undeveloped landscapes, the national parks offer some of the few locations where such experiments in wildlife conservation can occur without arousing insurmountable political opposition.

The national park as a wildlife reserve accords with the park's roles as a tourist destination and as a laboratory. As wildlife have thrived in the national park setting, park visitors have been rewarded with unique wildlife viewing opportunities, perhaps none more thrilling than Yellowstone's reintroduced wolves. The presence of abundant wildlife, like Yellowstone's elk and bison, have enabled park visitors to see and marvel at nature in its living dimensions, an experience that only enhances the park's stunning scenic setting and its wilderness qualities. Moreover, the national parks offer scientists a unique opportunity to study wildlife in an undisturbed environment. Because the human presence is limited and controlled, the national parks provide a useful baseline for assessing how ecosystems function in the natural world, as illustrated by the evolving conditions on Yellowstone's northern range. In short, wildlife represent a visible testament that our national parks are living entities, places where visitors can reconnect with the nation's wilderness heritage and where scientists can better understand how nature operates.

As a protected sanctuary, the national parks have enabled the Park Service to experiment with different wildlife management policies. Those policies have evolved over the years, shifting from a static, "good" animal view of nature to an ecosystem-based, evolutionary perspective committed to conserving both native species and ecological processes. Unlike conventional wildlife management policy, which has traditionally focused on single species with obvious economic value (most often big-game or trophy animals), Park Service policy has moved toward a broader view based on biodiversity conservation and ecosystem preservation principles. Although still a work in progress, the idea that natural processes should be allowed to proceed unimpeded runs counter to our historical inclination to control nature, but this approach to wildlife conservation fits with the park's laboratory image and is also part of what makes the national parks unique as wildlife reserves.

Few wildlife management policies have proven as controversial as the Park Service's noninterventionist natural regulation policy, however. Nearly everywhere outside the national parks, wildlife are intensively managed, particularly elk and other big-game species that are maintained as a harvestable resource to benefit hunters. Outside the parks, wildlife—ranging from migratory ungulates to spotted owls, sage grouse, and other species—regularly encounter intensive development pressures that threaten to displace them, destroy key habitat, and fragment historical movement patterns. Because these same pressures are not present inside the national parks, agency officials have the unique opportunity to manage park wildlife without extensive intervention while also learning from it and adjusting as necessary. Of course, boundary lines and neighborly relations matter, so park officials must occasionally control certain species to prevent damage outside park borders. Moreover, the Park Service's commitment to maintaining and reestablishing historic ecological conditions also argues for occasional interventions, whether to remove nonnative mountain goats and other exotic species or to restore wolves and other native species so as to repair altered ecosystems. The decision to intervene generally turns, as it should, on whether the failure to act will impair park ecosystems and the processes that sustain them. Natural regulation is therefore not an absolute policy.

Much of the debate over the Park Service's wildlife conservation policies focuses on scientific disagreements, but science is not the only driving force in maintaining the parks as critical wildlife reserves. Whether or not Yellowstone's northern range is overgrazed by elk raises important scientific questions; including whether we should expect range conditions to fluctuate widely in response to dynamic ecosystem conditions and what the range looked like in presettlement times. It also raises difficult policy questions, however. What do we value most on the northern range, abundant elk or aspen groves or riparian habitat? Are we prepared to shoot excess elk to maintain or reestablish particular vegetative conditions? How should we manage species that migrate across administrative boundaries, especially when portions of the landscape are highly modified? Science can assist in answering these questions, but the general public—including park visitors, hunters, animal welfare advocates, and local officials—will also have an important say in any final wildlife management decision. Similarly, although ecological management principles may support reintroducing wolves and fire to the Yellowstone landscape, local politics will also figure into when and how these ecological restoration efforts move forward. In short, science can help us understand how park ecosystems are affected by various wildlife policies, but it cannot alone dictate the content of those policies or the strategies for implementing them.[67]

Perhaps nowhere is this tension between science and policy more evident than in the controversy that surrounds the Park Service's use of culling as a wildlife management strategy. As the mid-1960s Yellowstone elk-culling controversy demonstrated, the sight of park rangers shooting the animals they are charged with protecting arouses strong public passions. Similar reactions have accompanied the use of hunting to address Yellowstone's contemporary bison management controversy and lurk in the background as Rocky Mountain and Theodore Roosevelt cull their elk herds. Today's oversaturated, twenty-four-hour news cycle, along with the Internet, virtually ensures that such decisions will receive public scrutiny and provides a ready forum for opponents and critics. The paradox of killing park wildlife to preserve them is a difficult sell to a general public that views the national park as a wildlife sanctuary. It challenges agency officials to better educate the general populace about habitat constraints, population dynamics, and the like. It also begs the question whether national parks can be maintained in an unimpaired condition with unrestricted wildlife population growth when predators are absent and migration or dispersal opportunities are limited.

Indeed, as a wildlife reserve, the national parks are imperfect. The parks—although a near-ideal setting for experimental management policies like natural regulation and for difficult species reintroductions—are not immune from the larger world and its political pressures. Given the often grim realities of park boundaries and political power, the Park Service has sometimes found itself with few options but to endorse intensive management strategies to reduce friction with park neighbors and maintain peace across the landscape. As vividly illustrated by Yellowstone's current bison management policies and regional grizzly bear recovery strategies, park boundaries have not been drawn for wildlife or its needs. Thus, while providing refuge to wildlife, the national parks are still intensively connected to the larger landscape that surrounds them. The realization of just how these connections affect national park resource management policies is propelling another conception of the national park—as the vital core of larger ecosystems—to the fore.

CHAPTER 9

"A Vital Core"

Ecosystem-Scale Conservation

Neither grizzly bears nor wildfires recognize boundary lines. The sacrosanct borders that we have constructed as straight lines on maps to define our national parks are of no significance to the creatures and natural processes that give life and shape to park landscapes. Since the designation of Yellowstone and the inception of the national park concept, wildlife have wandered to and from park lands, responding to seasonal habitat needs and other urges. Fires have regularly burned across administrative boundaries, and the watersheds that often originate in our parks are connected like vital arteries to downstream human and natural communities. As obvious as these linkages are today, they have not always factored into how we understand national parks or how we manage them. Instead, the image of the park as an island—a separate enclave or sanctuary—has long prevailed, both in popular imagination and management circles, confirmed by powerful notions of property rights and bureaucratic autonomy. But the ecological sciences, now abetted by the overarching threat of global climate change, are breaking down this traditional enclave notion and giving the national park new meaning as the core component in a much larger ecosystem setting. Just as the national parks have long been connected with the communities and visitors that depend on them, the parks are also vital to the expansive natural systems that define their very essence.

ECOLOGY AND THE NATIONAL PARKS

Well before the national park system came into being, the military caretakers overseeing the early parks recognized how poorly these protected enclaves served the wildlife that called them home. In 1882, upon completing a second tour of Yellowstone, General Phillip Sheridan proposed nearly doubling that park's size to better protect the region's elk and other wildlife populations. A decade later, in 1891, Sheridan's plea was partially answered when President Benjamin Harrison designated the nation's first forest reserve—the Yellowstone Park Timber Land Reserve—just east of the park. Six years later, President Grover Cleveland added the Teton Forest Reserve south of the park. Although neither of these forest reserve designations formally enlarged the park, the practical effect of placing these critical areas under federal control was to provide additional sanctuary for the park's migratory wildlife populations. The same concerns were voiced a few years later when Yosemite's military commander called for additional protection for that park's wildlife when they moved seasonally onto adjoining forest reserve lands, where they were being shot by hunters. Elsewhere, at Sequoia and General Grant Grove (now part of Kings Canyon) in California, similar park expansion pleas were heard, all predicated upon the recognition that park wildlife and other vital resources were at risk without further boundary adjustments for these newly protected sanctuaries.[1]

The Problem with Enclaves

Once the Park Service assumed oversight responsibility for the nascent national park system in 1916, the drumbeat for enlarging park boundaries continued, both to better protect migratory wildlife and to preserve nearby scenic vistas. At Yellowstone, Park Service director Stephen Mather and park superintendent Horace Albright put forth expansion proposals reaching eastward and southward that sought to increase the park's size significantly to ensure important wildlife habitat as well as such scenic attractions as the Teton Range. Although intending to endorse the expansion proposal, writer Emerson Hough instead stirred local resistance in 1917 when he observed, "Give her Greater Yellowstone and she will inevitably become Greater Wyoming." Despite initial opposition from Jackson Hole ranchers and the Forest Service, Congress eventually expanded Yellowstone modestly and also created Grand Teton National Park as a separate entity. Not only did the expansion acknowledge on-the-ground ecological realities by encompassing critical winter range and by following hydrographic lines, but the Hough phrase introduced the "Greater Yellowstone" concept into the public consciousness, thus providing an arresting image for subsequent expansion proposals and wildlife conservation battles. The notion of an incomplete and inadequately protected park persisted, though, eventually finding even more cogent expression in the concept of the Greater Yellowstone Ecosystem.[2]

Beyond Yellowstone, most of the early national parks were subject to similar boundary expansion controversies, reflecting a growing recognition that, as originally conceived, they offered incomplete protection to wildlife, scenic vistas, and other resources. In the case of Mount Rainier, the park was originally defined in rectilinear terms (much like Yellowstone), prompting early boundary expansion proposals to afford wildlife more winter habitat and to provide better tourist access, proposals that Congress eventually validated in expansion legislation. At Sequoia, after a prolonged struggle, the original park boundaries were first extended eastward toward the Sierra crest and then northward to include the Kings Canyon country as a separate park that protected lower-elevation wildlife habitat as well as an entire watershed on the slope of the Sierra Nevadas. At Grand Canyon, the Park Service long sought and eventually secured a portion of the Kaibab Plateau, extending the park boundaries northward to include important wildlife habitat. At Glacier, though, repeated efforts to expand the park eastward to provide year-round habitat for the park's elk and other wildlife met sustained resistance from the restive Blackfeet tribe whose reservation abutted the park, and the proposals ultimately went nowhere.[3]

Although predicated on a general understanding of wildlife habitat needs, these early park expansion proposals were not firmly rooted in science. Likewise, early park management policies generally lacked sound scientific footing. That changed when biologist George Wright entered the scene and issued his initial *Faunal Survey* reports, reviewing the status of park wildlife populations and the agency's conservation efforts. Wright's reports bluntly recognized "the failure of parks as biological entities . . . [due to] their size and boundary location," noting that "when national parks were set aside as inviolate wildlife sanctuaries . . . it was assumed that the wildlife of these parks would find suitable refuge within them regardless of what happened outside." The 1920s Kaibab deer population explosion fiasco revealed otherwise, however, showing that

> things could not be done in the Kaibab without affecting Grand Canyon, and vice versa; cougars could not be killed without directly affecting the character, habits, and numbers of deer and indirectly affecting range;

> range plants could not disappear without affecting ground-dwelling birds and small mammals, and so on.[4]

One solution, according to Wright and his colleagues, was to erect "buffer areas . . . [that] would act as transformers to step-down the high, disruptive pressure against native forms coming from outside the parks." The proposed buffer areas would be several miles in width depending on the terrain, hunting and trapping would be limited, exotic species were prohibited, and predators would be both protected and restored. Another solution was to expand several existing parks, including Mesa Verde and Carlsbad Caverns, so that their boundaries would follow more natural features and encompass necessary wildlife habitat. Even at this early stage and even with the landscape surrounding most national parks still relatively undeveloped, Wright was concerned enough with the growing national park system's ecological shortcomings to advance several concrete proposals for addressing the problem.[5]

Wright's untimely death and the intercession of World War II, with its ramped up demands for minerals, timber, and other raw materials, quieted any discussion about remaking the national park system to address these boundary problems. After the war, Wright's ecological concerns still garnered little support among politicians, even as the national park system continued to grow in number. When Congress created Everglades National Park in 1934, it was for the area's wilderness and wildlife—rather than scenic—qualities, but the park's supporters failed in their efforts to enlarge the park's boundaries to embrace the greater sawgrass ecosystem, thus leaving the new park vulnerable to the water demands of Florida's growing coastal communities and upstream farmers, particularly during drought years. Beginning in the 1930s, local opposition stymied efforts to expand the Jackson Hole National Monument and to redesignate it a national park, leaving the stunning Tetons protected, but not the critically important lower-elevation lands that were used by most wildlife. Similar efforts to adequately safeguard northern California's coastal redwood trees were rebuffed, leaving only a few isolated tree stands protected in diminutive state park enclaves, where they were subject to periodic flooding events triggered by uncontrolled upstream logging. Although national park supporters were beginning to appreciate these ecological connections, the political world continued to deny them.[6]

During the 1960s, the notion that the national parks could not be managed as isolated enclaves found further expression in the landmark reports addressing the Park Service's seeming ignorance of science and its role in national park management. The aforementioned Leopold report acknowledged that "few . . . parks are large enough to be in fact self-regulatory ecological units; rather, most are ecological islands subject to direct or indirect modification by activities and conditions in the surrounding area," but did not make any specific recommendations on this point. The subsequent Robbins report, however, did directly address the problem, calling for "research . . . [on] significant changes in land use, in other natural resource use, or in other economic activities on areas adjacent to national parks, and likely to affect the parks." It also supported "cooperative planning . . . with other agencies which administer public and private lands devoted to conservation and to recreation." Interior secretary Stewart Udall endorsed these recommendations in a 1968 policy document that explained that "the responsibilities of the [Park] Service . . . cannot be achieved solely within the boundaries of the areas it administers" and then called for "close cooperation with all land-managing agencies, considering broad regional needs." Plainly, were science to be taken seriously, it had the power to transform how the Park Service and its constituents perceived and managed the enclaves under its care.[7]

External Threats to the Parks

That was the clear message from the 1970s Redwood National Park controversy, which vividly brought home the perils of treating the national parks as isolated entities without regard to the surrounding landscape. When originally created in 1968, Redwood National Park was designed to protect some of the few remaining groves of these majestic, long-lived trees that stretched along the northern California coastal mountains. The park, however, was an island engulfed by both public and private timberlands that faced liquidation by a lumber industry fueled by the postwar housing boom. Extensive upstream logging activities sent copious runoff, sediment, and debris coursing downstream, where park lands were eroding so badly that the iconic trees the Park Service was charged with protecting were beginning to topple, the shallow soil covering their root systems washed away. In 1978, after a bruising ten-year struggle, Congress agreed to expand the park by acquiring most of these upstream lands while directing the Park Service to restore them and providing retraining assistance for the displaced timber workers. The total for this work was estimated at more than $350 million, representing the most expensive park acquisition in the system's history thus far. The upshot was a park that more closely matched its watershed and was better protected from external activities that could destroy key features and disrupt critical ecological processes.[8]

Another result of the Redwood controversy was explicit recognition of what was being called the external threats problem. As part of the campaign to protect the park from upstream logging effects, Redwood's defenders had sued the Park Service and requested the court to order it to take action to protect the park. Relying on the Organic Act's mandate directing the Park Service to conserve park resources in an unimpaired condition and on an implicit public trust duty to protect these same resources, a federal judge directed the agency to seek congressional funding to acquire upstream lands and to take other steps to protect the park. Although this order produced few immediate results, the court's ruling established the principle that the Park Service's management responsibilities extended beyond the boundary line when necessary to protect its lands.

The entire controversy also convinced Congress to reinforce the agency's protective duties by adding the so-called Redwood Amendment to the Organic Act, directing that "the protection, management and administration of [national parks] shall be conducted in light of the high public value and integrity of the National Park System and shall not be exercised in derogation of the values and purposes for which these various areas have been established." Since then, consistent with the legislative history linking this amendment with the Redwood problem, the courts have ruled that the statute imposes an absolute duty on the Park Service to protect national park resources from threatening activities, but have then refrained from forcing agency officials to take any specific protective actions. The external threats problem has nonetheless found official expression in the law, reinforcing that national parks should be managed as part of the larger landscape.[9]

A series of reports issued during the 1970s further highlighted the external problems that confronted the national parks. In 1972, the Conservation Foundation recommended in *National Parks for the Future* that "national park boundaries should, whenever possible, include entire ecosystems. Neighboring political jurisdictions should be encouraged to conduct their land planning and regulatory activities in ways which support the purposes of a park unit." The report also suggested giving the Park Service authority "to implement protective land-use controls for inholdings and adjacent private lands that clearly affect the natural values of park lands." In the aftermath of the

Redwood controversy, the National Parks Conservation Association issued *Adjacent Lands Survey* in 1979, reporting that "nearly two thirds of the 203 [national park] respondents stated that their units suffer from a wide variety of incompatible activities on adjacent lands . . . [and] nearly 50 percent of the superintendents believe that they do not have sufficient authority or appropriate policy directive to respond to problems emanating from lands outside the boundaries of their units."[10]

Congress weighed in, too, directing the Park Service to prepare a systemwide state of the parks report. The ensuing seminal *State of the Parks, 1980* report represented the first time the agency had acknowledged the pervasive, systemic nature of the external threats problem. Its findings were truly worrisome: park resources were experiencing "significant and demonstrable damage" from air and water pollution, exotic plant invasions, adjacent industrial development, and visitor overuse that could prove "irreversible." More than half these threats emanated from outside the parks, but few of them were well documented because the agency lacked baseline scientific information. For the sixty-three large natural-area parks, the report noted that the once-pristine lands that had long surrounded and buffered the parks were disappearing, with the result being more external encroachment. The agency's underlying problems seemed to cut across the board: insufficient scientific documentation, inadequate legal authority, and little political will. Faced with this compelling call to action, park supporters in Congress sought to remedy the situation by introducing park protection legislation that would give the Park Service explicit authority outside its boundaries, but opposition from landowners and others prevented these rather bold proposals from ever advancing very far on the legislative agenda.[11]

The Emergence of Ecosystem Management

During the late twentieth century as the ecological sciences matured, the problematic enclave nature of the national parks drew even more attention, lending credibility and urgency to the external threats problem. Drawing on island biogeography theory and its compelling insights on extinction, scientists came to understand that changes in the structure of even large, mature ecosystems could have severe destabilizing effects, destroying or fragmenting habitat that left species isolated and unable to sustain viable populations. Applied to the national parks, it meant that development pressures on the periphery could leave park wildlife subject to possible extinction. In fact, one widely cited study by biologist William Newmark revealed that most of the western North American national parks had lost wildlife through extinction during the twentieth century because their legal boundaries did not include sufficient habitat to maintain viable populations. The emergent discipline of conservation biology put even finer points on these conclusions, showing that the aggregate impact of roads, human presence, oil wells, and other development on the periphery of protected areas endangered wildlife, particularly top-of-the-food-chain carnivores, which then had a cascading effect on other species within the ecosystem. It was all evidence, according to the scientists, of the need to connect national parks and other protected areas to facilitate species movement and genetic interchange.[12]

An array of controversies involving the national parks not only illustrated the potential cumulative effect of external threats, but also brought forth the concept of national parks as the vital core of larger ecosystems. Once again, Yellowstone was in the forefront, but this time it was not just the park but the much larger and recently denominated concept of the Greater Yellowstone Ecosystem that commanded attention. The Craighead brothers, as part of their groundbreaking grizzly bear studies during the 1960s, had first used the "greater ecosystem" terminology to explain that the park's grizzly bears ranged far beyond park boundaries and therefore needed secure habitat outside the national parks to survive. Biologists also explained that top-of-the-chain carnivores like grizzly bears commonly served as "surrogates" for the health of ecosystems. Conservationists thus seized on the Greater Yellowstone Ecosystem concept to describe the complex arrangement of public and private lands that surrounded Yellowstone and Grand Teton National Parks. Not only were these lands biologically connected to these parks, providing critical habitat and migratory corridors for park animals, but the headwaters of the all-important Columbia, Missouri, and Colorado Rivers—the lifeblood for the arid West—originated in the area. Having drawn these large-scale ecological connections, conservationists stood on firm ground when they asserted that uncontrolled timber harvesting, energy exploration, and subdivision development on the surrounding lands imperiled park wildlife, particularly the endangered grizzly bear that roamed broadly across the ecosystem.[13]

Similar concerns surfaced throughout the national park system. At Glacier, the threats were multiple: a proposed open-pit coal mine situated a couple miles north of the park in British Columbia on the North Fork of the Flathead River that defines the park's western boundary; oil and gas exploration projects sited near the park border on the two adjacent national forests, the Flathead and the Lewis and Clark; and timber harvesting near the park boundary in the adjacent national forests. At Canyonlands, the threat centered on a proposed nuclear storage facility to be located just outside the park; at Bryce Canyon, it was an adjacent strip-mining proposal. At North Cascades, Olympic, and Mount Rainier, unrestrained commercial logging was fragmenting the old-growth forests that spilled across park borders and provided important habitat for spotted owls and other old-growth–dependent species. At Grand Canyon, Bryce, and other southwestern parks, the overarching threat was mounting air pollution traced to nearby power plants and southern California auto exhausts. When asked to investigate the national parks' external threats problem, the General Accounting Office found that it was pervasive and had worsened since the *State of the Parks, 1980* report and that the Park Service had taken few effective steps to address it, a view other reports consistently confirmed.[14]

The obvious answer to these looming threats, at Yellowstone and other national parks, was to begin planning and managing at the ecosystem scale, with the national parks playing a central role in a regionally focused management effort. As the Conservation Foundation, in its *National Parks for the Next Generation* report, put it: "The tradition of park stewardship must gradually be extended beyond park boundaries, to domains where mainstream attitudes about private property and freedom of action still prevail today." The general idea—dubbed "ecosystem management" by its supporters—was to promote more collaboration between the various federal land management agencies and with property owners to ensure the ecological integrity of the shared landscape. That is exactly what the Park Service, in 1988 revisions to its management policies, sought to do by expressly acknowledging that "parks are integral parts of the larger regional environment" and by encouraging park managers to pursue cooperative regional planning opportunities. In the Yellowstone region, following a highly critical Congressional Research Service report that found the area's grizzly bear population at risk of extinction due to a lack of coordination among the responsible agencies, this effort took the form of a federally driven "vision document" that was intended to set new management priorities across the region's two national parks, seven national forests, and three national wildlife refuges.[15]

Viewed as an opportunity to create a "world-class model" for integrated natural resource management, the Greater Yellowstone vision initiative got off to a promising start. The draft document not only called for ecosystem management, but also envisioned "a landscape where natural processes are operating with little hindrance on a grand scale . . . and humans [are] moderating their activities so that they become a reasonable part of, rather than incumbrance upon, those processes." The project soon faltered, however, when disgruntled residents and industry representatives denounced it as an illegal federal land grab and enlisted local political officials to subvert the process, which they did. The final document represented a near total retreat; it discarded any reference to ecosystem management, dropped all the draft's visionary language, and emphasized the separate—rather than overlapping—missions of the various federal land management agencies. Although disappointing, the Greater Yellowstone vision process provided yet another lesson that scientific truth without political muscle would not be sufficient to knit together a fragmented landscape, even one as iconic as the Yellowstone country with its large assemblage of federally owned lands and charismatic species.[16]

This initial attempt proved to be just the opening salvo in the effort to promote a new ecosystem management approach to administering the federal lands. In 1992, the Park Service unveiled the "Vail Agenda" report, which stated that "the prevention of external and transboundary impairment of park resources and their attendant values should be a central objective of Park System policy" and recommended that "natural resources in the park system should be managed under ecological principles that prevent their impairment." With the election of Bill Clinton as president in 1992, the ecosystem management idea gained new life when it won his endorsement and was then embraced throughout the federal bureaucracy. Buoyed by laws like the Endangered Species Act and the National Environmental Policy Act, ecosystem management was given official recognition in the 1994 Northwest Forest Plan, a groundbreaking document devised to settle that region's long-standing timber wars. Besides espousing interagency coordination principles, the plan gave priority to protecting biodiversity at the ecosystem scale, doing so through the use of protected reserves and corridors designed to limit development on ecologically important federal lands. The broader principles that emerged from the Clinton administration's adoption of ecosystem management policies across the federal public lands boded well for the national parks, which stood to benefit from the restraining effect these policies had on development activity on their borders.[17]

Indeed, the ecological sciences garnered a new prominence in national park policy throughout the 1990s, providing further recognition to the vital role of the parks as the core of larger ecosystems. When the National Park Service undertook to revise its all-important management policies, it included specific provisions addressing the external threats problem; park officials were directed "to use all available authorities to protect park resources and values from potential harmful activities" and to engage with their neighbors to prevent harm to park resources. When Congress passed the 1998 National Parks Omnibus Management Act and gave the Park Service a new science mandate, it also expressly authorized scientific studies that viewed national parks as part of larger regions. Long-stalled plans to restore Everglades National Park's "sea of grass" ecosystem received new life, reconfirming that the park was a vital part of this larger watershed. At Yellowstone, the restoration of wolves served to reassemble the park's original wildlife contingent and is reshaping the entire ecosystem, while presidential intervention stopped the proposed New World Mine on the park's northeastern border, eliminating this major external threat. The Glen Canyon Dam water-release project restored some semblance of historical flooding patterns to the Colorado River where it flows through the Grand Canyon, revitalizing the river corridor within the park. A dam removal project on the Elwha River, much of which runs through Olympic National Park, was designed to restore native salmon runs into the park's interior. President Clinton, by designating large new national monuments on the flanks of Grand Canyon and Sequoia National Parks, introduced another strategy for safeguarding vulnerable parklands in the larger landscape setting. Moreover, despite the political recriminations that followed the Yellowstone and Cerro Grande conflagrations, the continued federal commitment to wildfire as an important process in shaping public land ecosystems further reinforced the advent of new policies sensitive to ecological realities and the need to view the landscape as a whole.[18]

Coordination as the Key

These ecological management initiatives, however, were mostly devised and implemented by the agencies themselves without congressional involvement, so it was no surprise that the ensuing George W. Bush administration took a different path without much regard for the national parks as critical parts of the larger landscape. Confronted with escalating energy prices, the Bush administration prioritized oil and gas development on the public lands, largely unconcerned where development activities took place or their environmental consequences. Extensive controversy and litigation ensued, culminating in 2008 with a major political firestorm over a proposed oil and gas lease sale that would have brought drilling rigs to the doorstep of Canyonlands, Arches, and Dinosaur National Parks. To the south, renewed interest in nuclear power prompted uranium speculators to file hundreds of new mining claims on the national forests surrounding Grand Canyon National Park, triggering a legislative proposal to withdraw the area from further mining claims. Add to this the overt—but ultimately unsuccessful—effort to eviscerate the Park Service's management policies by eliminating references to external threats and by revising the agency's primary resource conservation duties. Although neither the underlying science nor the vision of national parks as key core components of larger ecosystems had changed, political priorities had shifted, and with that, the commitment to safeguarding these reserves by tempering nearby development activities changed as well.[19]

The election of Barack Obama to the presidency signaled yet another turn in the long road toward implementing meaningful ecological management policies that would ensure real protection for the national parks. This time, though, the relevant science was not just about ecology; it also extended to climate and the pervasive threat that global warming represents not only to the national parks but to the entire web of life on Earth. Unlike the Bush administration, the Obama administration has taken global climate change seriously, acknowledging that it could have a devastating effect on the national parks. As we have seen, climate scientists predict that Glacier's namesake glaciers will be gone by 2030, that Joshua Tree's namesake cactus will disappear from that park, and that several coastal parks—Everglades, Dry Tortugas, Biscayne, and Virgin Islands—could end up under water. Although the potential magnitude of these effects is on a vastly different scale than more conventional external threat problems, the warming atmosphere serves as another reminder of how much the national parks are interconnected with the larger world surrounding them.[20]

The current antidotes to global climate change—mitigation and adaptation—have important implications for the national park system and park management policy. First, the national parks and other protected lands provide important carbon sinks that can help reduce the effects of carbon emissions into the atmosphere. Second, the higher-elevation parks will likely serve as important refuges for species driven from lower-elevation habitats by warming temperatures. Third, the southwestern parks could face extended drought conditions that make desert watercourses and riparian areas even more important for local species, whereas Florida coastal parks may find themselves inundated by rising seawater levels linked to the melting polar ice caps. Under any of these scenarios, better coordinated management across the landscape—whether denominated ecosystem management or something else—will be necessary to ensure that dispersal corridors are available to climate-displaced species and that other adaptation strategies can be implemented effectively. Moreover, new national parks strategically located on the landscape to protect ecosystem services could also help mitigate the worst warming effects. Put simply, the need for ecosystem-based conservation strategies to safeguard national parks as core components of larger, interconnected landscapes is more urgent than ever.[21]

GLACIER NATIONAL PARK AND THE CROWN OF THE CONTINENT ECOSYSTEM

According to the *State of the Parks, 1980* report, Glacier National Park in northwestern Montana was the most threatened park in the national park system. Myriad threats emanating from outside its boundaries put Glacier's wildlife, water, and air at risk: a proposed open-pit coal mine in southeastern British Columbia, oil and gas leasing proposals in the rugged Badger-Two Medicine region of the Lewis and Clark National Forest, clear-cut timber harvesting in the Flathead National Forest, and continued interest in paving North Fork Road on the park's western flank. Park officials were quite concerned about the prospect of industrial development and new roads in previously untouched areas that had effectively buffered the park over the years, but they had few ideas about how to respond other than to invoke the park's recent International Biosphere Reserve designation and to promote rather vague ideas about regional management. For Glacier's managers and admirers, the park was teetering on the verge of seeing its ecological integrity impaired and losing its essential wilderness character.[22]

A Threatened Park

It was difficult, though, to imagine Glacier as the most threatened national park in the system. Located in a remote and lightly populated region, Glacier is surrounded by publicly owned land: the expansive timber-rich Flathead National Forest to the west; the Lewis and Clark National Forest to the south, including the Badger-Two Medicine country on the iconic Rocky Mountain Front; the sprawling Blackfeet Indian reservation to the east; undeveloped and unpopulated British Columbia provincial lands to the northwest; and its sister Waterton Lakes National Park in Alberta to the northeast. Large segments of the adjacent national forests were protected as designated wilderness areas, including the renowned Bob Marshall Wilderness Area, and two designated Wild and Scenic Rivers—the North Fork and the Middle Fork of the Flathead River—formed the park's western and southern boundaries. A sizable grizzly bear population, legally protected in the United States under the Endangered Species Act, roamed across the region, and a pack of Canadian wolves chose the park's remote northwestern corner as the site of their return to the American West after a fifty-year absence. If Glacier was endangered, then what fate awaited other less-buffered national parks situated much nearer urban population centers?

Over time, however, the threats gradually diminished, staved off by determined legal counterattacks, tactical political maneuvers, and changing market forces. Timber harvesting in the Flathead National Forest was brought to a standstill by environmental lawsuits based largely on protecting grizzly bear habitat from incursions by new roads. In the Lewis and Clark National Forest, oil and gas development was blocked by endangered species litigation and by the Blackfeet's cultural heritage claims as well as by reduced market prices. The Cabin Creek coal mine proposal in southeastern British Columbia was stymied when the International Joint Commission intervened under the Boundary Waters Treaty, finding that the project posed unacceptable water-quality risks throughout the Flathead River basin and recommending that the North Fork area be converted into an international conservation area. The North Fork Road escaped asphalt surfacing when federal officials concluded that increased traffic in this remote area could jeopardize the protected grizzly bear. Although the Park Service helped pursue the International Joint Commission referral, it was otherwise reticent to intervene in these controversies, relying instead on third-party environmental advocates to help protect its interests. In addition, its efforts to convert Glacier's International Biosphere Reserve designation into meaningful regional-level management commitments soon faded, although the vision of better integrated management at the ecosystem scale did not.[23]

Before the twenty-first century was a decade old, however, Glacier once again found itself confronting another set of externally driven environmental challenges, some new and some old. As energy prices spiked upward, leaseholders in the Lewis and Clark National Forest sought permission to drill, while oil companies expressed renewed interest in leasing national forest lands along the Rocky Mountain Front. Northward in the Canadian Flathead Valley, several new project proposals surfaced—two coal mines, coal-bed methane exploration, and a gold mine—that cumulatively would convert the still-pristine upper reaches of the North Fork drainage into an industrial zone. Across the Flathead and the Lewis and Clark National Forests, recreational off-road vehicle activity had soared, bringing more people, illegal trails, and unremitting noise into what had been undisturbed wildlife habitat. Further, over the intervening years the scenic Flathead Valley had been discovered; subdivisions and second homes were pushing outward toward Glacier, not only raising water-quality concerns but also displacing wildlife from their traditional haunts. It was once again apparent that some sort of regional response was required to address these mounting and seemingly unrelenting external challenges to the park's ecological integrity.[24]

Regionalism Takes Hold

This time around, regional initiatives were under way in the Glacier area that offered potential forums in which external threat issues might be addressed. One was the state-created, multiagency Flathead Basin Commission, but it had only an advisory role and no managerial power. Another was the Yellowstone to Yukon, or Y2Y, project, which was conceived by environmental organizations to establish a network of protected nature reserves along the spine of the Rockies, but it had not generated much local support. Yet another was the Northern Rockies Ecosystem Protection Act, a congressional bill that envisions a sixteen-million-acre network of protected areas stretching across five states, but it, too, lacked any real political support. Although these initiatives each conceived of the region at vastly different scales, they all aligned with the idea of Glacier at the core of a larger, ecosystem setting. Each of them was also designed to break down the traditional jurisdictional boundaries that had long impeded rational management of wildlife and water systems, but none has yet produced a major breakthrough. It remains to be seen whether these organizational efforts will be central to the search for regional managerial integration.[25]

One promising initiative, though, is the Crown of the Continent Ecosystem Managers Partnership, which was pioneered by officials at Glacier National Park and Canada's Waterton Lakes National Park in an effort to improve management relations at a regional scale. The "Crown of the Continent" phrase—first coined by George Bird Grinnell in the 1890s to describe newly created Glacier National Park—is now being used to describe an international ecosystem that extends along the Rocky Mountains from the Bob Marshall wilderness complex in Montana to the Elk Valley in British Columbia, with the two sister national parks at its core. The namesake managers' group—composed of federal, state, tribal, provincial, First Nation, and university members—has endorsed the notion of collaborative ecosystem management and has adopted "an ecologically healthy Crown of the Continent Ecosystem" as its vision. Moreover, its mere presence has helped spawn a cadre of other regional initiatives bearing the Crown of the Continent name, including a science center, a conservation alliance, a roundtable for community dialogue, and a geotourism project, all committed to promoting some version of landscape-scale conservation.[26]

The Crown managers group, however, has disavowed any intent to involve itself in management decisions or processes, choosing instead to engage primarily in regional-level studies and mapping projects addressing wildlife movement patterns, road construction, recreational activities, and other notable land use trends. An early group effort to develop a joint cumulative effects model for the region was jettisoned after the Flathead forest supervisor objected to it, evidently concerned that the model might be used against the Flathead in environmental litigation over its timber and other forest management practices. That the group largely ignored the region's most pressing transboundary resource management problem—energy development in the Canadian Flathead—illustrates its apparent limitations for achieving truly meaningful ecosystem-level managerial coordination.[27] Whether that will change or not as the manager group's members gain more familiarity with one another and build more trust remains to be seen. Meanwhile, the various Crown groups are working diligently to highlight the region's extraordinary conservation values and rebrand the region in ecosystem conservation terms. The Obama administration has acknowledged these efforts by featuring the Crown in its America's Great Outdoors initiative as a model for landscape-scale conservation.[28]

As these initiatives have begun to take hold, there has been notable progress toward reducing Glacier's vulnerability to external threats on several fronts. Multiple factors have coalesced to help promote greater park protection and regionalism, not only between the park and its national forest neighbors, but also with its Canadian neighbors. One important factor has been the evolution of a local public that conceives its surroundings in regional terms and employs that sense of place to promote ecosystem-level protection. Another key factor has been reduced conflict between the missions of neighboring federal land managers as reflected in the Forest Service's transition away from its historic commodity production orientation and toward wildlife, recreation, and other amenities, which are more consistent with the Park Service's overall priorities. A third critical factor has been the presence and enforcement of powerful laws like the Endangered Species Act and the National Environmental Policy Act, which have played pivotal roles in promoting greater managerial harmony on the federal lands by essentially channeling incompatible uses toward environmentally benign locations. These forces are particularly pronounced on the southeastern side of the park, where the ecologically defined image of a Rocky Mountain Front has become a central reality. Not only has oil and gas development receded as a major concern there, but Lewis and Clark National Forest officials have prohibited off-road vehicle use in the Badger-Two Medicine area that adjoins the park.[29]

These same factors are also coming into play on Glacier's west side, in part because the realities of a uniquely pristine and interconnected landscape have resonated on both sides of the international border. Building on an original International Joint Commission recommendation from the 1980s to recognize the North Fork region's exceptional conservation values,[30] Montana's governor and British Columbia's premier entered into an unusual bilateral Memorandum of Understanding committing their respective governments to safeguard the watershed from mining or energy development.[31] On the Canadian side, that agreement has been translated into provincial legislation;[32] on the U.S. side, the North Fork Watershed Protection Act is pending in Congress, and Montana state officials have precluded any surface drilling in a strategically situated state forest.[33] Several oil companies have voluntarily surrendered their leases on Flathead National Forest lands, the Forest Service has implemented significant off-road vehicle limitations across the area, and the rampant subdivision of private lands in the Flathead Valley has eased during the recession. The county has shown no interest in adopting meaningful zoning laws or land use restrictions to protect environmental values, however, and the pace of development will almost certainly accelerate in the years ahead, leaving Glacier unprotected from further residential development on its west side with the attendant wildlife and water quality consequences.

Vigilance and Vision Going Forward

Glacier National Park faces much more imminent threats, however, and the region's ecological integrity remains at risk. The Blackfeet tribe has leased nearly its entire reservation along the park's eastern flank, and seismic activities are already occurring just outside the park border. Tribal politics, cultural site preservation concerns, and the federally protected grizzly bear could yet deter or mitigate any new oil field developments, but oil companies have leases and the evident right to explore and develop the area.[34] Although Lewis and Clark forest officials have prohibited off-road vehicles in the sensitive Badger-Two Medicine area, the area was dropped from a pending Rocky Mountain Heritage wilderness bill due to objections from the tribe and from motorized interests. Without permanent legal protection, the Badger-Two Medicine area is protected only by the Forest Service's administrative decisions, which could be changed during the next round of forest planning. Further, local conservationists worry about new mining and logging activity in the Elk Valley and Castle area on the Canadian side that could further fragment important wildlife migration corridors.[35] Thus, although important protections are in place that will benefit the park and its resources, the vision of an interconnected and secure Crown of the Continent landscape remains elusive.

Despite the challenges inherent in advancing a regional environmental protection agenda, the fate of Glacier and other national parks hinges on making progress toward thinking and acting on an ecosystem basis, with the parks at the core. Regular engagement with adjacent public land managers and other neighbors can pay dividends on some issues, as it did for Glacier in helping convince Lewis and Clark forest officials not to issue additional energy leases, even if a broader regional agenda is unattainable. Third-party advocates who are willing to file environmental lawsuits can play key roles in protecting national park resources, as occurred in bringing the Flathead's timber program under control and in stalling drilling in the Lewis and Clark forest. These same advocates can also be crucial allies in promoting a broader regional management approach as is unfolding under the Crown of the Continent banner. Even when formal regional institutions and endeavors pay few immediate dividends, carefully conceived efforts to promote a sense of regional identity and integration with the public can only help advance long-term park protection goals. In sum, the Glacier experience teaches that protecting park resources against external threats requires long-term visionary thinking at the ecosystem level as well as calculated short-term strategic moves and interventions to safeguard park resources from impairment.[36]

Restoring the Everglades Ecosystem

Far from Glacier country, a massive landscape-scale ecosystem restoration effort is under way at Everglades National Park in south Florida. Widely acknowledged to be "the most fully realized and best funded ecosystem restoration effort ever undertaken by humankind,"[37] the Comprehensive Everglades Restoration Program (CERP) pairs the federal government and the state of Florida together in an $8 billion effort to reestablish a semblance of the park's historic ecological conditions. The CERP project extends across eighteen thousand square miles and sixteen counties, essentially covering the entire south Florida peninsula from the northern edge of Lake Okeechobee southward. Much of the greater Everglades area is devoted to conservation, including three other national park units (Big Cypress, Biscayne, and Dry Tortugas) and sixteen national wildlife refuges, all of which provide critical habitat for wading birds, alligators, panthers, and other species, many of which appear on the federal endangered species registry. The restoration effort is designed to revive the water that used to slowly flow as a sheet southward from the lake and thus restore the "sea of grass" that long characterized the region. That vital hydrological connection has been severely altered from nearly a century of breakneck agricultural and urban development pressures. Whether the park's ecological integrity can be salvaged depends on the outcome of the unprecedented reengineering effort that is taking place outside the park's boundaries.[38]

An Imperiled Ecosystem

Originally authorized in 1934, Everglades National Park was not established until 1947 when Congress finally provided the necessary funding. At 1.3 million acres, Everglades is the third largest national park outside of Alaska. According to its enabling act, the park is "permanently reserved as a wilderness . . . [for] the preservation intact of the unique flora and fauna and the essential primitive natural conditions now prevailing in this area."[39] Although scientists uniformly view the park and surrounding area as "the most ecologically important subtropical wetland in the United States,"[40] the park only encapsulates 20 percent of the original Everglades ecosystem. The park's wetlands were historically connected to Lake Okeechobee by a series of rivers, streams, and sloughs that allowed the lake's freshwater to flow southward in a perpetual sheet where it sustained a vast freshwater ecosystem shaped by the region's seasonal flooding events. The resulting marshes, sawgrass prairies, mangrove stands, and other wetland habitats sustained an abundance of animals and plants adapted to this unique ecosystem. These attributes have garnered the park international recognition as an International Biosphere Reserve, a World Heritage Site, and a Wetland of International Importance, all testaments to its extraordinary ecological significance.

Into the early twentieth century, south Florida remained largely undeveloped, primarily due to the flooding that rendered much of the landscape uninhabitable. Once coastal areas like Miami and West Palm Beach began attracting residents, however, it was only a matter of time before local development interests turned their attention inland, calling for government assistance to help drain the swamps and protect against flooding. In 1949, confronting mounting postwar in-migration pressures, the U.S. Army Corps of Engineers dramatically expanded its reclamation efforts after Congress authorized the massive Central and Southern Florida (C&SF) project to facilitate urban and agricultural development. To control flooding, the fifteen-year C&SF project channelized the Kissimmee River that flowed south to Lake Okeechobee, diked the lake itself to create the 750,000-acre Everglades Agricultural Area immediately south of it, constructed a series of drainage canals extending to the Atlantic Ocean to safeguard the rapidly growing south Florida coastal communities, and assembled an array of new freshwater storage facilities. In total, more than fourteen hundred miles of canals, levees, spillways, and other structures were built to alter and control the natural processes that had sustained the area for centuries. As a result, south Florida's population boomed to more than five million people by 2000,[41] the coastal cities expanded ever farther inland toward the park, and a major new and politically powerful sugarcane industry took hold in the northern Everglades.

It soon became evident, however, that this extensive replumbing imperiled the vast Everglades ecosystem and most notably the park. The Everglades rescue effort is generally traced to 1947, when Marjory Stoneman Douglas—widely hailed as "Grande Dame of the Glades"—published *The Everglades: River of Grass,* which chronicled the extensive environmental changes afoot across the ecosystem. These changes only mounted in subsequent years once some of the individual C&SF projects were completed, affecting the quantity, quality, and timing of water that reached the park and introducing new phosphorus contaminants into the system as runoff from the Everglades Agricultural Area. Chemical fertilizers from the farms coursed downstream, not only contaminating the freshwater that entered the park but also spurring the growth of cattails that displaced the native sawgrass. The C&SF project's interlocked series of dikes, levees, canals, and pumping stations diverted huge quantities of freshwater from the system, depriving the park's marshes of a dependable water flow and altering sensitive habitat. In addition, as the region's booming housing market and growing agricultural sector sprawled ever more deeply into the Everglades, developers called for even more protective levees and canals. The effect on the park's native species was devastating: the wading bird population crashed by more than 90 percent, and sixty-eight different plants and animals soon found themselves on the federal endangered species list, including the Florida panther, American crocodile, manatee, and Cape Sable seaside sparrow.[42]

In response, environmental organizations banded together in 1968 to form the Everglades Coalition. A year later, Douglas formed Friends of the Everglades to also assist in repairing the ecosystem. Over time, their organized efforts began to pay dividends. At the state level, Florida adopted the Water Resources Act of 1972, which mandated minimum flows and water levels, and it later passed the Water Management Improvement Act of 1987, which required water districts to clean up their systems. A landmark federal lawsuit charging that phosphorus runoff from the sugarcane farms was polluting Everglades National Park's freshwater system was eventually settled with an unprecedented (and as yet uncompleted) cleanup agreement,[43] prompting passage of the state's Everglades Forever Act, which funded construction of new storm-water treatment areas to filter these contaminants from the runoff. At the federal level, Congress first intervened in 1970 to establish minimum water flow levels to the park, but the mandated level was set too low to make much difference.[44] In 1978, the Park Service established the South Florida Natural Resources Center, whose mission is to "conduct and communicate science for the preservation and restoration of the south Florida ecosystem."[45] Then, in 1989, Congress intervened again, not only expanding the park by 107,600 acres in an effort to safeguard critical habitat, but also requiring the Corps of Engineers, through the so-called Mod Waters program, to deliver more water to the area to reestablish more natural flows. These changes, however, were piecemeal and implementation has been delayed, resulting in little progress toward restoring even a semblance of the original dynamic system. Everglades proponents responded by renewing their call for a more comprehensive plan, which prompted the Clinton administration to establish a multiagency South Florida Ecosystem Restoration Task Force, a group that soon provided a solid scientific basis for a more comprehensive approach to reversing the environmental damage.

Restoring an Ecosystem

In 2000, drawing upon the task force's work, Congress endorsed the Comprehensive Ecosystem Restoration Plan designed to reengineer the system and restore historic ecological conditions. Although predicated on the numerous scientific studies documenting the system's decline, CERP is ultimately a political solution that addresses multiple interests, including an assortment of federal, state, tribal, and local governmental entities, developers, utilities, farmers, and dozens of environmental groups. As one astute observer put it, "There is something for everybody in the plan."[46] CERP's goals are to "restore, preserve, and protect the South Florida ecosystem while providing for other water related needs of the region, including water supply and flood protection," all designed to ensure that the right quantity and quality of water are delivered and properly distributed throughout the area at the right time.[47] What this plan should ultimately mean for the Everglades is the annual addition of 320 billion gallons of water to the ecosystem.

To achieve these goals, the plan contemplates sixty-eight different projects. They include eighteen new surface-water storage reservoirs covering 217,000 acres, 330 new water storage wells, two new reclaimed rock quarry water-storage sites, 35,000 acres of new storm-water treatment areas, the removal of more than 240 miles of canals and levees within the remaining Everglades and the addition of more than 500 miles of protective levees and canals on the periphery, two new wastewater treatment plants for the Miami area, and modified water delivery schedules to better mimic natural flows. Recognizing the complexity of the ecosystem and the undertaking, the plan incorporates adaptive management requirements, which oblige the agencies to monitor and adjust the projects while implementing them. Further, the plan established a science coordination team to oversee and evaluate implementation, mandated periodic reviews by independent National Academy of Sciences panels, established a review and comment process to ensure public involvement in project decisions, and required regular interagency coordination. Original estimates put the plan's costs at a whopping $7.8 billion, to be shared equally by the federal government and the state.[48]

If fully implemented, CERP, along with Mod Waters and other earlier restoration projects, should reverse the trajectory of ecological decline in the park. Despite ongoing project delays and inadequate funding, some progress is nonetheless evident. The Kissimmee River north of Lake Okeechobee has been realigned from a straight-line canal to reestablish more than forty miles of the historic meandering river channel, and the amount of phosphorus reaching the park—although not consistently meeting legal standards—has decreased, the result of land purchases in the Everglades Agricultural Area and improved farming practices. Plans are moving forward to elevate 6.5 miles of the Tamiami Trail highway that transects the south Florida peninsula to enhance natural water flows moving southward from Lake Okeechobee. After a lengthy political and legal battle, Congress has appropriated the necessary funds to buy out property owners in what is called the 8.5 Square Mile Area, which scientists deemed crucial to restore key park lands and historic flow patterns from the park to Florida Bay.[49] Under a program dubbed Acceler8, the state of Florida has begun work on several engineering projects designed to enhance natural flows to the park while also addressing urban water supply and flooding concerns. Throughout this lengthy process, the park's science center, which is actually mandated "to address the impacts of activities taking place outside park boundaries," has played an important role in evaluating project proposals and securing modifications to improve ecological benefits.[50] In addition, park officials have been fully engaged with their federal and state counterparts in an impressive although often contentious interagency coordination effort.

The jury on whether the effort will succeed is still out, however. Critics cite numerous flaws with the Everglades restoration program, questioning whether this massive reengineering effort will work or whether a more natural approach favored by environmental interests would have been more effective. Despite reductions in phosphorus levels, invasive cattails continue to displace native sawgrass; funding delays have slowed the pace of land acquisition, which has meant the loss of critical wetlands areas to development as well as increased land purchase costs; political compromises have given priority to urban water-supply and flood-control projects at the expense of ecologically beneficial projects; and federal funding has lagged seriously behind state funding. Moreover, even with these modifications to the Everglades hydrology, the specter of climate change could not only alter precipitation and water flow patterns but also trigger a rise in sea levels that would allow salt water to intrude into the freshwater system, radically disrupting its fragile ecology.[51]

The Everglades restoration project nonetheless represents an extraordinary commitment to restoring ecological health to the national park, demonstrating its vital position in the larger Everglades ecosystem. Indeed, the presence of Everglades National Park and public concern over its health were plainly the main catalyst for this multi-billion-dollar joint federal-state restoration effort. That other federally protected national parks and wildlife refuges situated in the same south Florida ecosystem will also benefit from the project is further testament to the cornerstone role of a park like Everglades in promoting landscape-scale conservation initiatives. As Everglades restoration has unfolded, the Park Service has been a pivotal player in the effort, particularly by providing important scientific information to ensure that environmental concerns are being addressed and by serving as a counterweight to the Corps of Engineers and development interests during project negotiations.[52] An energized and well-informed environmental constituency has also been crucial to the effort, serving to mobilize public opinion, litigate when necessary, and help build the necessary political consensus for such an expansive project. In the end, as in the Glacier–Crown of the Continent case, knitting the Everglades National Park together with the surrounding landscape is yielding important environmental benefits that should eventually enable the park to meet once again its nonimpairment resource conservation obligation.

SUSTAINING NATIONAL PARK ECOSYSTEMS

The experiences at Glacier and Everglades National Parks demonstrate just how challenging it can be to promote and establish an effective ecological management approach that appropriately acknowledges the national parks as the vital core in larger landscapes. Clearly, the view that national parks can be sustained and managed in an unimpaired condition without regard to the broader setting in which they are located is no longer tenable in a world where intensified industrial activity and second-home growth threaten even a park as remote and seemingly well-buffered as Glacier. The manifold challenges involved in restoring the hydrology of the Everglades demonstrate just how expansive our view must be to ensure ecological integrity for individual parks, to say nothing about the air pollutants that emanate from even more distant sources or the specter of climate change. To ensure that vital park resources are not impaired from beyond the boundary line, the national park idea must be enlarged to embrace the broader landscape.

The scientific underpinnings for this ecosystem view of the national parks are firmly established. Conservation biologists and other scientists have convincingly demonstrated, in location after location, that the long-term survival of wildlife species requires an extended, interconnected network of protected areas and that national parks are often at the core of these ecosystem complexes. It is not sufficient to merely protect these core areas, however; many species migrate seasonally, and others must disperse large distances, both to meet their daily food requirements and to interbreed to maintain their genetic viability. The presence of secure corridors or migratory routes promotes connectivity between parks and other protected areas, not only facilitating species movement but also expanding the effective scale of any conservation effort. When the anticipated effects of climate change are factored into the future needs of native wildlife and plant species, the case for an interconnected mosaic of protected areas and coordinated management across the landscape becomes even more compelling. That is why the national parks alone are inadequate as wildlife reserves and why the Park Service must possess the scientific knowledge needed to make a persuasive case for ecosystem management.

Any vision of the national parks at the protected core of larger landscapes cannot be brought to fruition solely on the basis of science. As is evident by the Yellowstone, Glacier, and Everglades experiences, external threat problems and ecosystem-based regional management proposals pose very difficult—but not insurmountable—legal and political challenges. When a national park is surrounded primarily by federally owned lands, which is the case with many of the large western parks, the Park Service and its neighboring land managers operate under a common set of overarching federal laws. These laws establish uniform Endangered Species Act and National Environmental Policy Act requirements, and they contain inter-agency coordination requirements that are found in the planning legislation governing each federal land management agency. Moreover, the Glacier experience suggests that the Forest Service's traditional, single-minded focus on commodity activities is fading, with forest managers giving greater emphasis to recreation, wildlife, and the like. That many communities located adjacent to national parks are now pursuing tourism, recreation, and related activities for their economic sustenance, as seen in chapter 5, should also help promote more compatible management between the national parks and nearby federal land managers.

Clear divisions remain between the national parks and their federal neighbors, however. No land manager is prepared to relinquish his or her discretionary management authority to a rival agency, nor will the law allow such an arrangement. Deeply instilled philosophical views about the role of humans on the landscape continue to divide the federal land management agencies, most often setting the Park Service, with its preservationist mandate and noninterventionist policies, apart from its sister agencies. Similar ideological divides are evident within local communities, where some residents still adhere to a strong commodity-oriented view of national forest and Bureau of Land Management lands. Even as some communities have embraced what observers describe as a "new West economy" based on scenic and recreational values, new high-tech business ventures, and off-site income sources, others continue to cling to tradition and stridently resist these new economic realities. These differences make it all the more remarkable that a strong regional consensus built around the Crown of the Continent and the Rocky Mountain Front in the Glacier country is beginning to emerge and pay dividends in the form of on-the-ground conservation accomplishments. Of course, this evolution in public attitudes confirms the critical importance of developing and pursuing a long-term regional strategy to lay the political seeds for eventual ecosystem-level protection.

Any effort to coordinate land use priorities and management strategies between the national parks and nearby private landowners presents even more daunting challenges. As reflected in the ongoing private land development problems that Glacier confronts in the Flathead Valley, local officials and property owners are generally reluctant to acknowledge any obligation to a national park neighbor, even as they share the same landscape and even as property values are typically enhanced by proximity to a national park. The same holds true in the Everglades, where sugarcane farmers and subdivision developers have fought to prioritize their water-control projects over ecologically oriented projects in the restoration effort. In the United States, property ownership rights are rooted in the U.S. Constitution and come with a strong tradition of autonomy, which holds that landowners may ordinarily use their property as they see fit. Although this tradition may be eroding, undercut by an emerging consensus that property ownership also entails responsibility for the shared landscape, an antipathy toward zoning and land use regulation still prevails in many locations, particularly in rural areas across the interior West and the South. Nonetheless, various tools are available to promote harmony between national parks and adjacent private lands, ranging from incentive-based strategies, such as land exchanges, conservation easement purchases, and tax breaks, to more rigorous regulatory strategies, such as wildlife zoning requirements and subdivision limitations. With careful planning and sensitivity to local sentiment, Glacier and other national parks can help support these strategies to protect park lands and resources from continued encroachment.

Whatever setting surrounds a national park, neither the Park Service nor the public can continue to view that park solely in terms of its defined boundaries. The national parks are essential parts of larger landscapes that provide sustenance for wildlife and that contain shared watersheds and airsheds critical to nearby communities. Just as we have long understood that the national parks are a key element in the regional economy, we also now understand that healthy ecosystems are important for the parks' overall well-being. Quite simply, the parks serve both as the wild heart of complex ecosystems and as an anchor tenant for local economies. Difficult as it may be, sensitive and coordinated management on a regional scale is imperative to ensure the long-term ecological integrity of the national parks as well as sustainable local economic growth. This enlarged ecosystem-level view of the national parks also underscores the need to explore opportunities to expand the national park system, either through the establishment of new parks or the expansion of existing ones.

CHAPTER 10

"Growing the System"

New Parks and New Strategies

The American national park system, as currently constituted and managed, does not reflect the dramatic evolution we have witnessed in the national park idea. Indeed, the national park system fails to fully capture either the critical ecosystem science principles or the contemporary social values highlighted in previous chapters, a fact reflected in the myriad controversies examined throughout these pages. Most notably, the system fails to encapsulate our modern understanding that national parks must be conceived and managed in a broader ecological context. Important ecosystem types remain unrepresented, key wildlife habitat still lies outside park boundaries, and adjacent development proposals regularly imperil park lands and resources. Sites interpreting the nation's racial and ethnic heritage are also poorly represented in the system. Moreover, as profound social and demographic changes further alter the relationship between people and nature, the need for new opportunities to expose citizens to our natural heritage and to safeguard ecologically sensitive areas is more apparent than ever.

The central question for the future is not whether but, rather, how to expand the national park system, if not through formal new legal designations, then at least through new management arrangements on the ground. It is plainly time to move beyond the haphazard growth that has been the hallmark of the past hundred years and to embrace a new, more comprehensive commitment to conservation for the next hundred, one that embodies our evolved understanding of the national park idea by focusing on landscapes to pursue new park designations as well as other protected areas and strategies.

EVOLUTION OF THE NATIONAL PARK SYSTEM

The American national park system, despite its prominence and esteem, is hardly a monument to visionary planning. Nearly a hundred years after its inception, the system has grown in size and stature well beyond what any of its founders could have imagined. At first just a handful of western parks set aside to protect a few spectacular places, the national park system now boasts close to four hundred units extending across forty-nine states and several territories, and it covers a bit less than eighty-five-million acres, most of which is located in Alaska. Although legally referred to as a system, the national parks are actually a diverse collection of natural, recreational, historical, cultural, and other sites that have been melded together without much foresight, more the result of tough-minded political calculations and attractive scenic features than any purposeful commitment to preserving diverse ecosystems or key biological specimens. The current approach has nonetheless preserved many important sites, ranging from large natural parks like Yosemite to hallowed battlefields like Gettysburg, each of which captures an aspect of the nation's heritage and most of which can legitimately be described as nationally significant. This random approach does not ensure that all such meritorious sites have received federal protection, however, nor does it suggest any evident commitment to the unifying themes embodied in the national park idea today. Instead, the national park system has evolved piecemeal, park by park, either by congressional legislation or presidential edict. And as the system has evolved, despite its best efforts to shape the system it manages, the Park Service has not always been a central player in the larger political drama of new park creation.

The Park Creation Process

From the beginning, Congress has assumed the primary responsibility for establishing new national parks. The first parks were designated to protect spectacular mountainous western landscapes, often to ensure that these areas did not fall into private hands under the ubiquitous disposal laws that governed public land policy at the time. Congress, acting under its expansive property clause power, initially created Yellowstone National Park in 1872, followed by a succession of other parks, including Yosemite, Sequoia, Mount Rainier, Crater Lake, Glacier, and Rocky Mountain National Parks during the twenty-five-year period from 1890 to 1915. In each case, a dedicated group of citizens, sometimes driven by a single visionary, saw the value in preserving an area in an undeveloped state to be enjoyed by everyone, not just an elite few or an entrepreneurial coterie anxious to exploit the scenery for personal profit. In the case of Glacier National Park, George Bird Grinnell, a widely respected eastern conservationist who cofounded the Boone and Crockett Club and edited the popular *Forest and Stream* magazine, mounted a concerted campaign to safeguard the "Crown of the Continent" as a national park. In Rocky Mountain's case, a local naturalist named Enos Mills, sometimes referred to as "the John Muir of the Rockies," campaigned tirelessly for the new park, penning an endless series of promotional articles for the *Saturday Evening Post* and other national publications designed to attract national support for the park proposal. The result, by 1916, was a modest but significant collection of legally protected parks that lacked any central management direction or much in the way of services for the intrepid tourists who were beginning to visit them.[1]

As the twentieth century got under way, a parallel preservation effort emerged to protect the rapidly disappearing archaeological relics scattered across the Southwest. In 1906, Congress adopted the Antiquities Act and delegated to the president the authority to create new national monuments, several of which have eventually been redesignated by Congress as national parks. Under this legislation, the president was empowered to protect "historic landmarks, historic and prehistoric structures, and other objects of historic or scientific interest," but was directed to limit his designations to "the smallest area compatible with the proper care and management of the objects to be protected."[2] Never shy about exercising power, President Theodore Roosevelt promptly put his new authority to work, christening the Grand Canyon, Mount Olympus, Lassen Peak, Carlsbad Caverns, Petrified Forest, and several other sites as national monuments and thus consecrating a president's important role in the establishment of new parks.

Although local residents and politicians regularly railed against Roosevelt's enthusiasm for new national monuments, the U.S. Supreme Court read the legislation as granting the president nearly unlimited discretion and issued a landmark ruling that sustained his decision to set aside more than eight hundred thousand acres as Grand Canyon National Monument. With the passage of time, once the initial local uproar over a national monument designation has subsided, Congress has frequently converted these presidentially decreed monuments into national parks, including such icons as the Grand Canyon, Olympic, Zion, Grand Teton, and Death Valley.[3]

The National Parks Organic Act of 1916 consolidated these early individual parks and monuments into a single national park system and established a uniform management standard for them. But the Organic Act only hinted at how the new system might be expanded and which lands might qualify and was silent on what role the new National Park Service might play in the expansion process. Because the act referred to "such other national parks and reservations of like character that may be hereafter created by Congress," however, it was clear that Congress would have the final word over any new parks.[4] As for what sites might qualify for future national park status, the statutory language makes specific reference to preexisting national parks and monuments "of like character," which provides at best only limited direction. Future Congresses, of course, were unlikely to feel constrained by such language, particularly when no one in 1916 had any clear idea of what a national park system actually meant or might become.

The Park Service was quick to assert a role for itself in the new park creation process. That role was explained in the 1918 Lane letter, which explicitly identified standards for evaluating new national park proposals:

> In studying new park projects, you should seek to find scenery of supreme and distinctive quality or some national feature so extraordinary or unique as to be of national interest and importance. . . . The national park system as now constituted should not be lowered in standard, dignity, and prestige by the inclusion of areas which express in less than the highest terms the particular class or kind of exhibit which they represent.[5]

These original standards have been carried forward over the years and still form the principal basis by which the Park Service evaluates new park proposals. The standards have also provoked a long-standing controversy, both within the Park Service and among its allies, about what constitutes a nationally significant site. Proponents of park system expansion have construed these terms rather loosely and have supported a wide array of new park proposals, whereas system "purists" have read the terms quite literally and regularly opposed various "unworthy" additions to the system.[6]

Over time, Congress has given the Park Service a modest role in the new park designation process and has developed more detailed criteria for evaluating areas for possible inclusion in the system. Under a 1976 amendment to the Organic Act, Congress directed the secretary of the Interior "to investigate, study, and continually monitor the welfare of areas whose resources exhibit qualities of national significance and which may have potential for inclusion in the National Park System."[7] The secretary was also instructed to submit to Congress an annual list of potential additions to the system. In 1998, Congress provided additional guidance in the form of specific factors the agency was to consider in evaluating whether an area merited national park protection: not only must it "meet the established criteria of national significance, suitability, and feasibility," but it should also contain "themes, sites, and resources not already adequately represented in the National Park System." Other factors to be considered included the rarity and integrity of the resources, existing threats to those resources, the potential for public use, the site's interpretive and educational potential, possible socioeconomic effects of any designation, the level of local and general public support, and whether the area can be adequately protected over the long term. At the same time, evidently to protect its own prerogatives, Congress prohibited the agency from initiating new park studies without specific legislative authorization.[8]

Building the System

Since 1916, the addition of new parks to the system has been an often-haphazard process, but one that has gradually expanded the system across the country while fostering several new protective designations. Once the Organic Act was on the books, the national park idea was soon extended beyond the mountain west and exported eastward, giving the new system a more diverse national character and a broader political base. In 1916, Congress established Hawaii Volcanoes National Park, followed one year later by Mount McKinley National Park in Alaska, protecting not only the nation's highest mountain but also the wildlife that inhabited the surrounding landscape. In the Southwest, national monument designations were attached to the lands that would eventually become Zion, Bryce Canyon, Carlsbad Caverns, and Arches National Parks, and the Grand Canyon was converted from national monument to national park status. In 1919, Congress extended the system eastward, designating Lafayette National Park along the Maine coast, later renamed Acadia, and in 1926, it authorized the establishment of Shenandoah, Great Smoky Mountains, and Mammoth Cave National Parks. However, these new Appalachian region park designations were contingent on the affected states and communities, most of whom had enthusiastically supported the designations, raising the funds to purchase the privately owned lands that would make up the new parks. The national park idea, originally associated solely with protecting scenic western settings, was taking hold in locations where the bulk of the populace lived, a testament to the growing popularity of the idea itself as well as the apparent economic benefits attached to it.[9]

It was just the beginning, however, as President Franklin D. Roosevelt's tenure in the White House marked an era of extraordinary expansion for the park system. In 1933, with a stroke of his pen, Roosevelt more than doubled the number of park system units when he signed an executive order transferring sixty-four national monuments, military parks, battlefield sites, memorials, and cemeteries to the Park Service. Although most of these new units came from the War Department, the Forest Service was also divested of its national monuments: Jewel Cave, Bandelier, Saguaro, and Mount Olympus among others. The transfer further deepened animosities between the two agencies, a rift that traced to the origin of the Park Service and prior decisions creating several of the early national parks out of national forest lands. Franklin Roosevelt, like his cousin Theodore Roosevelt, unabashedly used his Antiquities Act power, proclaiming eight new national monuments, including Joshua Tree, Capitol Reef, Channel Islands, and Jackson Hole, all of which were eventually redesignated national parks. However, when Roosevelt announced the new Jackson Hole National Monument in 1943, his decision unleashed a firestorm of local protest that saw Congress attempt to abolish the new monument and refuse to appropriate funds for its management. When the dust finally settled, Wyoming had managed to exempt itself from the Antiquities Act, and presidential use of the act slowed noticeably for the next two decades.[10]

During Franklin Roosevelt's presidency, Congress was also active in creating new parks and related protected areas. In 1934, Congress designated Florida's Everglades a new national park, deviating from its traditional view that only scenically spectacular locations, as reflected in the early western parks, merited national park status. With the Everglades designation, the system now included a tropical wilderness unit and reached into the far southeastern states for the first time. In 1936, Congress established the first National Recreation Area at the new reservoir named Lake Mead in southern Nevada, not only creating a new type of protected area, but also giving recreation an explicit priority that it did not enjoy in the traditional national parks. In 1937, Cape Hatteras National Seashore was created and brought into the system, further expanding the type of areas that merited protection.

Other new congressionally established national parks included Big Bend in west Texas, Theodore Roosevelt in central North Dakota, Olympic in western Washington (actually a national monument conversion), and Kings Canyon in California's southern Sierras, which incorporated and expanded the original General Grant park lands. Moreover, drawing on New Deal public works projects, Congress established the Blue Ridge and Natchez Trace Parkways as new park units. In sum, Franklin Roosevelt's era not only witnessed dramatic expansion and diversification of the system, but it also thrust conservation back into the political limelight and showed that expansion was politically feasible even in times of war and fiscal crisis.[11]

The postwar period—an era highlighted by the nation's newfound prosperity and an unprecedented population boom—witnessed a major makeover for the national parks as well as another surge in new park units. During the 1950s, as we have seen, the Park Service embarked on its ambitious Mission 66 initiative, designed to expand and upgrade national park facilities to accommodate the record number of visitors to the parks. The Mission 66 building spree, along with the new interstate highway system, transformed the national parks into more attractive and accessible venues, which only served to increase visitation pressures. With the advent of the 1960s, a sympathetic Congress and the ebullient new administration of John F. Kennedy, spurred on by the beginnings of the environmental movement, was eager to take up an aggressive conservation agenda, one that included park system expansion and several new laws that would extend federal nature conservation efforts in new directions. In what proved to be inspired moves, Kennedy appointed Arizona congressman Stewart Udall as his secretary of the Interior, and Udall soon installed George Hartzog, a hard-driving lawyer turned park superintendent, at the helm of the Park Service. As the decade unfolded, these two men emerged as major national park proponents and astute political tacticians who seized the opportunities presented them.[12]

Indeed, the years that extended from the 1960s into the early 1970s are widely regarded as a golden era within Park Service circles. Director Hartzog, besides working with Udall, also served under two Nixon-appointed Interior secretaries—Wally Hickel and Rogers Morton—both of whom also turned out to be supporters of park expansion. Prodded by these officials, a sympathetic Congress added several new parks to the system, including Virgin Islands, Canyonlands, North Cascades, Redwood, and Guadalupe Mountains in west Texas, and it redesignated Arches and Capitol Reef from national monument to national park status. Congress also approved an array of new designations that included national seashores on Cape Cod, Point Reyes, and Padre Island; national lakeshores at Indiana Dunes, Sleeping Bear Dunes, and Apostle Islands; national recreation areas at Ross Lake, Lake Chelan, and Delaware Water Gap; the Ozark National Scenic Riverways; and the Appalachian National Scenic Trail. By the time Hartzog's eight-year tenure as director came to an end in 1972, the park system had grown by sixty-eight new units, which included a plethora of new designations that extended the national park idea to urban areas, river corridors, and trails.[13]

During the 1960s, besides adding new units to the park system, Congress also put into place several other landmark laws that extended federal nature conservation efforts into new realms. In 1964, Congress passed the Wilderness Act, creating an even more protective designation for undeveloped federal lands than found in the national park legislation. Under the Wilderness Act, the Park Service, Forest Service, and U.S. Fish and Wildlife Service—but not the Bureau of Land Management (BLM)—were vested with rigorous new preservation responsibilities once Congress designated a segment of their lands as a wilderness area, a prospect that also gave the Forest Service a useful tool to fend off new national park proposals targeting its lands. In 1965, Congress passed the Land and Water Conservation Fund Act, which established a dedicated fund, derived from offshore oil and gas lease sale revenues, to purchase lands for recreational and conservation purposes. With this new fund, at least in theory, the federal government would no longer have to depend on state or local financial capacity to acquire private lands for national park purposes, as had occurred to establish Shenandoah and Great Smoky Mountains National Parks. In addition, Congress passed the National Wild and Scenic Rivers Act and the National Trails Act in 1968, both of which created new types of protected areas that could be overseen by any of the federal agencies. In sum, this 1960s-era legislative cornucopia represented a significant federal expansion into land and water preservation, one that extended to national parks and involved the other principal land management agencies. Now, national park status was just one of several ways to protect areas from development.[14]

Increasingly concerned about the direction in which the system might be headed and its own role in the park creation process, the Park Service released its first National Park System Plan in 1972. Under Secretary Udall's instructions, Director Hartzog and his colleagues had embraced the idea of developing a systematic plan with clear standards to guide future growth of the national park system. Besides asserting its own role in this process, the agency's objectives were to expand the type of areas that might be included in the system and to constrain the growing congressional enthusiasm for new parks, no matter their significance or uniqueness. The plan's standards, in large measure, mirrored those originally set forth in the Lane letter: the area or region must be significant from a natural history perspective (that is, ecologically significant, as the idea would be expressed today), and it must not already be adequately represented in the system. Employing these standards, the plan identified gaps in the natural and cultural themes represented in the system, intending that new park proposals should be designed to fill these gaps. According to the agency's analysis, places like the Great Plains, Great Basin, Wyoming Basin, and Gulf Coastal Plain were not adequately represented in the system, whereas the Virgin Islands, Cascade Range, Sierra Nevada, and Northern Rocky Mountains were well represented.[15] The emerging vision, although not expressed in these terms, was to create an ecologically representative park system that stretched across the nation. Impending and inevitable changes in scientific knowledge and cultural values meant that it would be an ongoing and evolutionary process, however.

The Park Service's system plan, although it may have helped channel Congress's expansionist impulses, did not significantly temper them. Over the decade that stretched through Jimmy Carter's presidency, Congress strayed from its traditional one-park-at-a-time legislative approach and approved three new omnibus park bills as well as several other individual new park bills. First, in 1972, Congress assembled six small historic site proposals into one bill, thus setting a new precedent for omnibus park legislation. A few years later, under the guidance of California congressman Philip Burton, legendary chairman of the House subcommittee on parks and an unsurpassed legislative strategist, Congress passed the National Parks and Recreation Act of 1978, which approved fifteen additions to the system, including Santa Monica Mountains National Recreation Area, New River Gorge National River, four national historic trails, and boundary adjustments to thirty-nine existing park units. These new areas complemented an array of diverse designations that Congress had already approved during the intervening six years: the Golden Gate National Recreation Area, Gateway National Recreation Area, Big Cypress National Preserve, Big Thicket National Preserve, Cuyahoga Valley National Recreation Area, and Congaree Swamp National

Monument, to name a few. With the system growing at a record pace, critics decried this new legislative packaging approach to national park system expansion, branding it "park barrel politics."[16]

Undeterred by its critics, Congress followed the same omnibus course two years later when it finally ended a decade of bickering over Alaska's public lands by adopting the Alaska National Interest Lands Act (ANILCA). The ANILCA legislation more than doubled the size of the national park system and also dramatically increased the national wilderness and wildlife refuge systems. The park system gained 43.6 million new acres and ten new units, including Gates of the Arctic, Wrangell–St. Elias (now the largest park in the system at 13.2 million acres), Lake Clark, and Kenai Fjords National Parks. Congress also greatly expanded Denali, Katmai, and Glacier Bay National Parks, but did so by putting part of the new acreage in national preserve status, which is historically less protective than park status because it allows activities such as off-road vehicular access, hunting, and even mining and energy leasing on occasion. In addition, Congress expressly authorized subsistence uses inside most Alaskan parks, preserves, and monuments, defined as the taking of renewable resources by hunting, fishing, trapping, and otherwise for personal consumption or barter. This omnibus Alaska legislation culminated an extraordinary decade of park system expansion that also addressed several of the system's ecological shortcomings as identified in the 1972 system plan.[17]

Since then, Congress has slowed its pace, however, and the system has seen only a few major additions. In 1986, Congress added Great Basin National Park in Nevada to the system, thus providing much-needed representation for the Great Basin region. In 1994, at the behest of the California congressional delegation, Congress moved to resolve long-festering controversies over California's desert lands. The California Desert Protection Act converted Death Valley and Joshua Tree national monuments to national parks and expanded their boundaries, while also creating a new Mojave National Preserve under Park Service management. In total, more than five million acres were either renamed or changed hands from the BLM to the Park Service, creating the largest complex of park lands in the contiguous states and establishing Death Valley as the largest park outside of Alaska. In 1996, a small Tall Grass Prairie National Preserve in Kansas was incorporated into the system, but, to address local political concerns, it was placed under the aegis of the National Park Trust rather than the Park Service. The preserve finally protected an important remnant of the Great Plains' once-abundant tallgrass prairie ecosystem, much of which was already lost to the plow. Several other new park units were also added along the way, including Niobrara River National Scenic River in northwestern Nebraska, Little River Canyon National Preserve in Alabama, Great Sand Dunes National Preserve in southwestern Colorado, and Craters of the Moon National Preserve in Idaho. Although slowed by politics and other factors, the national park system was nonetheless still on a growth trajectory and plainly not being regarded as a complete system.[18]

The Park Service's own enthusiasm for new parks has waxed and waned since the 1980s. Agency officials have consistently maintained that new park proposals must meet the "national significance" standard and must be both suitable and feasible for inclusion in the system. They are well aware that new parks cost money and fear that these funds will come from the existing budget, a troublesome prospect because Congress has chronically underfunded the agency. When opposed to a proposed addition, the Park Service has generally recommended that the site be protected by another federal agency or by state or local authorities. But Congress, driven by its own motives, has not consistently respected these concerns or recommendations, as in the cases of the Presidio in San Francisco and Steamtown National Historical Site in Pennsylvania. Exasperated by this "thinning of the blood" tendency, former Park Service director James Ridenour has asserted that "members of Congress have blatantly disregarded the standards that have been traditionally used in evaluating the creation of new national park units." These same "national significance" concerns have periodically prompted members of Congress to call for decommissioning some parks, perhaps most notably when Utah congressman James Hansen chaired the House subcommittee on national parks during the 1990s and proposed eliminating Great Basin National Park. For the most part, however, Congress has been undeterred by such complaints and has taken a broad view of the type of lands and waters that might be added to the system.[19]

Viewed through a historical prism, several themes emerge from this excursion through the new park designation process. First, whereas the original national park system primarily consisted of a few large western natural parks and national monuments, it has been dramatically expanded and diversified over the years with the addition of national recreation areas, national seashores, national preserves, national trails, and the like. Second, the notion that only scenically spectacular locations merit national park protection has fallen by the wayside. Beginning with the addition of Everglades National Park in 1934, several national parks have been designated or expanded as much for their ecological or wilderness values as for their scenic splendor. That was true in the case of the Alaska parks and the North Cascades complex, and it is reflected in later additions to Grand Canyon, Redwood, Death Valley, and Joshua Tree. Third, the advent of such new designations as national recreation areas, national lakeshores, and gateway parks stand as proof that the "national significance" criteria has been diluted, if not abandoned, in several instances, partly to meet the growing demand for close-to-home recreational opportunities. In sum, these historical changes reflect an evolving national park system, one that both tracks and confirms the evolution of the national park idea.

Diverse Players and Competing Concerns

Figuring out exactly how the national park system has evolved to its present shape offers a fascinating glimpse into democracy in action. The identification and creation of new national parks is inherently a political process. The Park Service, despite its obvious interest in shaping the system, has not consistently played a major role in the new parks designation process. Instead, the vast majority of new parks have come into being through the vision and persistence of a single individual or group of citizens dedicated to protecting a treasured local landscape or waterway. For example, naturalist Enos Mills gets a large measure of the credit for championing Rocky Mountain National Park, and publisher George Bird Grinnell played that same role for Glacier National Park. Others who deserve similar credit include William Gladstone Steele, who first learned of Crater Lake as a Kansas schoolboy and then later dedicated his life and savings to protecting it as a national park; Horace Kephart and George Masa, who helped bring the Great Smoky Mountains National Park proposal to eventual fruition through their relentless writing, photography, and lobbying efforts; and Ernest Coe, a Miami developer who tirelessly championed creation of Everglades National Park and was known as "Father of the Everglades." Even though people like Mills, Grinnell, Steele, Coe, and others may have spearheaded these park campaigns, they enjoyed substantial grassroots support, both at the local and national levels, where organizations like the Sierra Club and National Parks Association were usually ready allies. Moreover, these new park campaigns were until recently generally nonpartisan in nature, another measure of their remarkable democratic character.[20]

Private philanthropy has also played a major role in the creation of new national parks. The early marriage between the national parks and private capital was cemented under the leadership and personal generosity of Park Service director Stephen Mather, who used his own private funds to purchase the old Tioga Road in Yosemite as well as private lands outside Glacier for Park Service offices and who personally paid Robert Sterling Yard's publicist salary. All too familiar with the obstacles that privately owned lands regularly posed for the creation of new parks and with the problems that private inholdings created for park managers, Mather and his successor, Horace Albright, sought support from some of America's wealthiest individuals in an effort to acquire these lands. They found an early ally in John D. Rockefeller Jr., an unlikely but passionate conservationist prepared to expend some of his family's vast fortune on the nascent national park system. Rockefeller's clandestine land purchases in Jackson Hole, Wyoming, during the late 1920s helped save much of what became Grand Teton National Park's frontcountry from subdivision and development. The Rockefeller family also helped acquire privately owned lands in Maine's Acadia National Park and assisted North Carolina and Tennessee in their efforts to purchase the lands that became part of Great Smoky Mountains National Park. At Acadia, wealthy New Englanders, most notably George Dorr from Boston, joined the Rockefellers in funding the acquisition of private lands on Mount Desert Island that were then conveyed to the federal government for the new park. Later, the Mellon family made its contribution to the national park system, playing key roles in acquiring lands at Cape Hatteras and on Cumberland Island. These early commitments by wealthy individuals not only contributed to the current shape of the national park system, but also provided a model for the role that private philanthropy might play in safeguarding sensitive lands for conservation purposes, a role that has more recently been assumed by The Nature Conservancy, the Trust for Public Land, the Conservation Foundation, and other nonprofit land trust organizations.[21]

In several instances, states have also provided crucial financial support to underwrite new national parks. During the early part of the twentieth century, several of the major eastern national parks, while being patched together from privately owned lands, were underwritten by their home states, whose political leaders willingly committed tax revenues to acquire the necessary property. That was the case with Great Smoky Mountains National Park, where North Carolina and Tennessee both agreed, as part of the political deal establishing the park, to fund the necessary land purchases. Virginia likewise agreed to raise the funds necessary to buy out rural landowners ensconced in the Shenandoah Valley to create a new Shenandoah National Park in the Blue Ridge Mountains. In the case of Everglades National Park, the state of Florida did the same. The truth is that Congress was neither willing nor able during these early years to underwrite widespread land acquisition to establish new parks, especially once the nation lapsed into the Great Depression. That eventually changed during the 1960s when Congress created Cape Cod National Seashore and appropriated federal funds to purchase the privately owned lands necessary to establish this new coastal park unit in an already developed area. Since then, following establishment of the Land and Water Conservation Fund, Congress has been much more willing to use federal funds to acquire key private parcels necessary to establish new parks or to complete existing ones.[22]

One constant tension underlying the growth of the national park system has been the presence of the rival U.S. Forest Service and other federal land management agencies. At least since the creation of Rocky Mountain National Park in 1917, when Congress transferred highly regarded national forest lands to the Park Service for the new park, the Forest Service has regularly opposed relinquishing its lands for this purpose. This tradition is reflected in the Forest Service's current opposition to transferring its Mount St. Helens National Monument in southern Washington to the Park Service. In fact, not long after the national park system came into being, the Forest Service bestowed a new "primitive area" designation on some of its prime scenic and recreational lands, largely to keep them away from the Park Service. Since the Wilderness Act passed in 1964, the Forest Service has regularly argued that its wilderness lands are much better protected than national park lands, noting that no roads, tourist facilities, motorized recreation, or other such intrusions are permitted in federally designated wilderness areas. Following passage of the Federal Land Policy and Management Act in 1976, the BLM has likewise contended that its wilderness-eligible lands are well protected and should not be given to the Park Service, as had also been the practice. Moreover, laws like the Wild and Scenic Rivers Act, the National Trails Act, and the Endangered Species Act have further expanded the protectionist management responsibilities of these other agencies. And with the emergence of state and local park systems, nonfederal protection options are also available.[23]

This growth in federal protected land systems means that national park status is no longer the only option for protecting scenically appealing, ecologically important, or recreationally attractive public lands. All the principal federal land management agencies now boast important land protection responsibilities: the Forest Service for its wilderness lands and Research Natural Areas; the BLM for its National Landscape Conservation System, which includes wilderness lands and Research Natural Areas; and the U.S. Fish and Wildlife Service for the National Wildlife Refuge System and its wilderness lands. Since the 1960s, these systems have been on a clear growth trajectory. The national forest wilderness system now totals more than 35 million acres, with another 60 million acres mostly protected as roadless lands; the BLM's National Landscape Conservation System—consisting of wilderness areas, wilderness study areas, national conservation areas, and other protected lands—exceeds 27 million acres; and the National Wildlife Refuge system embraces more than 93 million acres. In total, nearly 250 million acres, or 40 percent of the federal public lands, are in some form of protected status. Although not part of the national park system, these alternative protective designations provide an opportunity to link together protected areas—such as national parks and adjacent wilderness lands—to conserve much larger and more ecologically intact landscapes. As we have seen, this ecosystem-based approach to nature conservation has become the common rallying cry for scientists, preservationists, and others concerned about accelerating biodiversity losses and potential climate change effects.[24]

Although many of the large national parks were created from existing federal lands, this has not invariably been the case. The major eastern parks were stitched together from privately owned lands, an arrangement that required various acquisition strategies. To create Great Smoky Mountains, Shenandoah, and Everglades National Parks, Congress initially authorized the new parks conditioned upon the home states acquiring designated private lands, but then eventually supplemented these state efforts with federal funding. In other instances, private lands have been integrated into a new national park unit through different means. At Cape Cod National Seashore, for example, the Park Service was vested with condemnation authority and also charged with developing zoning standards to ensure compatibility between local land use decisions and the adjacent preserved landscape. More recently, Congress has employed a new National Heritage Area designation that does not disturb land-ownership patterns but gives the Park Service a partnership role in conserving and interpreting a community landscape deemed to have historical, natural, and recreational values. The acquisition of strategically located conservation or scenic easements, either by the Park Service or by a land trust organization, represents yet another method to effectively extend park-like protections without altering existing ownership or boundaries. The lesson is clear: with sufficient funding, several options are available to piece together nonfederal lands into a new or expanded national park unit.[25]

TOWARD AN ECOSYSTEM APPROACH

What does the future portend for growth of the national park system? Given the history of haphazard and unplanned system expansion and the inherently political nature of this process, the answer to this question is far from clear, but the evolution of the national park idea provides some guidance. The same concerns that have moved us to alter our conception of a national park are also in play on the system-wide level, helping reshape our image of the national park system in the years ahead. Today, we no longer view the parks as isolated scenic wonders or as mere playgrounds or tourist destinations. Instead, we understand that national parks are vital parts of larger landscapes, that they are interconnected with surrounding ecosystems and communities, and that they are unparalleled natural laboratories for understanding nature and educating ourselves about its wonders and threats. In a world where climate change may soon alter the natural order in alarming ways, where nature deficit disorder is now an identified malady among children, and where most people live in an urban environment far removed from nature, the national park system of the future will have to take on new roles and additional responsibilities.[26]

Putting the National Park Idea to Work

To meet tomorrow's needs, system expansion proposals must rest on our evolving view of the national park idea and the future role of parks in our ongoing nature conservation efforts. Although new park proposals must comport with the legally binding "national significance" criteria, our understanding of what constitutes a "nationally significant" addition is evolving, too, in response to the same forces that have altered our view of the parks themselves. The most prominent new national park idea gives priority to promoting ecological integrity and related biodiversity conservation objectives with a clear eye toward the potential effects that climate change will have on extant ecosystems. It holds that national parks should be large enough to allow nature to exist on its own terms and to protect the various species and ecosystem services connected to the site. It acknowledges that few of our current national parks are large enough to accomplish these goals and that they are not secure from outside development pressures, which have disrupted wildlife travel corridors, fouled park waters, polluted regional airsheds, and altered surrounding landscapes.[27] It recognizes that climate-induced changes to the natural world and the accelerating rate of biodiversity loss are becoming stark realities in a world that looks quite different from the early years of the twentieth century and where pending changes in the natural order could alter our approach to nature conservation. It understands that demographic shifts in the nation's population and living arrangements have converted us into an urban society with growing minority communities, creating new generations of children with little connection to nature or the outdoors. In addition, it acknowledges the Park Service's unique long-term engagement in nature education as well as the agency's unparalleled potential to enhance these educational efforts and thus reconnect us with the natural world. At the same time, of course, this new national park idea must be sensitive to the indisputable fact that political support is crucial in any park creation or expansion campaign.

Science, Politics, and New Parks

Congress holds the key to park system expansion, whether in the form of new national park designations or additions to existing park units. Over time, Congress has occasionally acknowledged the lessons of ecosystem science and conservation biology and has legislated in landscape-scale terms. The proof is in the legislation that has gradually increased the number and size of national parks and has expanded several existing parks to enhance their ecological integrity. With passage of ANILCA in 1980, Congress signaled that it understood the need to view parks in an ecosystem context. Not only did ANILCA dramatically expand Denali and other Alaskan national parks, but it also created several large, new national parks, such as Wrangell–St. Elias, Lake Clark, and Gates of the Arctic. Before ANILCA, Congress designed the North Cascades complex—which includes North Cascades National Park, the abutting Lake Chelan National Recreation Area, Ross Lake National Recreation Area, and Glacier Peak Wilderness Area—with an increasing degree of ecological forethought. The 1978 congressional decision to expand Redwood National Park to eliminate damaging upstream logging operations represents another example of ecologically conscious expansion, as does the 1974 establishment of the Big Cypress National Preserve adjacent to Everglades National Park. As another example, the 1994 legislation expanding the California desert parks—Death Valley and Joshua Tree National Parks and the Mojave National Preserve—along with nearby BLM wilderness area designations manifests a similar sensitivity to reconnecting a fragmented landscape. In addition, congressional wilderness area designations on national forest and BLM lands adjacent to several western national parks serve as a type of de facto park expansion that promotes landscape-scale conservation objectives.[28]

Whether and where Congress might be persuaded to take similar actions is hard to predict, but linking the designation of new parks and the expansion of existing ones to ecosystem preservation is a vital step toward reenvisioning the national park system and its role in nature conservation. Several existing proposals are designed to advance large-scale ecological conservation objectives. One calls for establishing a new Maine Woods national park on acquired corporate timberlands in central Maine, situated adjacent to the state-owned Baxter State Park and Mount Katahdin, the Appalachian Trail's northern terminus. As occurred in the case of Acadia, a lone individual—in this case, Roxanne Quimby, cofounder of Burt's Bees—has purchased key parcels with the intention of donating them to the Park Service, although local opposition has slowed this initiative. Another proposal that has local congressional support would transfer the Valles Caldera federal trust lands in north-central New Mexico from the Forest Service to the Park Service, which would ensure compatible management of these ecologically significant lands with the adjacent Bandelier National Monument and would enable the Park Service to interpret the area's rich Native American and Hispanic American history. A third proposal supports incorporating adjacent BLM lands into Canyonlands National Park in Utah to create a geologically complete park and better protect the fragile desert landscape from energy development and unregulated off-road vehicle use. Yet other large-scale proposals envision a new Park Service–administered San Gabriel Mountains national recreation area carved out of the Angeles National Forest in southern California, a new High Allegheny Plateau national park patched together from national forest lands in northeastern West Virginia, and transfer of the Mount St. Helens National Monument in southern Washington from the Forest Service to the Park

Service. Each of these proposals has evident ecological merit, but the politics of national park system expansion are complex and will turn on much more than scientific merit.[29]

To ensure success, most expansion campaigns have enlisted support within nearby communities and among key leaders by demonstrating how a new or expanded park unit will generate local economic and other benefits. Those communities that have embraced nearby national parks and other protected lands have generally prospered, both from an influx of new arrivals with their own financial resources and from growth in the regional tourism economy. Other options for enlisting local support include incorporating local citizen councils into the park management scheme; packaging a park proposal with community development opportunities, such as strategic land exchanges or infrastructure improvements; and providing local employment opportunities in ecotourism or other park-related jobs. In any event, making these connections explicit can only bolster the political case for ecologically based park creation or expansion efforts.[30]

The National Restoration Area Concept

Another approach to expanding the national park system is to target damaged lands for inclusion in the system as national restoration areas. In today's world, where the human presence is ubiquitous nearly everywhere, some degraded and developed lands can nonetheless have real ecological value, especially if they can be restored to an ecologically functional condition and are connected to nearby national parks or other protected lands. This idea is not new, either in theory or practice. The Conservation Foundation's 1972 *National Parks for the Future* report endorsed a "restoration reserves" strategy to expand the national parks portfolio. Several of the major eastern national parks, including Great Smoky Mountains and Shenandoah, were created from previously disturbed private lands. After the states acquired the lands during the Great Depression and transferred them to federal ownership, the Park Service restored the logged-over areas, which now contain mature forests. Similar restoration efforts have reduced the scars of farming, grazing, and other human activities, while the mere passage of time has allowed some disturbed sites to heal naturally. The result is two revered national parks located near the eastern seaboard and thus readily accessible to millions of people seeking to experience wild nature or just escape the travails of modern urban life.[31]

Federal restoration efforts elsewhere have helped transform damaged lands into ecologically valuable landscapes. In the case of Redwood National Park, confronted with destructive upstream logging and extensive flooding, Congress expanded the original park boundaries to encompass the entire watershed by acquiring privately owned timberlands in the upper drainage and by directing the Park Service to restore these lands to ensure the ecological integrity of the newly expanded park. Outside the national parks, the Weeks Act of 1911 authorized the fledgling Forest Service to acquire cutover and devastated eastern, southern, and midwestern timberlands, which it then proceeded to restore. These acquired national forest lands now provide myriad wilderness, wildlife habitat, recreational, and other benefits to a large segment of the nation's populace. The U.S. Fish and Wildlife Service has regularly restored depleted agricultural lands to reestablish native habitat for waterfowl, birds, and other species in national wildlife refuges. In short, the notion of federal acquisition and restoration of damaged lands for conservation purposes is an old idea that has long been deployed with impressive results.[32]

Adding damaged but restorable lands to the national park system will require a new long-term, science-based perspective on the system's goals and purposes. In their current state, disturbed landscapes offer neither outstanding scenic nor recreational settings and thus may not meet the "national significance" standard for new park designations. When restored, however, these lands may hold important ecological benefits as new wildlife habitat or new ecosystem types, as connective corridors to existing park lands, as ecosystem services reservoirs, as urban nature preserves, or as mitigation for climate change effects. In a restored condition, lower-elevation Pacific Northwest lands that were previously logged could serve as an important complement to the higher-elevation lands already protected at Mount Rainier and Olympic National Parks, providing wildlife additional habitat or movement routes to mitigate the effects of warming temperatures on regional ecosystems. Similar ecological benefits might accrue by adding the heavily logged national forest lands on Yellowstone's western border to the park. When restored, these lower-elevation lands could facilitate wildlife migration out of the park during harsh winter months, provide additional sanctuary for the region's grizzly bear population, and afford visitors new recreational opportunities. In addition, the restoration process itself holds the promise of new jobs and business opportunities in often-depressed rural areas.

The national restoration area concept presents a complex legal question, namely whether the concept satisfies the current "national significance" statutory criteria for new park designations. This standard, according to the Park Service, requires that proposed new areas must be "a true, accurate, and relatively unspoiled example of a resource."[33] As a federal agency, however, the Park Service has the clear legal authority to reinterpret its statutory mandates as long as it provides an explanation and adheres to congressional intent.[34] Tellingly, such iconic parks as Great Smoky Mountains, Shenandoah, and Redwood would not have initially met the current "national significance" standard given their condition when incorporated into the system. The Park Service should therefore redefine the "national significance" standard to treat it as a two-step process for proposed restoration areas. At the initial restoration stage, the question would simply be whether the site has the potential to enhance the national park system. Only at the second stage, after the area has been restored, would the question of "national significance" be relevant. At this point, if the area has demonstrable long-term ecological value, then the criteria would be satisfied. Thus, the "national significance" legal standard should not deter us from vesting the Park Service with restoration responsibilities designed to eventually incorporate ecologically significant lands into the system.

The national restoration area concept also poses daunting but not insurmountable political challenges. Other federal land management agencies and private landowners may well oppose any ownership transfer, even of damaged lands, but the degraded condition of so much of our landscape is a powerful argument against leaving select sites in their current status, at least when the site can be restored to meet the redefined "national significance" criteria. The owners of damaged private lands, whether located in rural or urban settings, may actually be amenable to disposing of them at a reasonable price, and local opposition may be minimized if it is clear that the targeted lands will eventually be restored to a productive ecological state and will afford new recreational, educational, and other opportunities. Application of this idea in both urban and rural settings could also provide much-needed jobs, which should make it even more politically attractive. As a bonus, creating such areas in urban locations where young people can engage with them would also help address the nature deficit disorder problem. Moreover, the proposed two-step approach fits comfortably with the now-well-accepted political tradition of converting national monuments to national park status over time. A new national restoration area strategy therefore presents a unique opportunity to significantly expand and strengthen the park system, taking the long-range view that dynamic ecological processes, climate change, and other concerns have now injected into our conservation efforts.

Alternative Designations and Strategies

With far fewer opportunities available today to convert large swaths of public land to national park status, the conservation objectives underlying national park system expansion might still be achieved de facto through alternative protective strategies. Indeed, the goal of knitting the landscape together into a more coherent ecological entity can also be accomplished through better-coordinated management arrangements, wildlife dispersal and migration corridors, strategic conservation easements, and other such approaches. Although these strategies may not change the actual acreage under Park Service management, the overall tenor and direction of resource management across the landscape would be better aligned with national park protectionist policies and resource management objectives.

One strategy involves establishing a new landscape-scale overlay designation to protect select landscapes or "greater ecosystems" for conservation purposes. Congress could do so legislatively by devising a special designation—perhaps calling it a national ecological reserve or a wildlife heritage area—to overlay an array of contiguous federal lands that extend across a particularly sensitive, vital, or treasured landscape, such as the Greater Yellowstone area, the Crown of the Continent Ecosystem, or the Greater Grand Canyon region. Without changing ownership or administrative responsibilities in the newly designated national ecological reserve or wildlife heritage area, new resource management and interagency coordination standards would be devised to better protect the larger ecosystem and thus ensure the area's biological integrity. Despite the different terminology and scale, the Wildlands Network, the Northern Rockies Ecosystem Protection Act proposal, and the Yellowstone to Yukon initiative are all based on this type of landscape-scale strategy and built around existing national parks and wilderness areas to promote more effective regional conservation. Linking such a carefully designed "greater ecosystem" legislative proposal with climate change mitigation provisions and local economic concerns might make such a proposal politically attractive enough to at least secure congressional consideration.[35]

In the face of potential political obstacles, nonlegislative measures can be pursued to achieve the same objectives. The president has the authority under the Antiquities Act to create new ecosystem-based—or landscape-scale—national monuments that transcend existing boundary lines. The president could, for example, designate a new Crown of the Continent National Monument or a Greater Grand Canyon National Monument that would overlay the public lands surrounding Glacier or Grand Canyon National Parks and establish more consistent and coordinated management standards for each landscape. During the Clinton presidency, Interior secretary Bruce Babbitt intentionally designed new large-scale national monuments with ecological conservation goals in mind, yet he left management with the existing agency but under new national monument guidelines, an approach that has been sustained by the courts.[36] Clinton endorsed this approach by designating several new national monuments on public lands adjacent to existing national parks: Giant Sequoia National Monument, situated adjacent to Sequoia National Park, curtailed logging, road building, and other industrial activities on the national forest lands incorporated into the new monument; and the Grand Canyon–Parashant National Monument and the Vermillion Cliffs National Monument designations imposed new conservation-oriented management restraints on BLM lands abutting Grand Canyon National Park.[37] Although this approach does not expand the national park system, it would effectively link existing national parks with adjacent public lands under a resource management framework designed to ensure ecological integrity at the regional scale. Alternatively, the president has the apparent authority under the Antiquities Act to transfer existing monuments from one agency to another. The president might, for example, shift administration of the Giant Sequoia National Monument from the Forest Service to the Park Service, citing the monument's proximity to Sequoia National Park and the ecological connections between the areas. [38]

Another strategy contemplates establishing formal wildlife corridors that extend outward from core national park areas into the surrounding landscape, designed to enable park wildlife to migrate seasonally or to disperse in response to climate change impacts. Scientists universally recognize the value of migration corridors to meet the basic habitat needs of migratory species and to facilitate genetic interchange for biodiversity conservation purposes. Scientists also agree that dispersal corridors are critical to enable species to respond effectively to warming temperatures, enabling them to move to more suitable terrain as their traditional habitats are altered. Many of the lands surrounding national parks, however, face significant development pressures that make safe passage treacherous at best and lethal at worst.

The problem is exemplified in the Upper Green River Valley south of Grand Teton National Park in Wyoming, where extensive natural gas development and subdivision pressure has constricted the seasonal migration route for the region's signature pronghorn herds. Confronted with potential loss of this natural spectacle, federal land managers, state officials, and local landowners set about creating a "Path of the Pronghorn" migration corridor to ensure the pronghorn safe passage over the 170-mile route from their summer habitat to their winter habitat. The Park Service obtained a key parcel along the route through a land exchange with the state; the Bridger-Teton National Forest amended its forest plan to designate a first-of-its-kind pronghorn migration corridor across its lands; and the BLM, state of Wyoming, and local landowners also took steps to safeguard the migration route. This unprecedented landscape-scale arrangement—creatively constructed out of existing legal authority—is certainly noteworthy, but whether such coordinated administrative actions can be duplicated elsewhere remains to be seen.[39]

Drawing on this experience, Congress might consider new federal wildlife corridor legislation. The Western Governors' Association has endorsed the concept of protected wildlife corridors, and it is engaged in identifying potential corridors and designing a process to protect them while also cooperating with federal agencies to develop the necessary biological data. New federal wildlife corridor legislation could be modeled after the amended National Trails System Act, which has designated and funded several trail corridors and established a process for future designations. To create a wildlife corridor system, Congress might direct federal land managers and state wildlife officials to collaboratively determine optimal corridor locations and dimensions. On federal public lands, the new corridor designations would simply overlay the existing landscape, imposing some new management standards and planning obligations to ensure adequate protection. On private lands, federal funds or tax incentives could encourage landowners to participate in the corridor program. As has proven true with the national trail system, it should be possible to design a national wildlife corridor program that will address wildlife needs without disrupting land-ownership patterns.[40]

Yet another strategy for strengthening and protecting the national park system is the alluring but still ill-defined concept of ecosystem management. During the Clinton administration, all the federal land management agencies endorsed ecosystem management, and these principles remain part of their basic policies. The Park Service's current management policies direct park managers to "maintain all the components and processes of naturally evolving park ecosystems" and to engage in "cooperative conservation beyond park boundaries . . . to preserve the natural and cultural resources of parks."[41] At its core, ecosystem management means that agency planning and project decisions must take account of all affected ecosystem components and processes and must ensure meaningful coordination among the various agencies and entities responsible for them. Despite much lip service and some progress, however, the goal of meaningful interagency coordination still remains elusive, as illustrated by the high-profile controversy that erupted in 2008 over the BLM's proposal to lease lands near several Utah national parks for oil and gas exploration.[42]

Congress might therefore consider legislatively encouraging more-effective management coordination to promote landscape-scale conservation. Within the federal agencies, the options include amending the National Environmental Policy Act to require a new interagency coordination statement as part of the environmental assessment process, adopting a new formal consultation requirement whenever an agency action might adversely affect national park resources, or mandating such cooperation through new statutory consistency requirements based on a similar provision in the Coastal Zone Management Act. Outside the federal agencies, a new model for coordinating natural resource conservation efforts might be derived from the cultural preservation laws, which establish an interlocking series of federal and state entities responsible for overseeing these resources, whether located on public or private lands. Any legislation endorsing meaningful and enforceable coordination policies would help promote more ecologically sound management practices on landscapes shared by the national parks, other federal land management agencies, and state, tribal, and private owners.[43]

Making Nature Accessible

National park system expansion must also take account of the nation's changing demographics. American society looks quite different today than it did at the system's outset in 1916. Besides doubling in size since the 1950s, the U.S. populace has become significantly more diverse, with nonwhite minority group members currently accounting for a third of the total population.[44] Civil rights and social justice concerns are now mainstay issues in our political culture. In addition, unlike a century ago when most people lived in a rural or semirural setting, Americans now reside primarily in urban communities where wild nature is not an integral part of their daily lives. In the face of these changes, visitation to the national parks has risen dramatically during most of the past half century, due in part to the advent of new, urban-focused parks located closer to the nation's population centers and the seacoasts. Minority citizens, however, are not frequent park visitors, and many children are no longer exposed to nature on a regular basis. Engaging these constituencies presents a paramount challenge if the national park system is to remain relevant in our changing world and maintain broad-based political support.

New parks established near urban centers represent one option for introducing children to the natural world and revitalizing the spirit of conservation in mainstream American life. These new urban-based parks might be managed by the Park Service or by a state or local park authority. The Park Service, with its well-established interpretation and education programs, is uniquely positioned to educate our urban populace about natural history and critical environmental issues, such as climate change, endangered species, and sustainability. Although not perfect models, the urban-based parks at Santa Monica Mountains, Golden Gate, and New York City's Gateway offer a glimpse of how such a system featuring nature education, wildlife conservation, and recreation could be structured, while the so-called Cape Cod Formula might be employed for acquisition and management purposes. An alternative model is found in the national heritage areas that have proliferated over recent years, giving the Park Service a role in preserving and interpreting local natural and historic properties without transferring their ownership to the federal government. Another model is the aforementioned national restoration area concept, which could be applied to restore damaged landscapes near urban areas for eventual management as park sites and which could provide employment and educational opportunities during the restoration process. These models would not only begin to address concerns about nature deficit disorder for the next generation, but would also provide urban residents an opportunity to experience nature close to home while encouraging them to visit other national parks.[45]

Whether due to socioeconomic, cultural, or other factors, the nonwhite population does not currently have a strong attachment to the national parks. Park visitation figures are telling: one survey found that 36 percent of whites visited a national park within the prior two years compared with only 13 percent of the African American population. The challenge, which the Park

Service has begun to address, is to make the national parks relevant and accessible to the nation's extraordinarily diverse communities of color. Our urban-based parks, whether new or existing, should be designed to meet diverse cultural needs by including, for example, large picnic sites where extended families might gather. New historical and cultural sites should be established to honor the experience of specific ethnic groups while providing visitors an opportunity to enjoy and learn about nature, as now occurs at several battlefield sites. Park interpretation programs should incorporate the minority experience, as Yosemite is doing by highlighting the important role of African American buffalo soldiers in safeguarding the park during its early days. Given the powerfully democratic origins of the national parks, such efforts to connect with our changing population base can only strengthen the future position of the parks.[46]

Besides responding to the needs of a growing urban and minority population, the Park Service cannot overlook the legitimate concerns of another minority group, Native Americans. As we have seen, many American Indian tribes and their members have a unique relationship with the national parks derived from a history of dispossession and neglect. In this case, the goal is not about reintroducing tribal members to nature; rather, it is about addressing social and economic injustice claims and cultural preservation concerns in ways that also safeguard park resources. Significantly, the Park Service's relationship with individual tribes is evolving in creative new directions that acknowledge legitimate tribal claims and endorse new comanagement relationships designed to further nature conservation and cultural heritage objectives. As these efforts gather additional momentum, the opportunity exists to reconnect the parks with their original inhabitants.[47]

RETHINKING THE NATIONAL SIGNIFICANCE STANDARD

Of course, any formal effort to expand the national park system must meet the agency's long-standing "national significance" standard. Ever since the Lane letter, the Park Service has viewed "national significance" as the primary criteria for designating new national park units, whereas organizations like the National Parks Association regularly invoked the standard to avoid any effort to devalue the park system by adding unworthy units.[48] In 1976, Congress affirmed the standard in amendments to the National Parks Organic Act, which also incorporated "'suitability" and "feasibility" as additional criteria, and then reaffirmed these criteria in the National Parks Omnibus Management

Act of 1998.[49] In turn, the agency has elaborated on these criteria for proposed new natural areas. A new area must (1) be an outstanding example of a particular type of resource, (2) possess exceptional value or quality illustrating or interpreting the nation's natural heritage, (3) offer superlative opportunities for public enjoyment or scientific study, and (4) retain a high degree of integrity as an accurate and relatively unspoiled example of a specific resource.[50] Although few will quarrel with these lofty objectives, Congress has honored them in the breach. Not only has it transformed the national park idea into a smorgasbord of new designations, but it has begun to shift the original focus from grandiose scenery to new ecological concerns as reflected in the establishment of Everglades and other wildlife-rich units, creation of the expansive Alaskan parks, and the California desert park additions.

Put simply, the concept of "national significance"—like beauty, justice, and other majestic terms—is in the eyes of the beholder. The concept has evolved over time and has proven inherently malleable to meet the perceived needs of the day, largely tracking the same evolutionary trajectory of the national park idea. Today, science rather than sublime scenery or recreational potential has become a primary touchstone for the nation's conservation efforts, and this commitment to science will only intensify in the years ahead as we confront the challenges of climate change and biodiversity protection. One of the lessons derived from the science of ecology is the need for redundancy. It is not sufficient to protect just one representative species population or ecosystem type; rather, we must protect several to guard against unexpected events that could entirely destroy the protected area or species.[51] Another lesson is the need to protect large areas to preserve ecological integrity and ensure resiliency in the face of change. Applying these lessons to the national park system will go far toward protecting park resources against impairment, whether the proposal involves creating new units, adding onto existing ones, designating new national restoration areas, establishing secure migration corridors, or extending conservation efforts to adjacent lands. Meeting these new conservation challenges is plainly in the national interest; thus, park system expansion proposals and strategies that do so should, by definition, meet the "national significance" standard. By the same logic, the creation or expansion of park units near urban population centers to address new demographic and social justice pressures is also in the national interest.

Just as our concept of a national park has evolved over the years, the nature of the national park system has also changed in response to new circumstances, knowledge, and values. American society looks and thinks quite differently today than it did when the national park system was created. Better informed by science about ecological imperatives, species conservation, restoration requirements, and climate change needs, we have the opportunity to reassess the purpose of the national parks and continue redesigning the system to meet these emerging challenges. Although it may be difficult to implement a new vision for the national park system on our fragmented and contentious landscapes, this challenge must be confronted so as to strengthen and grow the system to ensure a sustainable future. Although each presents its own political challenges, the proposals outlined above—ecosystem-based expansions, national restoration areas, multiagency landscape-based national monuments, federal wildlife corridor legislation, better-coordinated ecosystem management arrangements, and minority-focused urban park units—can help meet tomorrow's conservation demands. Without creative thinking and courageous initiatives, however, we risk diminishing the national park idea and the extraordinary conservation legacy it represents.

CHAPTER 11

Nature Conservation in a Changing World

Still widely heralded as "America's best idea," the national parks actually represent an assortment of ideas that have evolved over time. As the Park Service's founding director, Stephen Mather poured his enormous energy, passion, and personal wealth into making the national parks one of the country's most cherished institutions, and, judging by the extraordinary growth of the national park system and in visitor numbers, he largely succeeded. The national parks inevitably evoke powerful positive images of unsullied landscapes; majestic mountain peaks; free-roaming wildlife; clear, flowing rivers; ranger-led campfire talks; and carefree family vacations. The public rarely contemplates other prevalent but less savory images: car-clogged roads, a cacophony of two-cycle engines, degraded ecosystems, the pervasive taint of commercialism, and unrelenting local development and political pressures. In the midst of such awesome beauty, it is hard to acknowledge that the national park idea is still far from settled, much less that it is often shrouded in controversy. But that has been the reality from the beginning, and it is no less true today.

THE NATIONAL PARK IDEA REVISITED

The national park idea embodies our commitment to nature conservation, itself a matter of ongoing controversy. Forged at a time when the nation's principal goal was to subdue nature and populate the continent, the national park idea ran counter to these goals; it held that our natural heritage was important enough to preserve intact for the benefit of present and future generations. Congress soon translated this sentiment into the Organic Act, employing the language of conservation, promotion, enjoyment, and nonimpairment, a terminology that has set the standard for our nature conservation efforts ever since.[1] To give further meaning to this new and imprecise language, Director Mather and his early associates conceived the Lane letter, which fixed a management course for the new agency that emphasized park visitation, scenic preservation, recreational opportunities, and strategic partnerships to foster visitation and infrastructure development. The letter said little or nothing about wilderness, Native Americans, the role of science, wildlife conservation, or ecosystem integrity. It was a document of its time, geared to introduce the American public to the nascent national park system and the principal ideas behind it.[2]

Drawing on these sources, the new National Park Service set about opening its nature reserves to the public. To attract visitors to these out-of-the-way venues and to accommodate the arrival of the automobile, the Park Service was intent on taming, not preserving, the wilderness, which meant building new roads, lodges, campgrounds, and the like. In short order, the Park Service transformed these wilderness settings—in roughly equal measures—into inviting tourist destinations and outdoor playgrounds, places where people came to view stunning natural wonders and to play among them. Wild nature was tamed, rendered accessible, and put on display. Paradoxically, just as the public was being invited into the wilderness to witness nature's splendor, the nature they encountered was being disassembled into a destination vacation site and a recreational paradise. Any idea of the park as a wilderness enclave soon lost any real currency.[3]

This early evolution in the national park idea was abetted by a growing connection between the parks and commercial enterprise. To help create an accommodating environment where nature was on display and where visitors would feel comfortable, the Park Service enlisted the private sector—initially private concessioners and then later the gateway communities—to provide lodging and other services for park visitors. By promoting recreation as part of the national park experience, the Park Service helped foster the eventual emergence of a mass recreation culture and captured the attention of a growing recreation industry. Inevitably, these relationships between the national parks and private enterprise have been largely commercial in character, which has significantly affected park resource management policy over the years. The private sector's market-driven interest in profits does not readily square with the public interest in nature conservation. The early conception of a national park as a tourist destination and recreational playground not only belied any wilderness sanctuary notions, but it actually went far toward transmuting these new nature reserves into commercial commodities.[4]

The national park idea has another aspect, however, one that has over time assumed much greater importance. This aspect is tied to the ecological sciences, species conservation, ecosystem integrity and restoration, social justice, and civic education. As we have seen, science initially took a backseat to scenery with the Park Service, which concentrated its early management efforts on protecting scenic values to maintain a pretty façade. Wildlife conservation did not originally mean protecting all species, nor did it necessarily mean treating park wildlife as wild animals rather than zoo residents, let alone ensuring that their seasonal habitat needs were met. The notion that a national park was part of a larger ecological complex or that ecological processes played a critical role in sustaining the natural scene that was on display was likewise absent from any management calculus. Nor was there any recognition that the Native Americans who originally occupied many of the lands that became national parks might have a legitimate, ongoing connection to them. To the extent that education was part of the original mission, it focused on natural history and museum displays. Moreover, in a nation that was largely of European origin, there was little need to see the visiting public in terms of a diversifying population.[5]

As time has passed, however, these dimensions of the national park idea have emerged as central to our understanding of that idea. To meet our nature conservation responsibilities and to pass these magnificent places onto future generations in an "unimpaired" condition, we have expanded on the national park idea, incorporating contemporary values and knowledge into the concept. It has meant acknowledging (1) that science must be integrated into park resource management policy, (2) that effective wildlife management entails maintaining biodiversity at all levels, (3) that park wildlife needs and ecological processes transcend park boundaries and require an ecosystem perspective, (4) that Native Americans have valid treaty-based claims and cultural concerns that must be addressed through more sensitive management policies, (5) that more expansive public education efforts are essential to promote popular engagement in nature conservation, and (6) that nature conservation must be brought closer to where people live and introduced into the urban setting and minority communities. Some of this new understanding has already been translated into law and policy as reflected in the Park Service's new science mandate, the Endangered Species Act's species recovery goals, and the agency's efforts to promote ecosystem-wide conservation strategies in Glacier, the Everglades, and elsewhere.

That the national park idea is not a single idea but rather an amalgam of evolving ideas is not surprising. The national park concept has been on an evolutionary trajectory from the beginning. A beguilingly simple idea on the surface—just set aside a block of land to safeguard wild nature from humanity—the national park idea has generated plenty of perplexity when we examine how it has changed over time. In fact, conservation has never been a simple matter, even in the protected, island-like confines of a national park. As the controversies we have reviewed suggest, the notion that we can conserve nature in an unimpaired condition inevitably runs up against competing human interests that derive from park visitors as well as from commercial and other forces outside park boundaries. Moving forward, then, we can no longer view the parks in isolation; rather, to have any hope of achieving our nature conservation goals, we must understand and manage them as part of the larger landscape.

THE ENDURING ORGANIC ACT

The durable centerpiece of the national park idea is the 1916 Organic Act, which has guided national park policy from the system's inception. The act mandates that national parks be managed to conserve the scenery, natural and historic objects, and wildlife in an unimpaired condition for the benefit of present and future generations. This statutory language, unlike the evolutionary character of the national park idea, has endured unchanged since 1916, and Congress has even expressly reconfirmed it as the predominant purpose of the national parks. Consequently, the Organic Act's spare language has taken on near biblical qualities, although it masks major complexities over how the Park Service should go about accomplishing its conservation mission. That the Organic Act also calls for promoting the parks and public enjoyment adds more layers of complexity to the national park idea and the agency's fundamental conservation responsibilities.[6] It also helps explain why the Park Service, for much of its history, has focused on meeting visitor demands, often at the expense of its explicit conservation obligations. Of course, Mather and his colleagues, charged with bringing a new and unfamiliar nature conservation initiative to the nation, viewed public support as crucial to their task.

Two years after the Organic Act was enacted, the seminal Lane letter offered an original interpretation of the act's language and purpose that still resonates today. The letter laid down three aforementioned general principles to guide national park management: the parks must be maintained in "absolutely unimpaired form" for present and future generations; they are for "the use, observation, health, and pleasure of the people;" and the "the national interest" must govern the agency's management decisions. Elaborating on the first point, the letter admonished: "Every activity of the Service is subordinate to the duties imposed upon it to faithfully preserve the parks for posterity in essentially their natural condition."[7] This understanding of the Organic Act and the paramount conservation duties it imposes on the Park Service plainly captures what Congress intended at the inception.[8] It also explains why the federal courts, invoking this same congressional intent, have ruled consistently that resource conservation takes priority over other uses or concerns.[9] Although the Park Service has wavered from this conservation priority over the years, it now officially endorses this hierarchy of park purposes and priorities, as reflected in its 2006 *Management Policies*.[10]

Indeed, the Organic Act sets a high conservation standard that governs management of the national parks. More than once, the courts have reminded park officials that they have a duty to meet the Organic Act's nonimpairment standard. At the same time, though, the act allows the Park Service some flexibility in determining how to go about fulfilling its nature conservation responsibilities. Under the banner of the Organic Act, the agency has moved from policies that conflated the national park idea with scenic preservation, high-volume tourism, mass recreation, and concession-driven facilities and entertainment to policies that incorporated science into resource management policy, treated the parks as natural laboratories, revised conventional wildlife management practices, restored wolves and other extirpated species, and are beginning to engage neighbors in broader ecosystem-level management efforts. The act has also been sufficiently flexible to allow agency officials to address complex human-nature conundrums, as in the case of Native American cultural claims, bison management, and other wildlife issues. In short, the Organic Act's conserve unimpaired language has proven adaptable enough to accommodate our changing view of the national park idea.

Moreover, Congress has adopted other laws strengthening the nation's commitment to nature conservation that apply to the national parks. Where the Organic Act once stood preeminent as the most powerful federal law promoting nature conservation, laws like the Wilderness Act and the Endangered Species Act extend additional—and, in some cases, even greater—protection to the national parks and their resources. The Wilderness Act of 1964 created a new protective federal designation that precludes any permanent human presence on designated wilderness lands, making the management standards for wilderness areas even more restrictive than those governing the parks. A formal wilderness designation can overlay and enhance the level of protection enjoyed by park lands, as witnessed by the Grand Canyon controversy, where the presence of motorized rafting on the Colorado River has blocked formal wilderness protection for park lands and has fomented related management conflicts.[11] The Endangered Species Act of 1973 provides an extraordinary level of protection to federally "listed" species that face extinction and takes precedence over conflicting national park management policies.[12] A key provision in that act provided the legal foundation for reintroducing wolves to Yellowstone, one of the Park Service's major ecological restoration accomplishments.[13] The Wild and Scenic Rivers Act extends protection to designated river corridors even to the point of limiting human recreational impacts, a reality that has kept Yosemite's Merced River management plan in the courts and has forced the Park Service to reexamine its approach to auto access into Yosemite Valley.[14] These laws not only reinforce the Organic Act's conservation mandate, but they strengthen it, putting additional teeth into the meaning of nonimpairment in the national park setting.

Under this comprehensive legal framework, the national park idea has evolved to reflect new and different management goals and strategies, but the abiding standard remains one of safeguarding national park resources from impairment. The Organic Act, as supplemented by other related laws, has enabled the Park Service to confront an array of thorny resource management controversies—including the use of snowmobiles and off-highway vehicles in the parks, the ill-located giant sequoia complex, overdevelopment in Yosemite Valley, overabundant ungulate populations, and nonnative fish species—with a clear sense of priorities, even if park officials have not always selected the most conservation-oriented option. When political realities have intervened and constrained the Park Service's ability to resolve thorny resource management issues, Congress has occasionally weighed in with targeted amendments to address the problem, as seen in the case of the Redwood Amendment, concessions reform, air tour overflights, the new science mandate, and Elwha River restoration. The net result has not always been as sensitive to nature conservation priorities as it might have been, but no law can provide ironclad protection against politically powerful competing demands. The Organic Act, at least, has stood the test of time and is widely understood to put conservation first when other interests collide over national park policy.

TOWARD A NEW AGE

This evolution in the national park idea can be viewed as the advent of a third era in the history of national park conservation policy. During the early days, the Park Service concentrated on protecting scenery and promoting park visitation, employing intensive resource management practices so as to attract and entertain visitors. In the aftermath of World War II, however, a second era unfolded, one driven by an emerging appreciation for science and the demands of industrial tourism. The 1963 Leopold report prompted the Park Service to rethink the role of science in nature conservation and its longstanding interventionist resource management policies, while the onslaught of visitors during the postwar period made clear that the parks could not accommodate everyone's personal recreational preferences. As science has assumed a more prominent role in resource management, it has helped reveal how connected the parks are to the larger landscape and the diverse natural and human forces that affect park resources, including climate change.

Recognizing that these distant forces present ever more complex conservation challenges, national park policy may be entering a new post-Leopold era that will see further policy changes. The hallmarks of this new era will include an increased reliance on conservation biology science to guide management, less reluctance to intervene with nature so as to maintain and restore critical species and ecological processes, engagement in collaborative planning and management efforts that extend beyond park boundaries and align with on-the-ground ecological realities, a greater appreciation for public involvement in conservation efforts (including an enhanced Park Service role in public education), and a heightened awareness of social justice and diversity concerns.[15] These impending shifts in conservation policy reflect an evolving national park idea and the type of management strategies necessary to achieve sustainable nature conservation goals in this fast-changing world.[16]

A paramount lesson derived from this excursion through the ongoing evolution of the national park idea is that our parks are—and always have been—integral parts of a larger natural and human landscape. From the beginning, the national parks and their conservation policies have been deeply influenced by the human communities that surround them. The railroads certainly helped shape the early western parks, and gateway communities like Gatlinburg and Estes Park are credited with helping bring Great Smoky Mountains and Rocky Mountain National Parks, respectively, into existence.[17] For the most part, these early connections between the parks and surrounding towns were economic in nature, abetted by the political interests of local congressional delegations. These same economic and political forces are still quite evident today, taking the form of local resistance, for example, to any effort to eliminate snowmobiling from Yellowstone, to revise dam operations on the Colorado River upstream of the Grand Canyon, or to replumb the Everglades' life-sustaining river of grass. Simply put, we have long understood the economic connections that the national parks have with the surrounding landscape, and we have deferred to them.

At the same time, we thought of the national parks as islands, places where we could practice nature conservation without regard to what was occurring outside park boundaries. This island—or enclave—view was easy to indulge when most parks were in remote locations and surrounded by an undeveloped landscape.[18] Those days are over, however, halted by the plethora of controversies and reports that have revealed just how ecologically connected the parks are to their surrounding environs. Whether measured in terms of wildlife habitat needs, shared watersheds, air-quality concerns, wildfire patterns, or climate change, the national parks are not isolated enclaves, and any effort to manage them as such is doomed to fail. Civilization has found the national parks, many of which are today bordered by fast-growing communities and face recurrent development proposals on adjacent federal and private lands. The ecological and atmospheric sciences have brought home the message that the parks are part of larger ecosystems and must be managed in this context.

This new reality conjures up several new roles for the national parks that further reshape the national park idea. As an institution devoted to nature conservation, the national parks are—and must be treated as—the vital core of larger ecosystems, places where wild nature enjoys a special status unlike elsewhere in our ever more developed world. This new image of the national park calls for restraint on the periphery, a greater sensitivity to how development activities affect wildlife and other park resources, and recognition of just how far afield some of these threats are. It also calls for ecosystem-level planning, with the Park Service more directly engaged with its neighbors. To identify, develop, and implement appropriate conservation strategies in this expanded landscape, the national parks will have to assume a greater role as nature's laboratory, a place where we can test the new theories and policies that will enable us to live more sustainably on the landscape and to respond to a warming atmosphere. As the principal overseer of our conservation heritage and warden of these core nature reserves, the Park Service will have to increase its education efforts to inform the general public about contemporary conservation challenges, the need for experimental management initiatives, and the role of the national parks in responding to these challenges. It will have to do so by engaging a populace that is itself changing and has fewer connections with the natural world than was true with prior generations.

Viewing the national parks as the cornerstone of larger natural and human landscapes also has profound implications for the future shape of the national park system. Where necessary to meet nature conservation needs, we need to explore strategic park expansions. Over time, the original boundaries of several parks have been extended to better protect critical resources, and our improved ecological knowledge now gives us a much clearer picture of what healthy wildlife populations and other park resources require to thrive. We should also explore new national park designations, other types of protective designations, and new wildlife corridor designations or other ways to better connect our already extensive network of national parks, wilderness areas, wildlife reserves, and other nature preserves. In addition, we should explore opportunities to restore damaged landscapes to promote ecological integrity in a world where few landscapes have not been significantly altered by humankind. We must insist upon much more integrated planning and decision-making processes across the landscape to better harmonize both conservation and human needs with an eye toward a more sustainable landscape. In short, there is no reason to view the present park system as complete.[19]

This expanded vision of the national park idea, however, cannot ignore the abiding political realities that have long shaped the national park system and related nature conservation policies. Of course, social values and economic concerns have always figured prominently in the nation's conservation efforts as revealed by the evolution of the national park idea itself. To make the political case for this broader view of the national park idea and for system expansion proposals that incorporate these new ideas, we must begin educating the general public about the vital role that parks play in preserving the nation's dwindling natural heritage, about new ecological management opportunities, and about shifting economic trends on the broader landscape. Doing so will call for more effective engagement by the Park Service with its neighbors and new strategic alliances to underscore the case for this new view of the national park idea. In the interest of social justice and cultural diversity, it will also call for more-sensitive involvement with Native American neighbors, urban communities, and minority populations to not only address past wrongs but to strengthen the base of popular support for nature conservation more generally.

The American national park idea has evolved over time, reflecting the prevalent social norms and felt necessities of the day as well as contemporary advances in scientific knowledge. This evolution of the national park idea is not at an end; rather, it is still a work in progress. Our understanding of the natural world and the imperatives of nature conservation continue to evolve, just as our cultural norms and values remain in flux. The national parks were an aspiration in the beginning, an effort to redefine our relationship to the natural world and to acknowledge an obligation to future generations. The unvarnished truth is that the parks will always be confronted with new demands and threats, testing our commitment to the fundamental principles underlying the hallowed notion of conserving nature in an unimpaired condition.

NOTES

Chapter 1

1. For background on the Yellowstone snowmobile controversy, see Michael J. Yocim, "Snow Machines in the Gardens: The History of Snowmobiles in Glacier and Yellowstone National Parks," *Montana Western History Magazine* (Autumn 2003): 2; David G. Havlick, *No Place Distant: Roads and Motorized Recreation on America's Public Lands* (Washington, DC: Island Press, 2002), 166–68.
2. The controversy is recounted in two court cases: Fund for Animals v. Norton, 294 F. Supp. 2d 92 (D.D.C. 2003); and International Snowmobile Manufacturers Association v. Norton, 304 F. Supp. 2d 1278 (D. Wyo. 2004). See also National Park Service, *Winter Use Plans Final Environmental Impact Statement: Yellowstone and Grand Teton National Parks* (2007).
3. The Bandelier fire controversy is recounted in Roger G. Kennedy, *Wildfire and Americans: How to Save Lives, Property, and Your Tax Dollars* (New York: Hill and Wang, 2006), 89–101.
4. The Leopold report is found at A.S. Leopold, S.A. Cain, C.M. Cottam, I.N. Gabrielson, and T.L. Kimball, "Wildlife Management in the National Parks" (July 10, 1964), in *America's National Park System: The Critical Documents*, ed. Lary M. Dilsaver (Lanham, MD: Rowman and Littlefield, 1994), 237–51. On the evolution of the National Park Service's approach to resource management, see Richard West Sellars, *Preserving Nature in the National Parks: A History* (New Haven: Yale University Press, 1997); Alfred Runte: *National Parks: The American Experience*, 2d ed. (Lincoln: University of Nebraska Press, 1987), 197–208.
5. For criticism of the Park Service's resource management policies, see Frederic H. Wagner, Ronald Foresta, Richard Bruce Gill, Dale Richard McCullough, Michael R. Pelton, William F. Porter, and Hal Salwasser, *Wildlife Policies in the U.S. National Parks* (Washington, DC: Island Press, 1995); Stephen Budiansky, *Nature's Keepers: The New Science of Nature Management* (New York: Free Press, 1995); Karl Hess Jr., *Rocky Times in Rocky Mountain National Park: An Unnatural History* (Niwot: University of Colorado Press, 1993); Alston Chase, *Playing God in Yellowstone: The Destruction of America's First National Park*

(Boston: Atlantic Monthly Press, 1986). The Park Service's ecological restoration efforts are described in Douglas W. Smith and Gary Ferguson, *Decade of the Wolf: Returning the Wild to Yellowstone* (Guilford, CT: Lyons Press, 2005); Robert W. Adler, *Restoring Colorado River Ecosystems: A Troubled Sense of Immensity* (Washington, DC: Island Press, 2007); Phillip M. Bender, "Restoring the Elwha, White Salmon, and Rogue Rivers: A Comparison of Dam Removal Proposals in the Pacific Northwest," *Journal of Land, Resources, and Environmental Law* 17 (1997): 189.

6. See Joseph L. Sax and Robert B. Keiter, "Glacier National Park and Its Neighbors: A Study of Federal Interagency Relations," *Ecology Law Quarterly* 14 (1987): 207–64; Joseph L. Sax and Robert B. Keiter, "The Realities of Regional Resource Management: Glacier National Park and Its Neighbors Revisited," *Ecology Law Quarterly* 33 (2006): 233–312; see also Tony Prato and Dan Fagre, eds., *Sustaining Rocky Mountain Landscapes: Science, Policy, and Management for the Crown of the Continent Ecosystem* (Washington, DC: Resources for the Future, 2007).
7. On the Greater Yellowstone Ecosystem and the controversies surrounding it, see Robert B. Keiter and Mark S. Boyce, eds., *The Greater Yellowstone Ecosystem: Redefining America's Wilderness Heritage* (New Haven: Yale University Press, 1991); Susan G. Clark, *Ensuring Greater Yellowstone's Future: Choices for Leaders and Citizens* (New Haven: Yale University Press, 2008).
8. The history of Grand Teton National Park is described in Robert W. Righter, *Crucible for Conservation: The Creation of Grand Teton National Park* (Boulder: Colorado Associated University Press, 1982); see also Runte, *National Parks,* 118–27.
9. On the economic value of national parks, see National Parks Conservation Association, *Gateway to Glacier: The Emerging Economy of Flathead County* (Washington, DC: 2003); Sonoran Institute, *Prosperity in the 21st Century West: The Role of Protected Public Lands* (Tucson, AZ; 2004); Dennis J. Stynes, *Economic Benefits to Local Communities from National Park Visitation and Payroll* (Washington, DC: National Park Service, 2011); see also references cited in chapter 5, note 21. On the California desert protection campaign, see U.S. Congress, California Desert Protection Act of 1994: House Report 103-832 (to Accompany S.21) (Washington, DC: Government Printing Office, 1994). For a discussion and analysis of the roles Congress and the National Park Service play in the designation of new national parks, see James M. Ridenour, *The National Parks Compromised: Pork Barrel Politics and America's Treasures* (Merrillville, IN: ICS Books, 1994); George B. Hartzog, *Battling for the National Parks* (Mt. Kisco, NY: Moyer Bell, 1988); Ronald A. Foresta, *America's National Parks and Their Keepers* (Washington, DC: Resources for the Future, 1984).
10. For a description and analysis of the Canyonlands expansion proposal, see Robert B. Keiter, "Completing Canyonlands," *National Parks* (March/April 2000): 25–30.

11. The Bryce quotation can be found in Thomas R. Vale, *The American Wilderness: Reflections on Nature Protection in the United States* (Charlottesville: University of Virginia Press, 2005), 90. Wallace Stegner also frequently used this quote, which has been attributed to him, too. See Wallace Stegner, *Marking the Sparrow's Fall: The Making of the American West* (New York: Henry Holt, 1998), 135, 137.
12. The National Parks Organic Act provisions noted here are found at 16 U.S.C. §§ 1, 1a–1. For an analysis of the original legislation and congressional intent, see Robin W. Winks, "The National Park Service Act of 1916: 'A Contradictory Mandate'?" *Denver University Law Review* 74 (1997): 575.
13. See "Secretary Lane's Letter on National Park Service Management" (May 13, 1918), in Dilsaver, *America's National Park System*, 48–52; National Park Service, *Management Policies* (2006). For biographies of these two key early Park Service officials, see Robert Shankland, *Steve Mather of the National Parks* (New York: Alfred A. Knopf, 1970); Donald C. Swain, *Wilderness Defender: Horace M. Albright and Conservation* (Chicago: University of Chicago Press, 1970).

Chapter 2

1. Ferdinand V. Hayden, *Sixth Annual Report of the United States Geological Survey of the Territories* (Washington, DC: Government Printing Office, 1873), 32.
2. For examples of early park enabling acts, see 16 U.S.C. § 43 (Sequoia resources to be retained in their "natural condition"); 16 U.S.C. § 92 (Mount Rainier resources to be retained in their "natural condition"); 16 U.S.C. § 162 (Glacier to be preserved "in a state of nature"); 16 U.S.C. § 202 (Lassen to be preserved "in a state of nature"). For a discussion and analysis of industrial assaults on the early national parks, see John Ise, *Our National Park Policy: A Critical History* (Baltimore: Johns Hopkins University Press, 1961).
3. The National Parks Organic Act is found at 16 U.S.C. §§ 1–18f; the Yellowstone enabling act is found at 16 U.S.C. §§ 21–22.
4. "Secretary Lane's Letter on National Park Service Management" (May 13, 1918), in *America's National Park System: The Critical Documents*, ed. Lary M. Dilsaver (Lanham, MD: Rowman and Littlefield, 1994), 48–52.
5. A brief description and analysis of the early Mather strategy for safeguarding the new national park system can be found at Ronald A. Foresta, *America's National Parks and Their Keepers* (Washington, DC: Resources for the Future, 1984), 19–30. For a comprehensive treatment of the National Park Service and wilderness, see John C. Miles, *Wilderness in National Parks: Playground or Preserve* (Seattle: University of Washington Press, 2009).
6. The quotations are from "Superintendent's Resolution on Overdevelopment" (Nov. 13–17, 1922), in Dilsaver, *America's National Park System*, 57–58; see also, Miles, *Wilderness*, 47–8.
7. The incident is recounted in Alfred Runte, *National Parks: The American*

Experience, 2d ed. (Lincoln: University of Nebraska Press, 1987), 120–24; the initial quotation is from Albright, characterizing his meeting with the dude ranchers, and the subsequent quotations are from Mather.

8. See Alice Wondrak Biel, *Do (Not) Feed the Bears: The Fitful History of Wildlife and Tourists in Yellowstone* (Lawrence: University Press of Kansas, 2006), 7–27; Runte, *National Parks*, 197–208.
9. See John C. Paige, *The Civilian Conservation Corps and the National Park Service, 1933–1942: An Administrative History* (Washington, DC: National Park Service, 1985), 103–120, 216–39; see also Conrad L. Wirth, *Parks, Politics, and the People* (Norman: University of Oklahoma Press, 1980), 94–157.
10. Rosalie Edge, "Roads and More Roads in the National Parks and National Forests," in Dilsaver, *America's National Park System*, 138–39.
11. 16 U.S.C. § 410(b), 48 Stat. 817 (1934); see also Runte, *National Parks*, 134.
12. Richard West Sellars, *Preserving Nature in the National Parks: A History* (New Haven: Yale University Press, 1997), 151–53; see generally Ben W. Twight, *Organizational Values and Political Power: The Forest Service versus the Olympic National Park* (University Park: Pennsylvania State University Press, 1983).
13. Wallace Stegner, "The Marks of Human Passage," in *This Is Dinosaur: Echo Park and Its Magic Rivers*, ed. Wallace Stegner (Boulder, CO: Roberts Rinehart, 1955), 14.
14. For a brief account of the Echo Park controversy, see Roderick Nash, *Wilderness and the American Mind*, 3d ed. (New Haven: Yale University Press, 1982), 209–19. See generally Mark W.T. Harvey, *A Symbol of Wilderness: Echo Park and the American Conservation Movement* (Albuquerque: University of New Mexico Press, 1994).
15. On the National Park Service's Mission 66 program, see Wirth, *Parks, Politics, and the People*, 237–84; Foresta, *America's National Parks*, 52–56.
16. For an interesting analysis of the Park Service's evolving road construction program, see Thomas R. Vale, *The American Wilderness: Reflections on Nature Protection in the United States* (Charlottesville: University of Virginia Press, 2005), 113–17. The Wirth quotation is from internal Park Service memos and is noted in Sellars, *Preserving Nature*, 192; the Brower quotation appeared in the magazine *National Parks* and is noted in Sellars, *Preserving Nature*, 188.
17. The Wilderness Act is found at 16 U.S.C. §§ 1131–36. The term *wilderness* is further defined in 16 U.S.C. § 1131(c) as "an area of undeveloped Federal land retaining its primeval character and influence, without permanent improvements or human habitation, which is protected and managed so as to preserve its natural conditions." For more on the Wilderness Act and its passage, see Michael McCloskey, "The Wilderness Act of 1964: Its Background and Meaning," *Oregon Law Review* 45 (1966): 288–321; Nash, *Wilderness and the American Mind*, 200–237.
18. On the Park Service –Forest Service conflict over wilderness, see generally Miles, *Wilderness*.

19. On this early Great Smoky Mountains wilderness designation controversy, see Michael Frome, *Battle for the Wilderness*, 2d ed. (Salt Lake City: University of Utah Press, 1997), 176–86; Miles, *Wilderness*, 174–76; see also Sellars, *Preserving Nature*, 212.
20. Jim Walters, "An Evaluation of the National Park Service's Wilderness Program on the 40th Anniversary of the Wilderness Act," *George Wright Forum* 21(3) (2004): 9; Matt Jenkins, "Park Service Wilderness in Disarray," *High Country News* (February 16, 2004); Miles, *Wilderness*, 268–9; see also National Park Service, *Management Policies* 6.2 et seq. (2006) (detailing the agency's wilderness assessment and recommendation process).
21. The Wilderness Society v. Norton, 434 F.3d 584 (D.C. Cir. 2006).
22. Author's phone interview with Garry Olds, Chief, National Park Service Wilderness Stewardship Division (November 28, 2012); National Park Service, *National Parks—Our Treasured Landscapes: America's Wilderness* (2009); National Park Service, *Wilderness Stewardship Division, 2010–2011 Wilderness Report* (2011). See also Department of the Interior, Department of Agriculture, Environmental Protection Agency, and Council on Environmental Quality, *America's Great Outdoors: A Promise to Future Generations* (2011), 64 (recommending identification of new national park areas for wilderness designation).
23. Wilderness Watch v. Mainella, 375 F.3d 1085 (11th Cir. 2004).
24. Olympic Park Associates v. Mainella, 2005 WL 1871114 (W.D. Wash. 2005); see also Matt Jenkins, "Park Service Wilderness in Disarray," *High Country News*, Feb. 16, 2004.
25. National Park Service, *Management Policies* 6.3 et seq. (2006).
26. On the addition of new park units and wilderness designations in Alaska, see G. Frank Williss, *"Do Things Right the First Time": Administrative History: The National Park Service and the Alaska National Interest Lands Conservation Act of 1980* (Washington, DC: National Park Service, 1985), www.nps.gov/history/history/online_books/williss/adhi.htm; see also Daniel Nelson, *Northern Lights: The Struggle for Wilderness Alaska* (Washington, DC: Resources for the Future, 2004). On the Leopold report and National Park Service resource management policy revisions, see A.S. Leopold, S.A. Cain, C.M. Cottam, I.N. Gabrielson, and T.L. Kimball, "Wildlife Management in the National Parks," in Dilsaver, *America's National Park System*, 237–51; Sellars, *Preserving Nature*, 214–66; Runte, *National Parks*, 197–208. The Park Service's wilderness resource management policies are found in National Park Service, *Management Policies* 6.3 et seq. (2006).
27. Quoted in William deBuys, ed., *Seeing Things Whole: The Essential John Wesley Powell* (Washington, DC: Island Press, 2001), 64–65. Powell's journey down the Colorado River is recounted in Wallace Stegner, *Beyond the Hundredth Meridian: John Wesley Powell and the Second Opening of the* West (Lincoln: University of Nebraska Press, 1953); Donald Worster, *A River Running West: The*

Life of John Wesley Powell (New York: Oxford University Press, 2001).

28. 16 U.S.C. § 431. On the Antiquities Act, see David Harmon, Francis P. McManamon, and Dwight T. Pitcaithley, eds., *The Antiquities Act: A Century of American Archaeology, Historic Preservation, and Nature Conservation* (Tucson: University of Arizona Press, 2006). See also chapter 10 for further discussion of the Antiquities Act.
29. For a concise history of Grand Canyon National Park, see Michael F. Anderson, *Polishing the Jewel: An Administrative History of Grand Canyon National Park* (Grand Canyon, AZ: Grand Canyon Association, 2000), www.nps.gov/grca/historyculture/adhigrca.htm; Stephen Pyne, *How the Canyon Became Grand: A Short History* (New York: Viking, 1998). The U.S. Supreme Court decision sustaining creation of the Grand Canyon National Monument is Cameron v. United States, 252 U.S. 450 (1920).
30. The Grand Canyon National Park enabling legislation is found at 40 Stat. 1175 (1919), codified at 16 U.S.C. §§ 221–28. On the controversial federal deer reduction program, which was sustained by the U.S. Supreme Court over the state of Arizona's objections, see Hunt v. United States, 278 U.S. 96 (1928).
31. The Grand Canyon Enlargement Act provision addressing the Havasupai rights is found at 16 U.S.C. § 228i (2008). For further discussion of the National Park Service–Native American relationship at Grand Canyon National Park, see chapter 6.
32. See Anderson, *Polishing the Jewel*, 73–74. For a detailed accounting of the number of rafters annually on the Colorado River between 1869 and 1980, see Nash, *Wilderness and the American Mind*, 331.
33. The raft permitting system is described in Wilderness Public Rights Fund v. Kleppe, 608 F.2d 1250 (9th Cir. 1979); see also Nash, *Wilderness and the American Mind*, 330–39.
34. That prompting came in the 1975 Grand Canyon Enlargement Act, where Congress instructed the secretary of the Interior to report to the president within two years on the suitability of areas within the park for wilderness designations. 16 U.S.C. § 228i-1.
35. For an overview of the Grand Canyon wilderness designation controversy, see Kim Crumbo, "Wilderness Management at Grand Canyon—'Waiting for Godot?'" *International Journal of Wilderness* 2(1) (1996): 19; Miles, *Wilderness*, 262–68. For an opinionated account of the controversy, see Jeff Ingram, *Hijacking a River: A Political History of the Colorado River in the Grand Canyon* (Flagstaff, AZ: Vishnu Temple Press, 2003). The National Park Service's *Management Policies* 6.2.2.1 (2006) defines "potential wilderness" as park "lands that are surrounded by or adjacent to lands proposed for wilderness designation but that do not themselves qualify for immediate designation due to temporary nonconforming or incompatible conditions." These "potential wilderness" lands are "to be managed as wilderness to the extent that existing nonconforming conditions allow." National Park Service, *Management*

Policies 6.3.1 (2006).

36. Pub. L. 96-514 § 112, 94 Stat. 2957, 2972 (1980); see Peter Jacques and David M. Ostergren, "The End of Wilderness: Conflict and Defeat of Wilderness in the Grand Canyon," *Review of Policy Research* 23 (2006): 573, 581–82.
37. Crumbo, "Wilderness Management," 22. In 1993, an updated Park Service wilderness recommendation expanded the recommended acreage modestly to slightly more than 1.1 million acres, but continued to treat the river corridor as merely "potential wilderness." See National Park Service, *Colorado River Management Plan Grand Canyon National Park Final Environmental Impact Statement* 3.8 (2006).
38. See National Park Service, *Grand Canyon National Park General Management Plan* (1995); National Park Service, *Grand Canyon Draft Wilderness Management Plan and Environmental Assessment* (1998). For a more detailed recounting of these events, see Ingram, *Hijacking a River*, 393–409.
39. See National Park Service, *Colorado River Management Plan Record of Decision* (2006), 2–3, 24; see generally National Park Service, *Colorado River Management Plan Final Environmental Impact Statement*. For the court decision, see River Runners for Wilderness v. Martin, 593 F.3d 1064 (9th Cir. 2010).
40. 16 U.S.C. § 1133(c).
41. The quotations are from National Park Service, *Colorado River Management Plan Record of Decision*, 21–22.
42. This point appears in National Park Service, *Colorado River Management Plan Key Changes in the FEIS from the DEIS* (2006), 1.
43. Kim Crumbo and Bethanie Walder, "Restoring Wilderness at Grand Canyon," *Wild Earth* (Winter 1998–1999), 65.
44. On the Glen Canyon Dam, see Russell Martin, *A Story That Stands Like a Dam: Glen Canyon and the Struggle for the Soul of the West* (New York: Henry Holt, 1989); Philip L. Fradkin, *A River No More: The Colorado River and the West* (Berkeley: University of California Press, 1995).
45. The altered ecology of the Colorado River is described and analyzed in Robert W. Adler, *Restoring Colorado River Ecosystems: A Troubled Sense of Immensity* (Washington, DC: Island Press, 2007), 26–103; see also U.S. Bureau of Reclamation, *Operation of Glen Canyon Dam Final Environmental Impact Statement* (1995).
46. The Grand Canyon Protection Act of 1992 is found at Pub. L. 102-575 §§ 1801–9, 106 Stat. 4600, 4669; see also Michael Connor, "Extracting the Monkey Wrench from Glen Canyon Dam: The Grand Canyon Protection Act: An Attempt at Balance," *Public Land Law Review* 15 (1994): 135. For further discussion and analysis of Colorado River restoration efforts and ongoing scientific studies, see chapter 7.
47. On the Grand Canyon National Park dam controversies, see Byron E. Pearson, *Still the Wild River: Congress, the Sierra Club, and the Fight to Save the Grand Canyon* (Tucson: University of Arizona Press, 2002), 12, 52–53, 63–66;

Nash, *Wilderness and the American Mind*, 227–37; Anderson, *Polishing the Jewel*, 66–67.

48. 16 U.S.C. § 1133(d)(4).
49. The 1975 legislation is found at 16 U.S.C. § 228g. The 1987 legislation is found at Pub. L. No. 100-91, 101 Stat. 674, codified at 16 U.S.C. § 1a-1 (note). The Park Service actually recommends noise limitations to the Federal Aviation Administration, which must set those limits unless the proposed limit would affect flight safety. Notably, the act excludes helicopters used to transport river raft customers from the lower canyon bottom on the Hualapai Reservation to the North Rim outside park boundaries. For additional background information, see Anderson, *Polishing the Jewel*, 54–55. In early 2011, the Park Service released a draft EIS that sets forth specific overflight limitation recommendations that would further restrict air tours in terms of timing and location. National Park Service, *Special Flight Rules Area in the Vicinity of Grand Canyon National Park: Actions to Substantially Restore Natural Quiet Draft Environmental Impact Statement* (2011); John Dougherty and Ray Ring, "Park Service Finally Drafts a Solution to Conflicts over Canyon Flights," *High Country News* 43(10) (2011): 18.
50. For background information on air quality at Grand Canyon National Park, see Western Governors' Association, *Grand Canyon Visibility Transport Commission, Recommendations for Improving Western Vistas* (Denver: 1996); pertinent Clean Air Act provisions include 42 U.S.C. § 7491 (visibility protection for federal class one airsheds) and 42 U.S.C. § 7492 (creating visibility transport commissions). On the proposed legislation banning uranium mining adjacent to the park, see H.R. 5583 (108th Cong., 2008); April Reese, "Mining: Arizona Lawmaker Floats Bill to Ban Uranium Projects Near Grand Canyon," *Land Letter*, Mar. 20, 2008; Manuel Quinones, "Grand Canyon: Obama Admin Limits Mining near National Park," *Greenwire*, Jan. 9, 2012; Matthew Daly, "New 20-Year Ban on Mining Near Grand Canyon Is Final," *USA Today*, Jan. 9, 2012.
51. In 2012, environmental groups proposed creation of a new 1.7-million-acre Grand Canyon Watershed National Monument on national forest and BLM lands situated north of the park and south of the Utah border. The site lies between the two Clinton-era national monuments. Felicia Fonseca, Associated Press, "Groups Propose National Monument near Grand Canyon," http://www.ksl.com/?nid=148&sid=19994467.
52. Wallace Stegner, "Wilderness Letter," in *Marking the Sparrow's Fall: The Making of the American West* (New York: Henry Holt, 1998), 114 (quoting Sherwood Anderson).
53. Joseph L. Sax, *Mountains without Handrails: Reflections on the National Parks* (Ann Arbor: University of Michigan Press, 1980), 42.
54. On wilderness and national park values, see Max Oelschaelger, *The Idea of Wilderness* (New Haven: Yale University Press, 1991); McCloskey, "The Wil-

derness Act of 1964"; Stegner, "Wilderness Letter," 111; Stegner, "The Best Idea We Ever Had," in *Marking the Sparrow's Fall*, 135. For more on the "land ethic," see Aldo Leopold, *A Sand County Almanac with Essays on Conservation from Round River* (New York: Ballantine, 1966), 237–64.

55. 16 U.S.C. § 1131(c) (defining "wilderness); 16 U.S.C. § 1133(c) (setting forth specific prohibitions). The Wilderness Act does permit the agencies to allow preexisting aircraft and motorboat use to continue, with appropriate restrictions. 16 U.S.C. § 1133(d)(1). It would be difficult, however, to classify the motorized rafts used on the Colorado River as "motorboats." See also High Sierra Hikers Association v. U.S. Dept. of the Interior, 848 F. Supp. 2d 1036 (N.D. Cal. 2012), which ruled that the Park Service violated the Wilderness Act by permitting commercial outfitters to use horses and mules in a designated wilderness area within Sequoia and Kings Canyon National Parks without finding that their use was a "necessity" as required by the act. The court's ruling, however, was promptly overturned by Congress. Sequoia and Kings Canyon National Parks Backcountry Access Act, Pub. L. 112-128, 126 Stat. 373 (2012).

Chapter 3

1. See Yosemite Valley Act of 1864, 38 Cong. Ch. 184, June 30, 1864, 13 Stat. 325; Frederick Law Olmstead, "The Yosemite Valley and the Mariposa Big Tree Grove," in *America's National Park System: The Critical Documents*, ed. Lary M. Dilsaver (Lanham, MD: Rowman and Littlefield, 1994), 12, 22–23, 25. For an overview of the history of Yosemite Valley and the 1864 legislation, see Alfred Runte, *Yosemite: The Embattled Wilderness* (Lincoln: University of Nebraska Press, 1990), 5–44.
2. See Yellowstone Park Act of 1872, 42 Cong. Ch. 24, March 1, 1872, 17 Stat. 32, codified at 16 U.S.C. § 22. On the role of the railroads in establishing and promoting the early national parks, see Alfred Runte, *Trains of Discovery: Western Railroads and the National Parks*, 4th ed. (Boulder, CO: Roberts Rinehart, 1998).
3. On the concept of heritage tourism and its evolution in the Grand Canyon, see Hal K. Rothman, *Devil's Bargains: Tourism in the Twentieth Century West* (Lawrence: University Press of Kansas, 1998), 23–28, 50–80.
4. John Muir, *Our National Parks* (Boston: Houghton Mifflin, 1901), 2.
5. On John Muir and the Hetch Hetchy controversy, see Donald Worster, *A Passion for Nature: The Life of John Muir* (New York: Oxford University Press, 2008), 418–39, 447–53; Roderick Nash, *Wilderness and the American Mind*, 3d ed. (New Haven: Yale University Press, 1982), 161–81.
6. On the arrival and early role of automobiles in the national parks, see Rothman, *Devil's Bargains*, 143–53, 157–69; Paul Sutter, *Driven Wild: How the Fight against Automobiles Launched the Modern Wilderness Movement* (Seattle: University of Washington Press, 2002), 100–141. The Muir quotation is found in

Worster, *A Passion for Nature*, 447; the Fisher quotation is found in "Auto Use in the National Parks: Proceedings of the National Park Conference Held at the Yosemite National Park" (1912) in Dilsaver, *America's National Park System*, 42–43.

7. The pertinent sections of the National Parks Organic Act are 16 U.S.C. § 1 (conservation and enjoyment) and § 3 (leases, contracts, etc. authorized). For a compelling argument that the Organic Act does not create a dual mission for the Park Service, see Robin W. Winks, "The National Park Service Act of 1916: 'A Contradictory Mandate'?" *Denver University Law Review* 74 (1997): 575–624. See also John J. Reynolds, "Whose America? Whose Idea? Making 'America's Best Idea' Reflect New American Realities," *George Wright Forum* 27(2) (2010): 125–34 (arguing that the Park Service and its allies, drawing on the Organic Act's "shall promote" language, must focus on encouraging the nation's diverse ethnic and economic groups to visit and embrace the national parks, much like Mather and his cohorts did at the beginning of the twentieth century, to ensure the parks' political relevancy).
8. The Olmstead quotation can be found in Olmstead, "The Yosemite Valley," 20. On the Forest Service's opposition to the national park legislation, see Richard West Sellars, *Preserving Nature in the National Parks: A History* (New Haven: Yale University Press, 1997), 35–37, 57–58. On Stephen Mather, see Robert Shankland, *Steve Mather of the National Parks* (New York: Alfred A. Knopf, 1970).
9. The Lane letter and the quotations can be found in "Secretary Lane's Letter on National Park Management" in Dilsaver, *America's National Park System*, 48, 50–51. On the shift to recreational tourism, see Rothman, *Devil's Bargains*, 23–25, 143–67.
10. The Mather quotation can be found in Sellars, *Preserving Nature*, 59. On the 1925 park superintendent's auto caravan, see Sellars, *Preserving Nature*, 56. Mather's park-to-park loop tour idea is explained in Sutter, *Driven Wild*, 108–10; Shankland, *Steve Mather*, 150. Early national park resource management policies geared toward promoting visitation are described in Sellars, *Preserving Nature*, 47–90; see also chapter 8.
11. The idea that the automobile played a key role in democratizing and commercializing the national parks is developed in Sutter, *Driven Wild*, 23–27; and Rothman, *Devil's Bargains*, 143–67. See also John Ise, *Our National Park Policy: A Critical History* (Baltimore: Johns Hopkins University Press, 1960), 430 (noting that the automobile led to a demand for different visitor services than were originally available in the parks). The road mileage figure and related facts are noted in Sellars, *Preserving Nature*, 59. The joint superintendent's resolution is found in "Superintendents' Resolution on Overdevelopment" in Dilsaver, *America's National Park System*, 57.
12. The park visitation figures are found in Ise, *Our National Park Policy*, 429. On the expansion of the national park system, see Sellars, *Preserving Nature*,

136–45. Information about the CCC's contribution to the national parks can be found in Sellars, *Preserving Nature,* 140–42; and Conrad L. Wirth, *Parks, Politics, and People* (Norman: University of Oklahoma Press, 1980), 128–57; see also John C. Paige, *The Civilian Conservation Corps and the National Park Service, 1933–1942: An Administrative History* (Washington, DC: National Park Service, 1985). The Park Service's entry into national recreation area management at Lake Mead is recounted in Sellars, *Preserving Nature,* 137–39, 174–76.

13. See Ise, *Our National Park Policy,* 429; see also Janet A. McDonnell, "World War II: Defending Park Values and Resources," *Public Historian* 29(4) (2007): 15–33; Sellars, *Preserving Nature,* 150–55.
14. The visitation figures noted here can be found in Wirth, *Parks, Politics, and People,* 238. On the postwar societal changes, see Samuel P. Hays, *Beauty, Health, and Permanence: Environmental Politics in the United States, 1955–1985* (New York: Cambridge University Press, 1987), 26–34.
15. See Bernard DeVoto, "Let's Close the National Parks," in Dilsaver, *America's National Park System,* 183–89. On the Park Service's Mission 66 program, see Sellars, *Preserving Nature,* 180–85; Wirth, *Parks, Politics, and People,* 262, 266–70; Ise, *Our National Park Policy,* 546–50.
16. The Mission 66 figures are from Wirth, *Parks, Politics, and People,* 262, 266–71; see also Sellars, *Preserving Nature,* 184. For a firsthand description of the Mission 66 program, see "Mission 66 Special Presentation to President Eisenhower and the Cabinet by Director Conrad Wirth" (Jan. 27, 1956) in Dilsaver, *America's National Park System,* 193–96. On the "industrial tourism" idea, see Edward Abbey, *Desert Solitaire: A Season in the Wilderness* (New York: Simon and Schuster, 1968), 39–59; see also Susan Sessions Rugh, "Branding Utah: Industrial Tourism in the Postwar American West," *Western History Quarterly* 37(4) (2006): 445.
17. On criticism of the Mission 66 program, see Sellars, *Preserving Nature,* 185–87; John C. Miles, *Guardians of the Parks: A History of the National Parks and Conservation Association* (Washington, DC: Taylor and Francis, 1995), 187–94; Michael Frome, *Regreening the National Parks* (Tucson: University of Arizona Press, 1992), 64–65, 212. On the emergence of the wilderness movement, passage of the Wilderness Act, and its effect on the Park Service, see Sellars, *Preserving Nature,* 191–94, 213–14; John C. Miles, *Wilderness in National Parks: Playground or Preserve* (Seattle: University of Washington Press, 2009), 123–58.
18. The Leopold report can be found in A.S. Leopold, S.A. Cain, C.M. Cottam, I.N. Gabrielson, and T.L. Kimball, "Wildlife Management in the National Parks" (Mar. 4, 1963), in Dilsaver, *America's National Park System,* 237–51, with the noted quotation on 242; the Udall quotation is found in "Secretary Udall's Letter on National Park Management" (July 10, 1964) in Dilsaver, *America's National Park System,* 274.

19. The pertinent National Environmental Policy Act EIS provision is found at 42 U.S.C. § 4332(2)(C); the Endangered Species Act is found at 16 U.S.C. §§ 1531–43. See Conservation Foundation, *National Parks for the Future* (Washington, DC: 1972), 12; National Park Service, "State of the Parks, May 1980: A Report to Congress," partially reprinted in Dilsaver, *America's National Park System*, 405–7.
20. On Watt's policy priorities, see "Secretary Watt's Letter on National Park Management" (July 6, 1981) in Dilsaver, *America's National Park System*, 411. The Fishing Bridge controversy is described and analyzed in chapter 5.
21. The quotations are found in "National Parks for the 21st Century: The Vail Agenda" in Dilsaver, *America's National Park System*, 436.
22. On park visitor numbers, see https://irma.nps.gov/Stats/. On park visitor trends, see Oliver R.W. Pergams and Patricia A. Zaradic, "Evidence for a Fundamental and Pervasive Shift Away from Nature-Based Recreation," *Proceedings of the National Academy of Sciences* 105 (2007): 2295–2300; chapter 10 (discussing the nature deficit disorder syndrome). On the Zion Canyon and Bryce Canyon shuttle bus systems, see www.nps.gov/zion/planyourvisit/shuttle-system.htm and www.nps.gov/brca/planyourvisit/shuttle.htm, respectively; on the Grand Canyon south rim railroad line proposal, see Blaine Harden, "At Grand Canyon, No Way to Run a Railroad," *New York Times*, Jan. 28, 2002; on the Yosemite Valley plan, see "The Yosemite Valley Controversy" in this chapter.
23. On Las Vegas–based air tour flights over Grand Canyon National Park, see chapter 2. The subject of gateway communities and the national parks is discussed and analyzed in chapter 5; see also Rothman, *Devil's Bargains*, 162–67.
24. The Muir quotations appear in National Park Service, *A Sense of Place: Design Guidelines for Yosemite Valley* (2004), 11.
25. For a more detailed description of the state's early management challenges in Yosemite Valley, including its retrocession to the federal government, see Runte, *Yosemite*, 13–86; the early visitation figures are from page 38. The Muir quotation can be found in National Park Service, *A Sense of Place*, 38.
26. See generally Runte, *Yosemite*, 119–59; the visitation figures are found on pages 121 and 143; the 1937 quotation is found on page 144.
27. See generally Runte, *Yosemite*, 143–59; Sellars, *Preserving Nature*, 62–63.
28. On the Park Service during the Great Depression and World War II, see generally Sellars, *Preserving Nature*, 140–64; on the CCC in Yosemite, see Runte, *Yosemite*, 173; on Drury's postwar development concerns, see Runte, *Yosemite*, 189–92.
29. The Adams quotations appear in Runte, *Yosemite*, 186; the Tioga Pass road controversy is described on pages 192–97.
30. See generally Runte, *Yosemite*, 201–6.
31. The quoted passage from the plan is found in William R. Lowry, *Repairing Paradise: The Restoration of Nature in America's National Parks* (Washington,

DC: Brookings Institution Press, 2009), 72. Lowry's book contains a detailed description and analysis of the Yosemite Valley controversy, focusing on efforts to remove autos from the valley to restore more natural conditions. See Lowry, *Repairing Paradise*, 63–106.

32. For a more detailed description of the Merced River commercial rafting matter, see Runte, *Yosemite*, 213–16; for more on the jail and courthouse construction, see Runte, *Yosemite*, 225.
33. The Wild and Scenic Rivers Act of 1968 is found at 16 U.S.C. §§ 1271–87; for the comprehensive management plan provision, with the "user capacity" language, see 16 U.S.C. § 1274(d)(1). The statutory designation process as well as planning and management requirements are explained in Friends of Yosemite Valley v. Kempthorne, 520 F.3d 1024, 1027–29 (9th Cir. 2008).
34. The district court decision ordering the Park Service to prepare and adopt a Merced River plan within the next twelve months is found at Sierra Club v. Babbitt, 69 F. Supp. 2d 1202 (E.D. Cal. 1999). The relationship between the park's general management plan, Merced River plan, and Yosemite Valley plan is described in National Park Service, *Merced Wild and Scenic River Revised Comprehensive Management Plan and Supplemental Environmental Impact Statement I-7 to I-10* (2005).
35. See National Park Service, *Merced Wild and Scenic River Comprehensive Management Plan* (2000); National Park Service, *Merced Wild and Scenic River Revised Comprehensive Management Plan and Supplemental Environmental Impact Statement* (2005); National Park Service, *Final Yosemite Valley Plan Supplemental Environmental Impact Statement* (2000).
36. See Lowry, *Repairing Paradise*, 77–78, 87, 101.
37. *Friends of Yosemite Valley*, 520 F.3d 1024, 1033n2, affirming, Friends of Yosemite Valley v. Scarlett, 439 F. Supp. 2d 1074 (E.D. Cal. 2006).
38. *Friends of Yosemite Valley*, 520 F.3d 1024, 1035n5.
39. The other court decisions addressing the Merced River plan include Friends of Yosemite Valley v. Norton, 348 F.3d 789 (9th Cir. 2003), affirming and reversing in part, 194 F. Supp. 2d 1066 (E.D. Cal. 2002).
40. The "heart of the park" characterization is from Runte, *Yosemite*, 217. News reports suggest that Yosemite officials remain reluctant to limit visitor numbers. See Laura Petersen, "Development, Visitor Capacity Frame Debate over Merced River Management Plan," *Land Letter*, Dec. 15, 2011. On the Park Service's incremental approach to restoring Yosemite Valley, see Lowry, *Repairing Paradise*.
41. On the Zion Canyon shuttle system, see National Park Service, "Zion Canyon Transportation System Background Information and Introduction," in *Zion Canyon Transportation System Technical Analysis, Executive Summary* (2008). Transportation planning for Grand Canyon's South Rim is described in National Park Service, *South Rim Visitor Transportation Plan Environmental Assessment/Assessment of Effect* (2008), 10–12.

42. See Noelle Straub, "Settlement to End Great Smoky Mountains 'Road to Nowhere,'" *Land Letter*, Feb. 4, 2010. For background on the Fontana Lake road controversy, see Michael Frome, *Battle for the Wilderness*, rev. ed. (Salt Lake City: University of Utah Press, 1997), 178–81.
43. The Everglades case is found at Organized Fishermen of Florida v. Watt, 590 F. Supp. 805 (S.D. Fla. 1984), affirmed, 775 F.2d 1544 (11th Cir. 1985). See also Mausolf v. Babbitt, 125 F.3d 661 (8th Cir. 1997) (sustaining a Voyageurs National Park decision closing snowmobile trails to protect denning wolves).
44. National Park Service, *Management Policies* 8.2, 8.2.1 (2006). The policies also direct superintendents to use available research and studies in making their written determination about adverse effects to park resources or values.

Chapter 4

1. The quotations are found, respectively, in John Muir, *Our National Parks* (Boston: Houghton Mifflin, 1901), 2; Alfred Runte, *Trains of Discovery: Western Railroads and the National Parks*, 4th ed. (Boulder, CO: Roberts Rinehart, 1998), 48, 70; Enos A. Mills, *Your National Parks* (Boston: Houghton Mifflin, 1917), 379.
2. For an account of the legislative history of the National Park Service Act, see Robin W. Winks, "The National Park Service Act of 1916: 'A Contradictory Mandate'?" *Denver University Law Review* 74 (1997): 583–611. The so-called Lane letter is found in "Secretary Lane's Letter on National Park Management" (May 13, 1918) in *America's National Park System: The Critical Documents*, ed. Lary M. Dilsaver (Lanham, MD: Rowman and Littlefield, 1994), 48.
3. The Lane letter expressly included "motoring" as a favored form of national park recreation. See Dilsaver, *America's National Park System*, 49. On Mather's endorsement of golf, tennis, and ski areas in the parks, see Richard West Sellars, *Preserving Nature in the National Parks: A History* (New Haven: Yale University Press, 1997), 62–63.
4. Henry S. Graves, "A Crisis in National Recreation," *American Forestry* 26 (1920): 391–400.
5. See Parks, Parkway, and Recreation Act of 1936, ch. 735 § 1, 49 Stat. 1894 (1936), codified at 16 U.S.C. §§ 17(k)–(n). For the 1941 report, see National Park Service, "A Study of the Park and Recreation Problem of the United States," in Dilsaver, *America's National Park System*, 151.
6. See, e.g., 16 U.S.C. § 460n-3 (Lake Mead National Recreation Area legislation); see also Sellars, *Preserving Nature*, 137–39 (describing the Park Service's early role in what became the Lake Mead National Recreation Area); Cameron Binkley, *The Creation and Establishment of Cape Hatteras National Seashore: The Great Depression through Mission 66* (Washington, DC: National Park Service, 2007), 32–41.
7. See Sellars, *Preserving Nature*, 282.

8. The Lake Mead and Cape Hatteras legislation is found at 16 U.S.C. § 460n et seq. (Lake Mead) and § 459a-1 et seq. (Cape Hatteras). On the origins of the national recreation area concept, see Sellars, *Preserving Nature,* 138–39; on opposition to the Park Service's move into national recreation area management, see Sellars, *Preserving Nature,* 174–77; Ronald A. Foresta, *America's National Parks and Their Keepers* (Washington, DC: Resources for the Future, 1984), 46–47.
9. On the postwar recreation boom, see Samuel T. Dana and Sally K. Fairfax, *Forest and Range Policy: Its Development in the United States,* 2d ed. (New York: McGraw-Hill, 1980), 190–99; Jan G. Laitos and Thomas A. Carr, "The Transformation on Public Lands," *Ecology Law Quarterly* 26 (1999): 161; on the postwar societal changes, see generally Samuel P. Hays, *Beauty, Health, and Permanence: Environmental Politics in the United States, 1955–1985* (New York: Cambridge University Press, 1987).
10. See U.S. Outdoor Recreation Resources Review Commission, *Outdoor Recreation for America* (Washington, DC: Government Printing Office, 1962); Robin W. Winks, *Laurance S. Rockefeller: Catalyst for Conservation* (Washington, DC: Island Press, 1997), 121–39; see also Dana and Fairfax, *Forest and Range Policy,* 196–97; Sellars, *Preserving Nature,* 194–95; Foresta, *America's National Parks,* 62–63. The Land and Water Conservation Fund Act of 1964 is found at 16 U.S.C. § 460l-4 et seq.
11. On Mission 66, see Sellars, *Preserving Nature,* 180–85; Conrad L. Wirth, *Parks, Politics, and People* (Norman: University of Oklahoma Press, 1980), 262, 266–70; John Ise, *Our National Park Policy: A Critical History* (Baltimore: Johns Hopkins University Press, 1960), 546–50; see also chapter 2. The park system expansion figures are from Dana and Fairfax, *Forest and Range Policy,* 216. The General Authorities Act is found at Pub. L. 91-383, 84 Stat. 825 (1970), codified at 16 U.S.C. § 1a-1 et seq.
12. See Sellars, *Preserving Nature,* 143; Conservation Foundation, *National Parks for the Future,* 2, 32, partially reprinted in "Preservation of National Park Values" in Dilsaver, *America's National Park System,* 386. For the White speech, see John R. White, "Atmosphere in the National Parks," in Dilsaver, *America's National Park System,* 142.
13. See *America's National Park System: The Critical Documents,* ed. Lary M. Dilsaver (Lanham, MD: Rowman and Littlefield, 1994), 198; "A Backcountry Management Plan for Sequoia and Kings Canyon National Parks" in Dilsaver, *America's National Park System,* 211; Lary M. Dilsaver and William C. Tweed, *Challenge of the Big Trees: A Resource History of Sequoia and Kings Canyon National Parks* (Three Rivers, CA: Sequoia Natural History Association, 1990), 264–78; Mitchel P. McLaran, "Recreational Pack Stock Management in Sequoia and Kings Canyon National Parks," *Rangelands* 11(1) (1989): 3–8.
14. On the Grand Canyon rafting permit controversy, see Roderick Nash, *Wilderness and the American Mind,* 3d ed. (New Haven: Yale University Press,

1982), 329–39; Wilderness Public Rights Fund v. Kleppe, 608 F.2d 1250 (9th Cir. 1979); River Runners for Wilderness v. Martin, 593 F.3d 1064 (9th Cir. 2010); see also chapter 2 for a more detailed discussion of this controversy.

15. On recreational development in Yosemite, see Alfred Runte, *Yosemite: The Embattled Wilderness* (Lincoln: University of Nebraska Press, 1990), 152–53. The Sequoia superintendent's statement is found in John R. White, "Atmosphere in the National Parks," in Dilsaver, *America's National Park System*, 142.
16. See Conservation Foundation, *National Parks for the Future* (Washington, DC: 1972), 21–22, 32; Nash, *Wilderness*, 328–29; John C. Miles, *Guardians of the Parks: A History of the National Parks and Conservation Association* (Washington, DC: Taylor and Francis, 1995), 187–94.
17. The 1918 Lane letter included "motoring" as a preferred recreational activity in the new national parks. See "Secretary Lane's Letter," 50. For an excellent analysis of the appropriate role for recreation in the national parks, see Joseph L. Sax, *Mountains without Handrails: Reflections on the National Parks* (Ann Arbor: University of Michigan Press, 1980).
18. National Park Service, *Management Policies* 8.1.1 (2006) and at 1.5 (characterizing "appropriate uses" as "often associated with the inspirational qualities of the parks . . . [and] uniquely suited to the superlative natural and cultural resources found in the parks and that (1) foster an understanding of and appreciation for park resources and values, or (2) promote enjoyment through a direct association with, interaction with, or relation to park resources").
19. See National Park Service, *Management Policies* 8.1, 8.2 (2006); see also Robert E. Manning, *Parks and Carrying Capacity: Commons without Tragedy* (Washington, DC: Island Press, 2007).
20. 36 C.F.R. § 1.5; see also 36 C.F.R. § 7.1 et seq. (setting forth individual park rules).
21. The noted cases are, respectively, National Rifle Association v. Potter, 628 F. Supp. 903, 909 (D.D.C. 1986); Fund for Animals v. Norton, 294 F. Supp. 2d 92, 105 (D.D.C. 2003); Bicycle Trails Council of Marin v. Babbitt, 82 F.3d 1445, 1453 (9th Cir. 1996); Mausolf v. Babbitt, 125 F.3d 661 (8th Cir. 1997); see also Greater Yellowstone Coalition v. Kempthorne, 577 F. Supp. 2d 183 (D.D.C. 2008); Southern Utah Wilderness Alliance v. National Park Service, 387 F. Supp. 2d 1178 (D. Utah 2005). But see Voyageurs National Park Association v. Norton, 381 F.3d 759 (8th Cir. 2004) (sustaining the park superintendent's decision reopening several areas to snowmobiles).
22. On the contrasting Glacier and Yellowstone snowmobile policies, see Michael J. Yochim, "Snow Machines in the Gardens: The History of Snowmobiles in Glacier and Yellowstone National Parks," *Montana Western History Magazine* (Autumn 2003): 2.
23. See National Park Service's Draft Management Policies, Hearing before the Subcommittee on National Parks of the Committee on Energy and Natural

Resources, U.S. Senate, 109th Cong., 1st Sess., Sen. Hearing 109-313, Nov. 1, 2005; Richard West Sellars, *Preserving Nature in the National Parks*, 2d ed. (New Haven: Yale University Press, 2010), 301–6; Dan Berman, "Recreation Industry Makes Final Push to Change NPS Rules," *Land Letter*, July 20, 2006. The Jarvis' quotation is found in Sellars, *Preserving Nature*, 2d ed., 303.

24. On the need for a coordinated national recreation policy, see U.S. Outdoor Recreation Resources Review Commission, *Outdoor Recreation for America*, 124–26; Public Land Law Review Commission, *One Third of the Nation's Land: A Report to the President and to Congress* (Washington, DC: Government Printing Office, 1970), 197–216; President's Commission on Americans Outdoors, *Americans Outdoors: The Legacy, The Challenge* (Washington, DC: Island Press, 1987), 189–98.
25. For a history of the Yellowstone snowmobile controversy, see Michael J. Yochim, *Yellowstone and the Snowmobile: Locking Horns over National Park Use* (Lawrence: University Press of Kansas, 2009); Michael J. Yochim, "The Development of Snowmobile Policy in Yellowstone National Park," *Yellowstone Science* 7(2) (1999): 2–10.
26. The presidential executive orders are Executive Order 11644 (1972) (calling on the federal land management agencies to zone their lands for off-road vehicle travel) and Executive Order 11989 (1977) (requiring federal land management agencies to ban off-road vehicle travel that causes "considerable adverse effects"). For a comparison of the Yellowstone and Glacier snowmobile experiences, see Yochim, "Snow Machines."
27. The Yellowstone bison management controversy is examined in chapter 7. The early bison litigation is described in Robert B. Keiter and Peter H. Froelicher, "Bison, Brucellosis, and Law in the Greater Yellowstone Ecosystem," *Land and Water Law Review* 28 (1993): 32–38.
28. National Park Service, Record of Decision; Winter Use Plans for the Yellowstone and Grand Teton National Parks and John D. Rockefeller Jr., Memorial Parkway, 65 Fed. Reg. 80908, 80917 (Dec. 22, 2000) (hereinafter NPS 2000 Winter Use ROD).
29. See National Park Service, *Winter Use Plans for the Yellowstone and Grand Teton National Parks and John D. Rockefeller, Jr., Memorial Parkway: Final Environmental Impact Statement* (2000); and NPS 2000 Winter Use ROD.
30. National Park Service, *Winter Use Plans: Final Supplemental Winter Use Plans for the Yellowstone and Grand Teton National Parks and John D. Rockefeller, Jr., Memorial Parkway* (2003); Department of the Interior, National Park Service, Winter Use Plans Final Rule for the Yellowstone and Grand Teton National Parks and John D. Rockefeller Jr., Memorial Parkway, 68 Fed. Reg. 69268 (Dec. 11, 2003) (hereinafter NPS 2003 Winter Use Final Rule).
31. Fund for Animals v. Norton, 294 F. Supp. 2d 92, 108n22, 109n14, 116 (D.D.C. 2003).
32. See International Snowmobile Manufacturers Association v. Norton, 304

F. Supp. 2d 1278 (D. Wyo. 2004); Department of the Interior, National Park Service, Special Regulations, Areas of the National Park System, 69 Fed. Reg. 65,348 (Nov. 10, 2004), codified at 36 C.F.R. part 7. Although the 2004 interim rule allowed eight hundred snowmobiles in the park daily, it was challenged unsuccessfully by Wyoming business interests. Wyoming Lodging and Restaurant Association v. U.S. Dept. of the Interior, 398 F. Supp. 2d 1197 (D. Wyo. 2005). For an analysis of these initial snowmobile court rulings, see Joanna M. Hooper, "Blowing Snow: The National Park Service's Disregard for Science, Law, and Public Opinion in Regulating Snowmobiling in Yellowstone National Park," *Environmental Law Reporter* 34 (2004): 10975.

33. *Greater Yellowstone Coalition v. Kempthorne*, 577 F. Supp. 2d 183.
34. See Wyoming v. U.S. Dept. of the Interior, 587 F.3d 1245, 1249–51 (10th Cir. 2009) (describing the Wyoming federal district court's 2008 decision and finding the case moot).
35. See, e.g., National Parks Snowmobile Restrictions Act, H.R. 1465, 107th Cong., 1st Sess. (2001); Yellowstone Protection Act, S. 965, 108th Cong., 1st Sess. (2003); National Park Service Winter Access Act, S. 365, 107th Cong., 1st Sess. (2001).
36. National Park Service, Yellowstone National Park, *Supplemental Winter Use Plan/Environmental Impact Statement* (2012), www.nps.gov/yell/parkmgmt/currentmgmt.htm.
37. See Hooper, "Blowing Snow," 10975n7.
38. See generally Phil Taylor, "Wyoming Governor Urges More Winter Access to Yellowstone," *Land Letter*, July 21, 2011. On the Sylvan Pass controversy, see Mark Heinz, "Groups Urge Winter Closure of Sylvan Pass," *Cody Enterprise*, July 21, 2011; Eric Bontrager, "Study Group Suggests Keeping Open Yellowstone Snowmobile Route," *Land Letter*, June 5, 2008. On the plethora of winter recreation options in the Greater Yellowstone area, see Greater Yellowstone Coordinating Committee, *Winter Visitor Use Management: A Multi-Agency Assessment* (1999).
39. See U.S. Government Accountability Office, *All Terrain Vehicles: How They Are Used, Crashes, and Sales of Adult-Sized Vehicles for Children's Use* (Washington, DC: 2010), 53; Bluewater Network, *Off-the-Track: America's National Parks under Siege* (1999), 19; see generally David G. Havlick, *No Place Distant: Roads and Motorized Recreation on America's Public Lands* (Washington, DC: Island Press, 2002), 84–128; U.S. Forest Service, *Off-Highway Vehicle Recreation in the United States and Its Regions and States* (2008).
40. Southern Utah Wilderness Alliance v. National Park Service, 387 F. Supp. 2d 1178, 1192 (D. Utah 2005); see also Southern Utah Wilderness Alliance v. National Park Service, 7 F. Supp. 2d 1205 (D. Utah 1998), reversed and remanded, 222 F.3d 819 (10th Cir. 2000). And see San Juan County, Utah v. United States, 2011 WL 2144762 (D. Utah 2011), rejecting the county's RS 2477 claim that the Salt Creek route is a county-owned road.

41. See Conservation Law Foundation v. Secretary of the Interior, 864 F.2d 954 (1st Cir. 1989); Cape Hatteras Preservation Alliance v. U.S. Dept. of the Interior, 344 F. Supp. 2d 108 (D.D.C. 2004), 731 F. Supp. 2d 15 (D.D.C. 2010); U.S. v. Matei, 2007 WL 2079874 (criminal conviction for operating a vehicle without due care on Cape Hatteras National Seashore).
42. See 36 C.F.R. §§ 212.55–57, §§ 261.55–56 (national forest road designation and use rules); 43 C.F.R. §§ 8340, 9268.3 (BLM off-road vehicle regulations). See generally John C. Adams and Stephen F. McCool, "Finite Recreation Opportunities: The Forest Service, the Bureau of Land Management, and Off-Road Vehicle Access," *Natural Resources Journal* 49 (2009): 45.
43. For many water-based park units, the use of personal watercraft (PWCs) has provoked controversy and raised motorized recreation concerns similar to those involving OHVs. As a general matter, the Park Service has by rule prohibited PWCs in most parks, but has allowed twenty-one individual parks to permit and specially regulate their use. 36 C.F.R. § 3.9. See Bluewater Network v. Salazar, 721 F. Supp. 2d 7 (D.D.C. 2010); M. Steven O'Neill, "The Appropriate Use and Enjoyment of National Parks: Personal Watercraft and the Organic Act of 1916's 'Enjoyment' Mandate," *Temple Political and Civil Rights Law Review* 21 (2011): 245.
44. For the most part, the Park Service enjoys immunity from tort liability for recreational accidents that occur in the national parks, which means that liability concerns and damage award costs are generally not principal concerns for agency officials when establishing recreation policy. The Federal Tort Claims Act grants federal agencies and employees immunity from liability when engaged in a discretionary function or duty, which would include most recreational policy decisions. The primary exception to this rule is when an agency fails to warn visitors about a known but not easily detectable hazard. See Johnson v. United States, 949 F.2d 332 (10th Cir. 1991); Tippett v. United States, 108 F.3d 1194 (10th Cir. 1997).
45. See Bicycle Trails of Marin v. Babbitt, 82 F.3d 1445 (9th Cir. 1996). For court cases sustaining convictions for illegal base jumping in national parks, see U.S. v. Albers, 226 F.3d 989 (9th Cir. 2000); U.S. v. Oxx, 127 F.3d 1277 (10th Cir. 1997). The new Park Service mountain biking rule is found at 36 C.F.R. § 4.30 (2012); see also National Park Service, Vehicles and Traffic Safety—Bicycles, 77 Fed. Reg. 39927 (July 6, 2012).
46. Yosemite's climbing regulations are at www.nps.gov/yose/planyourvisit/climbing.htm; Joshua Tree's fixed-anchor regulations are at www.nps.gov/jotr/planyourvisit/fixed_anchors.htm. See also Timothy Dolan, "Fixed Anchors and the Wilderness Act: Is the Adventure Over?" *University of San Francisco Law Review* 35 (2000): 355.
47. Indeed, controversy has recently flared over Park Service proposals to outlaw commercial guided climbs in the Black Canyon of the Gunnison and to further limit the number of daily hikers allowed to climb Yosemite's Half

Dome, where a permit system is already in place. See Jason Blevins, "Park Service Balks at Banning Guided Climbs in Colorado's Black Canyon," *Denver Post*, Apr. 19, 2012; Scott Streater, "Plan Limiting Access to Popular Yosemite Climb Sparks Dispute," *Greenwire*, Mar. 15, 2012.

48. See Michael J. Yochim, "Kayaking Playground or Nature Preserve? Whitewater Boating Conflicts in Yellowstone National Park," *Montana: The Magazine of Western History* 55(1) (2005): 52–64.
49. See Scott Streater, "NPS Pressured to Bend Rules for Colo. Monument Bike Race," *Land Letter*, Mar. 17, 2011; Scott Streater, "Pro Cycling Race Barred from Colorado National Monument," *Land Letter*, Mar. 24, 2011; Scott Streater, "Former Colo. Monument Chief Recognized for 'Courage' in Political Storm over Bike Race," *Land Letter*, Oct. 20, 2011. Not surprisingly, given the local political pressures, the issue of whether to permit the bike race inside the national monument has resurfaced, and the Park Service appears willing to revisit the matter. Nancy Lofholm, "Colorado National Monument 'Listening' Process Could Create U.S. Model," *Denver Post* (November 12, 2012).
50. For cases in which Endangered Species Act and other claims have been made, see *Voyageurs National Park Association*, 381 F.3d 759 (8th Cir. 2004) (rejecting an ESA consultation claim); *Cape Hatteras Access Preservation Alliance*, 344 F. Supp. 2d 108 (D.D.C. 2004) (finding flaws in the U.S. Fish and Wildlife Service's critical habitat designation); *Sierra Club v. Norton*, 74 Fed. Appx. 376 (5th Cir. 2003) (discussing the Park Service's obligation to consult with the U.S. Fish and Wildlife Service regarding oil and gas development at Padre Island National Seashore). For a Wilderness Act case, see *High Sierra Hikers Ass'n v. U.S. Dept. of the Interior*, 848 F. Supp. 2d 1036 (N.D. Cal. 2012) (finding a Wilderness Act violation when Sequoia National Park officials continued to allow horses into the park's wilderness areas without a finding of necessity).
51. Sax, *Mountains without Handrails*; National Park Service, *Management Policies* 1.4.7, 4.9, 4.10 (2006).
52. On the Forest Service and the development of ski areas and related real estate development, see Hal K. Rothman, *Devil's Bargains: Tourism in the Twentieth-Century West* (Lawrence: University Press of Kansas, 1998), 168–286. For more on the relationship between Rocky Mountain National Park and Estes Park, see chapter 5; for more on the relationship between Glacier National Park and Flathead Valley communities, see chapter 9.
53. The Lane letter quotation can be found in "Secretary Lane's Letter," 48. The 1970 and 1978 Organic Act amendments are set forth and analyzed in National Rifle Association v. Potter, 628 F. Supp. 903, 905–6 (D.D.C. 1986). On Forest Service and BLM national OHV policy, see Adams and McCool, "Finite Recreation Opportunities."

Chapter 5

1. On the early history of lodging and concessions in Yosemite, see Alfred Runte, *Yosemite: The Embattled Wilderness* (Lincoln: University of Nebraska Press, 1990), 16–27, 28–30. For the Yosemite legislation, see "An Act Authorizing a Grant to the State of California of the 'Yo-Semite Valley,' and of the Land Embracing the 'Mariposa Big Tree Grove' " (13 Stat. 325; 1864) in *America's National Park System: The Critical Documents*, ed. Lary M. Dilsaver (Lanham. MD: Rowman and Littlefield, 1994), 11.
2. The quoted Yellowstone legislative provision is found at 17 Stat. 32 (1872), 16 U.S.C. § 22. On the early history of concessions in Yellowstone, see Mary Shivers Culpin, *A History of Concession Development in Yellowstone National Park, 1872–1966* (Yellowstone National Park, WY: Yellowstone Center for Resources, 2003), 9–32; see also Paul Schullery, *Searching for Yellowstone: Ecology and Wonder in the Last Wilderness* (New York: Houghton Mifflin, 1997), 89–105. Although these early concessioners were permitted to cut park timber to construct the new lodges, they were prohibited from building near the park's geysers and other natural features as well as from interfering with visitors' access to these features.
3. See Hal K. Rothman, *Devil's Bargains: Tourism in the Twentieth-Century American West* (Lawrence: University Press of Kansas, 1998), 50–80; Donald H. Robinson, *Through the Years in Glacier National Park: An Administrative History* (West Glacier, MT: Glacier National History Association, 1960), chapter 3; Michael F. Anderson, *Polishing the Jewel: An Administrative History of Grand Canyon National Park* (Grand Canyon, AZ: Grand Canyon Association, 2000), 1–10, www.nps.gov/grca/historyculture/adhigrca.htm.
4. On the 1911 national park conference, see Culpin, *History of Concession Development*, 153; on the origins of *Sunset* magazine, see Michael Frome, *Regreening the National Parks* (Tucson: University of Arizona Press, 1992), 179.
5. On Mather and his leadership of the National Park Service, see Robert Shankland, *Steve Mather of the National Parks* (New York: Alfred A. Knopf, 1970). The story of Mather's hiring as the first director is recounted on pages 7–11. The statutory provision on concessions is found at 39 Stat. 535 (1916), codified at 16 U.S.C. § 3. In addition, the secretary was required to approve the transfer of any lease or contract, and he could authorize the lessees to execute mortgages if the purpose was to finance new facilities or improvements designed for public accommodations. 16 U.S.C. § 3. The "promotion" language is found in the Organic Act, 16 U.S.C. § 1, which directs the Park Service to "promote and regulate the use of the . . . national parks." See John J. Reynolds, "Whose America? Whose Idea? Making 'America's Best Idea' Reflect New American Realities," *George Wright Forum* 27 (2010), 125.
6. On the early idea that national parks were engines for local economic growth, see Theodore Catton, *Wonderland: An Administrative History of Mount*

Rainier National Park (Washington, DC: National Park Service, 1996), chapter 9 (quoting from Director Stephen Mather's 1925 annual report). The Lane letter is found in "Secretary Lane's Letter on National Park Management" (May 13, 1918), in Dilsaver, *America's National Park System*, 48.

7. On the origins of the regulated monopoly idea, see Culpin, *History of Concession Development*, 58. For an extensive critique of the regulated monopoly-like features of the concession system, see George Cameron Coggins and Robert L. Glicksman, "Concessions Law and Policy in the National Park System," *Denver University Law Review* 74 (1979): 729, 739–59.
8. The transport incident is recounted in Lloyd K. Musselman, *Rocky Mountain National Park: Administrative History, 1915–1965* (Washington, DC: National Park Service, 1971), chapter 3. The Mount Rainier park hotel incident is described in Catton, *Wonderland*, chapter 1; Shankland, *Steve Mather*, 77–78.
9. On the effect of the automobile on park visitation, railroad travel, and local businesses, see John Ise, *Our National Park Policy: A Critical History* (Baltimore: Johns Hopkins University Press, 1961), 609; Paul S. Sutter, *Driven Wild: How the Fight Against Automobiles Launched the Modern Wilderness Movement* (Seattle: University of Washington Press, 2002), 100–141; Rothman, *Devil's Bargains*, 143–67.
10. The White's City saga is recounted in Rothman, *Devil's Bargains*, 162–64.
11. See John R. White, "Atmosphere in the National Parks," reprinted in Dilsaver, *America's National Park System*, 142, 147.
12. See Culpin, *History of Concession Development*, 81–83; the Ickes' speech is quoted on page 81.
13. The Chapman quotation can be found in Culpin, *History of Concession Development*, 98–99. Wirth explained his Mission 66 program in a Jan. 1956, cabinet-level presentation that can be found in Dilsaver, *America's National Park System*, 194.
14. On Yosemite's Tioga Pass Road project, see Runte, *Yosemite*, 192–97; on Wyoming's Yellowstone concession legislation, see Aubrey L. Haines, *The Yellowstone Story: A History of Our First National Park* (Boulder: University Press of Colorado, 1996), 2:373; Richard A. Bartlett, *Yellowstone: A Wilderness Besieged* (Tucson: University of Arizona Press, 1985), 369; and on the Everglades' Flamingo project, see Frome, *Regreening the National Parks*, 183. The concluding quotation, although referring to Yellowstone, can readily be applied across the park system; it is found in Schullery, *Searching for Yellowstone*, 179.
15. For the Concession Policy Act of 1965, see Pub. L. 89-249, 79 Stat. 969 (1965), codified at 16 U.S.C. § 20-20(g), but repealed by Congress in the National Parks Omnibus Management Act of 1998, Pub. L. 105-391 § 415(a), 112 Stat. 3497, 3515 (1998), codified at 16 U.S.C. §§ 5951–66. For a discussion of the legislative history underlying the Concession Policy Act, see Lake Mojave Boat Owners Association v. National Park Service, 78 F.2d 1360, 1366–67 (9th Cir. 1996).

16. For a comprehensive discussion and analysis of the 1965 Concessions Policy Act, see George Cameron Coggins and Robert L. Glicksman, "Concessions Law and Policy in the National Park System," *Denver University Law Review* 74 (1997): 729. The Yellowstone Park Company buyout is described in Conservation Foundation, *National Parks for a New Generation: Visions, Realities, Prospects* (Washington, DC: 1985), 179–80, which notes that some Park Service officials believed General Host's "possessory interest" was worth less than $5 million. See also Bartlett, *Yellowstone*, 366–75.
17. On the Yosemite development proposals, see Frome, *Regreening the National Parks*, 191; Runte, *Yosemite*, 202–5. The other report conclusions and recommendations are found in Conservation Foundation, *National Parks for the Future* (Washington, DC, 1972), 21–22; General Accounting Office, *Concession Operations in the National Parks—Improvements Needed in Administration* (Washington, DC: 1975), 9, www.gao.gov/assets/120/115124.pdf. The "industrial tourism" metaphor is found in Edward Abbey, *Desert Solitaire: A Season in the Wilderness* (New York: Simon and Schuster, 1990), 39–59, and the "conglomerate concessionaires" phrase comes from Frome, *Regreening the National Parks*, 192.
18. See Conservation Foundation, *National Parks*, 12, 46, 86–91; National Park Service, *State of the Parks* (Washington, DC: 1980); Dale A. Hudson, "Sierra Club v. Dept. of the Interior: The Fight to Preserve Redwood National Park," *Ecology Law Quarterly* 7 (1978): 781; Robert B. Keiter, "On Protecting the National Parks from the External Threats Dilemma," *Land and Water Law Review* 20 (1985): 355.
19. For the Watt memorandum, see "Secretary Watt's Letter on National Park Management" (July 6, 1981), in Dilsaver, *America's National Park System*, 411, 413. The Fishing Bridge controversy is recounted later in this chapter; see also Schullery, *Searching for Yellowstone*, 86–91; Frome, *Regreening the National Parks*, 187. James Watt's tenure as secretary of the Interior and his park policies, as well as those of his successors during the Reagan years, are described and critiqued in William J. Lockhart, "External Park Threats and Interior's Limits: The Need for an Independent Park Service," in *Our Common Lands: Defending the National Parks*, ed. David Simon (Washington, DC: Island Press, 1988), 7–24; see also Bartlett, *Yellowstone*, 377–78; Frome, *Regreening the National Parks*, 175.
20. See Conservation Foundation, *National Parks*, 180–96; see also U.S. Congress, *National Park Service Policies Discourage Competition, Give Concessioners Too Great a Voice in Concession Management*, H.R. Rpt. No. 869, 94th Cong., 2d Sess. (1976).
21. The quotations are from "National Parks for the 21st Century: The Vail Agenda" (1992), partially reprinted in Dilsaver, *America's National Park System*, 440, 441. The case for refocusing local economic activity toward amenities based on the presence of national parks is made in Raymond Rasker, "A

New Look at Old Vistas: The Economic Role of Environmental Quality in Western Public Lands," *University of Colorado Law Review* 65 (1994): 369; Jim Howe, Ed McMahon, and Luther Propst, *Balancing Nature and Commerce in Gateway Communities* (Washington, DC: Island Press, 1997); National Parks Conservation Association, *Gateway to Glacier: The Emerging Economy of Flathead County* (Washington, DC: 2003); Sonoran Institute, *Prosperity in the 21st Century West: The Role of Protected Public Lands* (Tucson, AZ: 2004); National Parks Conservation Association, *Gateways to Yellowstone: Protecting the Wild Heart of Our Region's Thriving Economy* (Washington, DC: 2006); Dennis J. Stynes, *Economic Benefits to Local Communities from National Park Visitation and Payroll* (Washington, DC: National Park Service, 2011); National Parks Conservation Association, *Landscapes of Opportunity: The Economic Influence of Southeast Utah's National Parks* (Washington, DC: 2009); see also Thomas M. Power, *Lost Landscapes and Failed Economies: The Search for a Value of Place* (Washington, DC: Island Press, 1996); Thomas M. Power and Richard N. Barrett, *Post-Cowboy Economics: Pay and Prosperity in the New American West* (Washington, DC: Island Press, 2001); Rick S. Kurtz, "Public Lands Policy and Economic Trends in Gateway Communities," *Review of Policy Research* 27(1) (2010): 77.

22. The 1998 concession reforms are found in the National Parks Omnibus Management Act, Pub. L. 105-391 §§ 401–19, 112 Stat. 3497, 3503-19, codified at 16 U.S.C. §§ 5951–66; see also National Park Service, *Management Policies* (2006), 10.2 et seq.
23. The quotation, from Donald Hummel who chaired the Conference of National Park Concessioners during the 1970s, is found in Frome, *Regreening the National Parks*, 174. The principal case challenging the Park Service's new concession policy rules is Amfac Resorts v. U.S. Dept. of the Interior, 282 F.3d 818 (D.C. Cir. 2002). The Glacier concessions incident is reported in Richard J. Ansson Jr., "Protecting and Preserving Our National Parks in the Twenty First Century: Are Additional Reforms Needed Above and Beyond the Requirements for the National Parks Omnibus Management Act?" *Montana Law Review* 62 (2001): 213, 229–31.
24. The quotations are from Frome, *Regreening the National Parks*, 108, 109.
25. This account of Sequoia National Park's Giant Forest controversy is derived primarily from Lary M. Dilsaver and William C. Tweed, *Challenge of the Big Trees: A Resource History of Sequoia and Kings Canyon National Parks* (Three Rivers, CA: Sequoia Natural History Association, 1990); the quotations are found on pages 140–41, 143. See also Brett Wilkinson, "Giant Forest Restoration in Sequoia National Park Balances Tourism, Preservation," *Visalia Times-Delta*, Dec. 1, 2009; National Park Service, *Giant Forest Restoration Overview*, www.nps.gov/seki/historyculture/gfmain.htm.
26. Dilsaver and Tweed, *Challenge of the Big Trees*, 153.
27. The Grant Village–Fishing Bridge controversy is described and analyzed

in Sue Consolo Murphy and Beth Kaeding, "Fishing Bridge: 25 Years of Controversy regarding Grizzly Bear Management in Yellowstone National Park," *Ursus* 10 (1998): 385–93; Schullery, *Searching for Yellowstone*, 187–90; Alice Wondrak Biel, *Do (Not) Feed the Bears: The Fitful History of Wildlife and Tourists in Yellowstone* (Lawrence: University Press of Kansas, 2006), 125; David P. Sheldon, "A Threatening Turn for a Threatened Species: The Impact of National Wildlife Federation v. National Park Service," *Public Land Law Review* 10 (1989): 157; see also National Wildlife Federation v. National Park Service, 669 F. Supp. 384 (D. Wyo. 1987). For a more opinionated analysis of the controversy, see Alston Chase, *Playing God in Yellowstone: The Destruction of America's First National Park* (New York: Atlantic Monthly Press, 1986), 198–231.

28. See *National Wildlife Federation*, 669 F. Supp. 384 (D. Wyo. 1987); Sheldon, "Threatening Turn."
29. Schullery, *Searching for Yellowstone*, 180–83. But see National Park Service, *Lake Area Comprehensive Plan Environmental Assessment* (2012) (proposing to expand the recreational vehicle campground at Fishing Bridge by paving an additional seven acres, provoking the ire of some local conservation organizations).
30. This account of the evolution of Estes Park as a gateway community to Rocky Mountain National Park is derived largely from Lloyd K. Musselman, *Rocky Mountain National Park: Administrative History, 1915–1965* (Washington, DC: National Park Service, 1971); Rick S. Kurtz, "Gateway Communities, Economic Development, and Environmentalism," paper presented at the 2003 Western Political Science Association Conference, www.citation.allacademic.com/meta/p_mla_apa_research_citation/0/8/7/8/3/p87835_index.html; Howe, McMahon, and Propst, *Balancing Nature*, 101–5.
31. For more on the term "*glitter gulch*," see Kurtz, "Gateway Communities," 9; Kurtz, "Public Lands Policy," 85.
32. See Howe, McMahon, and Propst, *Balancing Nature*, 101–5.
33. On the park's new elk management plan, see National Park Service, *Final Environmental Impact Statement Elk and Vegetation Management Decision for Rocky Mountain National Park Record of Decision* (2008); see also chapter 7.
34. This account of the evolution of Gatlinburg and Pigeon Forge as gateway communities is taken largely from C. Brenden Martin, *Tourism in the Mountain South: A Double-Edged Sword* (Knoxville: University of Tennessee Press, 2007); Margaret Lynn Brown, *The Wild East: A Biography of the Great Smoky Mountains* (Gainesville: University Press of Florida, 2000); Howe, McMahon, and Propst, *Balancing Nature*, 32–34.
35. The Cade's Cove restoration incident is recounted in Martin, *Tourism in the Mountain South*, 151.
36. On Gatlinburg's hillbilly and Dixie thematic marketing efforts, see Martin, *Tourism in the Mountain South*, 155–58. The Martin quotation can be found on page 59.

37. The quotation and much of this description are taken from Howe, McMahon, and Propst, *Balancing Nature*, 32–34, and the economic figures are from Brown, *Wild East*, 302; see also Brown, *Wild East*, 304–5, for additional local employment statistics.
38. On the history and evolution of Pigeon Forge and Dollywood, see Martin, *Tourism in the Mountain South*, 128–32, 158–60; Brown, *Wild East*, 302–5.
39. On the park's air pollution problems, see Brown, *Wild East*, 329–38.
40. On the Canyon Forest Project, see Julie Cart, "Finding Common Ground in Plan for Grand Canyon's Future," *Los Angeles Times*, Mar. 21, 1999; Tom Westby, "Grand Canyon Development Sparks Debate," *High Country News*, Aug. 30, 1999; Doug Kreutz, "Canyon at the Crossroads," *Arizona Daily Star*, Dec. 25, 2000. For an update on this issue, including a description of a new upscale project by the same foreign firm, see John Dougherty, "Under the Flight Path," *High Country News*, June 13, 2011.
41. The 1998 legislation is discussed in more detail earlier in this chapter.
42. See Joseph L. Sax and Robert B. Keiter, "The Realities of Regional Resource Management: Glacier National Park and Its Neighbors Revisited," *Ecology Law Quarterly* 33:(2006): 233, 258–65; see also chapter 9.
43. Daniel J. Stynes, *Economic Benefits to Local Communities from National Park Visitation and Payroll, 2010* (Washington, DC: National Park Service, 2011), www.nature.nps.gov/socialscience/docs/NPSSystemEstimates2010.pdf.

Chapter 6

1. The quotation appears in Robert H. Keller and Michael F. Turek, *American Indians and National Parks* (Tucson: University of Arizona Press, 1998), xvi; the other data on tribal–national park connections are found on page xiii.
2. On Catlin's vision for a national park, see Mark David Spence, *Dispossessing the Wilderness: Indian Removal and the Making of the National Parks* (New York: Oxford University Press, 1999), 9–10.
3. On the history of Native Americans at Yosemite, Yellowstone, and Glacier, see Spence, *Dispossessing the Wilderness*. For a more general treatment of the Indian experience with the national parks, see Keller and Turek, *American Indians and National Parks*.
4. 16 U.S.C. §§ 431–33. See chapter 7 for a more detailed discussion of the Antiquities Act.
5. See Zach Zipfel, *Shared Boundaries: American Indian Tribes and the National Park Service* (Washington, DC: National Parks Conservation Association, 2009), 21–23.
6. For an overview of the Roosevelt administration's Indian policies and the termination era, see Charles Wilkinson, *Blood Struggle: The Rise of Modern Indian Nations* (New York: W.W. Norton, 2005), 57–86. For the history of federal Indian policy in general, see Nell Jessup Newton, ed., *Cohen's Handbook of Federal Indian Law* (Newark, NJ: LexisNexis, 2005), chapter 1.

7. This new era in tribal self-determination and Indian rights is described in Wilkinson, *Blood Struggle*, 177–268, with the role of Congress and the courts described on pages 241–68. See also Zipfel, *Shared Boundaries*, 6–8. For a history of the "Boldt decisions" and litigation over tribal fishing rights, see Rebecca Ulrich, *Empty Nets: Indians, Dams, and the Columbia River* (Corvallis: Oregon State University Press, 1999). The noted laws are found at Indian Civil Rights Act, Pub. L. No. 90-284, Title II, 82 Stat. 77 (1968), codified as amended at 25 U.S.C. §§ 1301–3, 1321; Indian Self Determination and Education Assistance Act of 1975, Pub. L. No. 93-638, 88 Stat. 2203 (1975), codified as amended at 25 U.S.C. § 450 et seq.; American Indian Religious Freedom Act of 1978, Pub. L. No. 95-341, 92 Stat. 469 (1978), codified as amended at 42 U.S.C. §§ 1996 and 1996a; Indian Child Welfare Act of 1978, Pub. L. No. 95-608, 92 Stat. 3069 (1978), codified at 25 U.S.C. § 1901 et seq.; Indian Gaming Regulatory Act of 1988, Pub. L. No. 100-497, 102 Stat. 2467 (1988), codified as amended at 25 U.S.C. §§ 2701–21; Native American Graves Protection and Repatriation Act, Pub. L. No. 101-601, 104 Stat. 3048 (1990), codified as amended at 18 U.S.C. § 1170, 25 U.S.C. §§ 3001–13; Tribal Self-Governance Act of 1994, Pub. L. No. 103-413, Title II, 108 Stat. 4270 (1994), codified as amended at 25 U.S.C. chapter 14, subchapter II (amending the Indian Self Determination and Education Assistance Act of 1975).
8. See Executive Order 13175, 65 Fed. Reg. 67249 (Nov. 6, 2000); see also Presidential Memorandum of Apr. 29, 1994; Executive Order 13007, 61 Fed. Reg. 26771 (May 24, 1996); Heather J. Tanana and John C. Ruple, "Energy Development in Indian Country: Working within the Realm of Indian Law and Moving towards Collaboration," *Utah Environmental Law Review* 32(1) (2011): 6. For the quoted management policies, see National Park Service, *Management Policies* 1.11, 1.11.1, 5.3.5.3.1, 5.3.5.3.2, 8.5 (2006). See also Zipfel, *Shared Boundaries*, 11–16, 21–23.
9. For a detailed account of the Grand Canyon–Havasupai relationship, see Keller and Turek, *American Indians and National Parks*, 156–84.
10. Keller and Turek, *American Indians and National Parks*, 170.
11. See Grand Canyon National Park Enlargement Act, Pub. L. 93-620, 88 Stat. 2089 (1975), codified at 16 U.S.C. § 228a et seq.
12. The 1895 Blackfeet treaty is found at Agreement with the Indians of the Blackfeet Indian Reservation in Montana, Sept. 26, 1895, ch. 398, § 9, 29 Stat. 321, 353–54 (1896); the Glacier legislation is codified at 16 U.S.C. § 161 et seq. On the Blackfeet relationship with Glacier National Park, see Keller and Turek, *American Indians and National Parks*, 43–64; Philip Burnham, *Indian Country, God's Country: Native Americans and the National Parks* (Washington, DC: Island Press, 2000), 105–17, 187–218; see also Spence, *Dispossessing the Wilderness*, 83–100, on the early history of the park.
13. These early park expansion efforts are recounted in Burnham, *Indian Country, God's Country*, 112–13; Keller and Turek, *American Indians and National*

Parks, 53–56.

14. For a more detailed discussion and analysis of these tribal economic development issues, see Burnham, *Indian Country, God's Country*, 57–59, 152–54, 204–8.
15. This emergent Crown of the Continent conservation vision is discussed further in chapter 9.

16. The American Indian Religious Freedom Act is found at 42 U.S.C. §§ 1996 and 1996a. Key court cases interpreting AIRFA include Wilson v. Block, 708 F.2d 735 (D.C. Cir. 1983); Lockhart v. Kenops, 927 F.2d 1028 (8th Cir. 1991). See generally Sandra B. Zellmer, "Sustaining Geographies of Hope: Cultural Resources on Public Lands," *University of Colorado Law Review* 73 (2002): 413.
17. For a detailed description and analysis of the Devil's Tower controversy, see Lloyd Burton, *Worship and Wilderness: Culture, Religion, and Law in Public Lands Management* (Madison: University of Wisconsin Press, 2002), 123–44. See also Keller and Turek, *American Indians and National Parks*, 195–99.
18. See Bear Lodge Multiple Use Association v. Babbitt, 175 F.3d 814 (10th Cir. 1999), affirming 2 F. Supp. 2d 1448 (D. Wyo. 1998); George Linge, "Ensuring the Full Freedom of Religion on Public Lands: Devils Tower and the Protection of Indian Sacred Sites," *Boston College Environmental Affairs Law Review* 27 (2000): 307, 312.
19. Another sacred site controversy has festered at Rainbow Bridge National Monument for several decades, not only prompting several lawsuits but also a similar settlement that "voluntarily" closes the base of the arch to visitors to address tribal concerns. See David Kent Sproul, *A Bridge between Cultures: An Administrative History of Rainbow Bridge National Monument* (Washington, DC: National Park Service, 2001); Friends of the Earth v. Armstrong, 485 F.2d 1 (10th Cir. 1974); Badoni v. Higginson, 638 F.2d 172 (10th Cir. 1980); Natural Arch and Bridge Society v. Alston, 209 F. Supp. 2d 1207 (D. Utah 2002), affirmed on other grounds, 98 Fed. Appx. 711 (10th Cir. 2004).
20. A detailed recounting of the Park Service–Timbisha Shoshone relationship can be found in Burnham, *Indian Country, God's Country*, 3–8, 89–105, 162–65, 294–308. See also Theodore Catton, *To Make a Better Nation: An Administrative History of the Timbisha Shoshone Homeland Act* (Missoula: University of Montana / Rocky Mountain Cooperative Ecosystems Study Unit, 2009).
21. The "slumlord" quotation comes from a Death Valley park ranger and is found in Burnham, *Indian Country, God's Country*, 164. The quoted California Desert Protection Act provision is found at 16 U.S.C. § 410aaa-75.
22. See Timbisha Shoshone Homeland Act, Pub. L. 106-423, 114 Stat. 1875 (2000).
23. See An Act to Authorize the President of the United States to Establish the Canyon De Chelly National Monument within the Navajo Indian Reservation, Arizona, Pub. L. No. 71-667, 46 Stat. 1161 (1931), codified at 16 U.S.C. § 445 et seq.; President, Proclamation, Establishing Canyon De Chelly Na-

tional Monument, Arizona, Proclamation 1945, 47 Stat. 2448 (1931); President, Proclamation, Canyon De Chelly National Monument, Arizona, Area Comprising, Proclamation 2035, 47 Stat. 2562 (1933). The Park Service's relationship with the Navajo and Canyon de Chelly is recounted in Keller and Turek, *American Indians and National Parks*, 205–12; David M. Brugge and Raymond Wilson, *Administrative History: Canyon de Chelly National Monument Arizona* (Washington, DC: National Park Service, 1976). See also National Park Service, Navajo Nation, and Bureau of Indian Affairs, *Joint Management Plan: Canyon de Chelly National Monument Arizona* (1989). The 1989 plan was never implemented, however. Instead, the Navajo Nation created Canyon de Chelly Tribal Park, which would coexist with the national monument. This move allowed the tribe to control tourism in the park, which it claimed it had the right to do under the 1931 authorizing amendment. Cindy Yurth Tséyi', "Tour Guides Angry at New Fees Regulations," *Navajo Times* (Dec. 16, 2010).

24. See Burnham, *Indian Country, God's Country*, 218–43, for an insightful description and analysis of the south unit controversy. The quotation is found on pages 232–33.
25. For an overview of the history of the south unit, see National Park Service and Oglala Sioux Tribe, *South Unit Badlands National Park: Final General Management Plan and Environmental Impact Statement* (2012), 6–18 (hereinafter *Badlands Final GMP and EIS*); see also Burnham, *Indian Country, God's Country*, 218–43.
26. The 1976 agreement is reproduced in National Park Service and Oglala Sioux Tribe, *South Unit Badlands National Park: Draft General Management Plan and Environmental Impact Statement* (2010), 217 (appendix).
27. See Burnham, *Indian Country, God's Country*, 232–33.
28. Bureau of Indian Affairs, *American Indian Population and Labor Force Report* (Washington, DC: 2005).
29. The Wounded Knee incident is recounted in Wilkinson, *Blood Struggle*, 143–49; see also Peter Matthiessen, *In the Spirit of Crazy Horse* (New York: Penguin, 1992).
30. See "Trouble over the Badlands: Oglala Lakota Sioux Fight for Control of Part of Badlands National Park," *High Country News*, Aug. 18, 2003.
31. See National Park Service and Oglala Sioux Tribe, *Badlands Final GMP and EIS*, 19.
32. For a full description of the Tribal National Park proposal, see National Park Service and Oglala Sioux Tribe, *Badlands Final GMP and EIS*, 38–39.
33. For additional discussion of the national parks as wilderness areas, see chapter 2.
34. Elwha River Ecosystem and Fisheries Restoration Act, Pub. L. 102-495, 106 Stat. 3173 (1992). For background information on the Elwha River restoration initiative, see Phillip M. Bender, "Restoring the Elwha, White Salmon,

and Rogue Rivers: A Comparison of Dam Removal Proposals in the Pacific Northwest," *Journal of Land, Resources, and Environmental Law* 17 (1997): 189; National Park Service, *Elwha River Ecosystem Restoration, Final Supplement to the Final Environmental Impact Statement* (2005); Tom Callis, "National Park Service Director Praises Tribe for Dam Removal Support," *Peninsula Daily News*, July 4, 2010. The Elwha River Restoration Project is described and analyzed in more detail in chapter 7.

Chapter 7

1. The Washburn quotation can be found in John Ise, *Our National Park Policy: A Critical History* (Baltimore: Johns Hopkins University Press, 1961), 15; the Hayden quotations are found in Ferdinand V. Hayden, *Preliminary Report of the United States Geological Survey of Montana and Portions of Adjacent Territories* (Washington, DC: Government Printing Office, 1872), 4, 162.
2. The Olmstead report is in "The Yosemite Valley and the Mariposa Big Tree Grove" in *America's National Park System: The Critical Documents*, ed. Lary M. Dilsaver (Lanham, MD: Rowman and Littlefield, 1994), 23.
3. The Antiquities Act is found at 16 U.S.C. §§ 431–33. On the Antiquities Act, see Ronald F. Lee, *The Antiquities Act of 1906* (Washington, DC: National Park Service, 1970); Hal Rothman, *Preserving Different Pasts: The American National Monuments* (Urbana: University of Illinois Press, 1989); David Harmon, Francis P. McManamon, and Dwight T. Pitcaithley, eds., *The Antiquities Act: A Century of American Archaeology, Historic Preservation, and Nature Conservation* (Tucson: University of Arizona Press, 2006).
4. The Lane letter quotation is found in "Secretary Lane's Letter on National Park Management" (May 13, 1918), in Dilsaver, *America's National Park System*, 50. The early schism between the yet-to-be-formed National Park Service and the Forest Service can be traced to the Hetch Hetchy dam controversy in Yosemite National Park and to the Forest Service's opposition, through chief Gifford Pinchot, to the bill creating the Park Service. See Robert W. Righter, *The Battle over Hetch Hetchy: America's Most Controversial Dam and the Birth of Modern Environmentalism* (New York: Oxford University Press, 2005).
5. See "Secretary Lane's Letter," 50. The 1922 letter was penned by Roger W. Toll, Rocky Mountain National Park superintendent, and issued in conjunction with "Superintendent's Resolution on Overdevelopment"; both documents are found in Dilsaver, *America's National Park System*, 57–61.
6. The early history of the Park Service's education efforts and programs is recounted in Harold C. Bryant and Wallace W. Atwood Jr., *Research and Education in the National Parks* (Washington, DC: Government Printing Office, 1932), part 2.
7. The quotation is from Bryant and Atwood, *Research and Education*, quoted in R. Gerald Wright, *Wildlife Research and Management in the National Parks*

(Urbana: University of Illinois Press, 1992), 13. The George M. Wright story and his role in national park history is recounted more fully in chapter 7. The statistics on park naturalists and museums are from Ise, *Our National Park Policy*, 446.

8. National Parks Association, "National Primeval Park Standards: A Declaration of Policy" (1945), in Dilsaver, *America's National Park System*, 174, 176–77.
9. The Mission 66 program and its effect on agency science are discussed in Richard West Sellars, *Preserving Nature in the National Parks: A History* (New Haven: Yale University Press, 1997), 180–214; see also Ethan Carr, *Mission 66: Modernism and the National Park Dilemma* (Amherst: University of Massachusetts Press, 2007), 184–92 (discussing how Mission 66 affected Park Service education and interpretation efforts). On Freeman Tilden and his role in National Park Service interpretation, see E. Bruce Craig, introduction to Freeman Tilden, *Interpreting Our Heritage*, 4th ed. (Chapel Hill: University of North Carolina Press, 2007), 1–25. The early history of the Park Service's interpretation efforts is recounted in C. Frank Brockman, "Park Naturalists and the Evolution of National Park Service Interpretation through World War II," *Journal of Forest History* 22(1) (1978): 24–43; see also Barry Mackintosh, *Interpretation in the National Park Service: A Historical Perspective* (Washington, DC: National Park Service, 1986), www.nps.gov/history/history/online_books/mackintosh2/index.htm. The Tilden quotation is from Tilden, *Interpreting Our History*, 163.
10. See R. Gerald Wright, "Wolf and Moose Populations in Isle Royale National Park," in *Science and Ecosystem Management in the National Parks*, eds. William H. Halvorson and Gary E. Davis (Tucson: University of Arizona Press, 1996), 74–95; Rolf O. Peterson, *Wolf Ecology and Prey Relationships on Isle Royale* (Washington DC: National Park Service, 1977); Rolf O. Peterson, "Wolf-Moose Interaction on Isle Royale: The End of Natural Regulation?" *Ecological Applications* 9(1) (1999): 10–16; Brian E. McLaren and Rolf O. Peterson, "Wolves, Moose, and Tree Rings on Isle Royale," *Science* 266 (1994): 555–58.
11. Stanley Cain was a prominent University of Michigan biologist with lengthy ties to the national parks. His criticisms were delivered in a talk entitled "Ecological Islands as Natural Laboratories," in Dilsaver, *America's National Park System*, 200, 204; the quotation is found on page 204.
12. See Sellars, *Preserving Nature*, 195–201, 214–17; Wright, *Wildlife Research*, 24–28; James A. Pritchard, *Preserving Yellowstone's Natural Conditions: Science and the Perception of Nature* (Lincoln: University of Nebraska Press, 1999), 207–20.
13. A.S. Leopold, S.A. Cain, C.M. Cottam, I.N. Gabrielson, and T.L. Kimball, "Wildlife Management in the National Parks," in Dilsaver, *America's National Park System*, 237–51. The quotations appear on pages 239 and 243.
14. National Research Council, "A Report by the Advisory Committee to the

National Park Service on Research" (1963), partially reprinted in Dilsaver, *America's National Park System*, 253–62. The Robbins report is discussed and analyzed in Wright, *Wildlife Research*, 26–27; Sellars, *Preserving Nature*, 215–17.

15. On the post–Robbins report era for agency scientists, see Ervin H. Zube, "Management in National Parks: From Scenery to Science," in Halvorson and Davis, *Science and Ecosystem Management*, 13, 17; Wright, *Wildlife Research*, 27–31; Sellars, *Preserving Nature*, 217–66.
16. On the Yellowstone-Craighead grizzly bear management controversy, see Sellars, *Preserving Nature*, 249–53; Thomas McNamee, *The Grizzly Bear* (New York: Alfred A. Knopf, 1984), 99–122. See also Frank C. Craighead Jr., *Track of the Grizzly* (San Francisco: Sierra Club Books, 1979).
17. See Conservation Foundation, *National Parks for the Future* (Washington, DC: 1972), 12, 17; National Park Service, "State of the Parks: A Report to Congress" (1980), in Dilsaver, *America's National Park System*, 405–8; General Accounting Office, "Limited Progress Made in Documenting and Mitigating Threats to the Parks" (1987), 16–37, partially reprinted in Dilsaver, *America's National Park System*, 414–17.
18. See William L. Halvorson and Gary E. Davis, "Lessons Learned from a Century of Applying Research Results to Management of National Parks," in Halvorson and Davis, *Science and Ecosystem Management*, 334–43; Wright, *Wildlife Research*, 31; National Park Service, *Management Policies* 4:1, 4:2, 4:4 (1988).
19. The quotations are from National Academy of Sciences, "Science and the National Parks" (1992), partially reprinted in Dilsaver, *America's National Park System*, 446–49; "National Parks for the 21st Century: The Vail Agenda," partially reprinted in Dilsaver, *America's National Park System*, 434, 443.
20. On the National Biological Survey, see National Research Council, *A Biological Survey for the Nation* (Washington, DC: National Academies Press, 1993); Frederic H. Wagner, "Whatever Happened to the National Biological Survey?" *BioScience* 49 (1999): 220.
21. The quotation is from Sellars, *Preserving Nature*, 290.
22. The relevant sections of National Parks Omnibus Management Act of 1998, Pub. L. 105-391, 112 Stat. 3497 (1998), are codified at 16 U.S.C. §§ 5931–37; the quoted text is found at 16 U.S.C. § 5932.
23. National Park Service, *Natural Resources Challenge: The National Park Service's Action Plan for Preserving Natural Resources* (Washington, DC: 1999). On funding for the Natural Resources Challenge, see National Park Service, *Natural Resource Challenge Funding History—Program Summary* (Washington, DC: 2004); for information on the cooperative ecosystem studies units, see www.cesu.psu.edu.
24. The quotations can be found in the noted reports: National Park Service Advisory Board, *Rethinking the National Parks for the 21st Century* (Washington,

DC: National Park Service, 2001); National Parks Second Century Commission, *Advancing the National Park Idea* (Washington, DC: National Parks Conservation Association, 2009).

25. See "National Parks for the 21st Century," 441–42; National Park Service, *Management Policies* 7.5.3, 7.5.4 (2006).
26. See Richard Louv, *Last Child in the Woods: Saving Our Children from Nature-Deficit Disorder* (Chapel Hill, NC: Algonquin Books, 2006); Richard Louv, "Leave No Child Inside," *Orion Magazine*, Mar. / Apr. 2007; see also chapter 10 for further discussion of nature deficit disorder and its implications for new national park designations.
27. See National Park Service, *A Call to Action: Preparing for a Second Century of Stewardship and Engagement* (Washington, DC: 2011), 13–15.
28. See National Parks Second Century Commission, *Advancing the National Park Idea*, 18, 30; see also www.epa.gov / air / peg / parks.html; www.grand .canyon.national-park.com / ranger.htm.
29. For a brief history of fire policy on the public lands, see Robert B. Keiter, *Keeping Faith with Nature: Ecosystems, Democracy, and America's Public Lands* (New Haven: Yale University Press, 2003), 136–45; see also Stephen J. Pyne, *Fire in America: A Cultural History of Wildland and Rural Fire* (Princeton, NJ: Princeton University Press, 1982); David Carle, *Burning Questions: America's Fight with Nature's Fire* (Westport, CT: Praeger, 2002); Stephen J. Pyne, *America's Fires: Management on Wildlands and Forests* (Durham, NC: Forest History Society, 1997).
30. See Alfred Runte, *National Parks: The American Experience*, 2d ed. (Lincoln: University of Nebraska Press, 1987), 201–8; Ashley Schiff, *Fire and Water: Scientific Heresy in the Forest Service* (Cambridge: Harvard University Press, 1962), 51–115; Larry Bancroft, Thomas Nichols, David Parsons, David Graber, Boyd Evison, and Jan van Wagtendonk, "Evolution of the Natural Fire Management Program at Sequoia and Kings Canyon National Parks," in *Proceedings of the Symposium and Workshop on Wilderness Fire, USDA Forest Service General Technical Report INT-182* (1985).
31. The Park Service's early fire management and research program is recounted in David J. Parsons and Jan W. van Wagtendonk, "Fire Research and Management in the Sierra Nevada National Parks," in Halvorson and Davis, *Science and Ecosystem Management*, 25–48; see also Runte, *National Parks*, 201–8.
32. On the 1988 Yellowstone fires and their aftermath, see Pritchard, *Preserving Yellowstone's Natural Conditions*, 281–89; Mary Ann Franke, *Yellowstone in the Afterglow: Lessons from the Fires* (Mammoth Hot Springs, WY: Yellowstone Center for Resources, 2000); Dennis H. Knight and Linda L. Wallace, "The Yellowstone Fires: Issues in Landscape Ecology," *BioScience* 39 (1989): 700; see also Rocky Barker, *Scorched Earth: How the Fires of Yellowstone Changed America* (Washington, DC: Island Press, 2005); Micah Morrison, *Fire in*

Paradise: The Yellowstone Fires and the Politics of Environmentalism (New York: HarperCollins, 1993).

33. On the 2002 Cerro Grande fire and its aftermath, see Barry T. Hill, *Fire Management Lessons Learned from the Cerro Grande (Los Alamos) Fire* (Washington, DC: General Accounting Office, 2000); Roger G. Kennedy, *Wildfire and Americans: How to Save Lives, Property, and Your Tax Dollars* (New York: Hill and Wang, 2006), 88–104.
34. Adolph Murie, *The Wolves of Mount McKinley National Park* (Seattle: University of Washington Press, 1985), xvii.
35. See Murie, *Wolves*, xvii; Aldo Leopold, "Review of *The Wolves of North America* by S.P. Young and E.A. Goldman," *Journal of Forestry* 42 (1944): 929. For an overview of the early science on wolves, see Hank Fischer, *Wolf Wars: The Remarkable Inside Story of the Restoration of Wolves to Yellowstone* (Helena, MT: Falcon Press, 1995), 24–34.
36. The wolf reintroduction saga is recounted in Fischer, *Wolf Wars*; Thomas McNamee, *The Return of the Wolf to Yellowstone* (New York: Henry Holt, 1997); Renee Askins, *Shadow Mountain: A Memoir of Wolves, a Woman, and the Wild* (New York: Anchor Books, 2002).
37. Section 10(j) of the Endangered Species Act is found at 16 U.S.C. § 1539(j) and is implemented under 50 C.F.R. §§ 17.80–84. The reintroduction was approved in the Record of Decision accompanying U.S. Fish and Wildlife Service, *The Reintroduction of Gray Wolves to Yellowstone National Park and Central Idaho: Final Environmental Impact Statement* (1994). The court cases are Wyoming Farm Bureau Federation v. Babbitt, 987 F. Supp. 1349 (D. Wyo. 1997), reversed, 199 F.3d 1224 (10th Cir. 2000).
38. On the ecological effects of wolf reintroduction, see William J. Ripple and Robert L. Beschta, "Restoring Yellowstone's Aspen with Wolves," *Biological Conservation* 138 (2007): 514; Robert Beschta, "Cottonwoods, Elk, and Wolves in the Lamar Valley of Yellowstone National Park," *Ecological Applications* 13 (2003): 1295; Robert L. Beschta and William J. Ripple, "Increased Willow Heights along Northern Yellowstone's Blacktail Deer Creek Following Wolf Reintroduction," *Western North American Naturalist* 67(4) (2007): 613–17; Patrick J. White and Robert A. Garrott, "Yellowstone's Ungulates after Wolves: Expectations, Realizations, and Predictions," *Biological Conservation* 67(4) (2005): 141; Daniel Fortin, Hawthorne L. Beyer, Mark S. Boyce, Douglas W. Smith, Thierry Duchesne, and Julie S. Mao, "Wolves Influence Elk Movements: Behavior Shapes a Trophic Cascade in Yellowstone National Park," *Ecology* 86(5) (2005): 1320; Kim Murray Berger and Mary M. Conner, "Recolonizing Wolves and Mesopredator Suppression of Coyotes: Impacts on Pronghorn Population Dynamics," *Ecological Applications* 18(3) (2008): 599–612; Douglas W. Smith, Rolf O. Peterson, and Douglas B. Houston, "Yellowstone after Wolves," *BioScience* 53(4) (2003): 330. On the economic implications of wolf reintroduction, see John Duffield, "An Economic Analysis of

Wolf Recovery in Yellowstone: Park Visitor Attitudes and Values," in *Wolves for Yellowstone? A Report to the United States Congress* (Washington, DC: National Park Service, 1990); Tom Reed, *Yellowstone Wolves Bring Estimated $7–10 Million in Annual Tourism Revenue* (Nov. 6, 2009), http://www.yellowstonepark.com/2011/06/yellowstone-wolves-bring-estimated-7-10-million-in-annual-tourism-revenue/; see also Martin A. Nie, *Beyond Wolves: The Politics of Wolf Recovery and Management* (Minneapolis: University of Minnesota Press, 2003), 132–38; Timothy W. Clark, A. Peyton Curlee, Steven C. Minta, and Peter Kareiva IV, eds., *Carnivores in Ecosystems: The Yellowstone Experience* (New Haven: Yale University Press, 1999).

39. On the aftermath and ramifications of delisting wolves under the Endangered Species Act, see Douglas H. Chadwick, "Wolf Wars," *National Geographic* (March 2010): 34–55.
40. See Phil Taylor, "Famed Yellowstone Research Pack Hit Hard by Hunters," *Land Letter*, Oct. 29, 2009; Defenders of Wildlife v. Salazar, 729 F. Supp. 2d 1207 (D. Mont. 2010); Pub. L. 112-10 at § 1713, 125 Stat. 38, 150 (2011); April Reese, "Idaho, Montana Hunts Suspended as Wyoming Digs in Heels," *Land Letter*, Aug. 12, 2010; Amanda Peterka, "Montana, Idaho Populations Lose ESA Protection Today," *Land Letter*, May 5, 2011; Laura Petersen, "FWS Delists Wyo. Gray Wolves," *Greenwire*, Aug. 31, 2012.
41. See U.S. Department of the Interior, National Park Service, *Final Environmental Impact Statement Elk and Vegetation Management Decision for Rocky Mountain National Park Record of Decision* (2008); Joshua Zaffos, "An Ecosystem Wanting for Wolves," *High Country News*, Jan. 23, 2006. See also Wildearth Guardians v. National Park Service, 804 F. Supp. 2d 1150 (D. Colo. 2011).
42. On national park wildlife restoration policy, see National Park Service, *Management Policies* 4.1.5, 4.4.2.2 (2006); on other national park wolf restoration proposals, see John T. Ratti, Mike Weinstein, J. Michael Scott, Patryce Aysharian Wiseman, Anne-Marie Gillesber, Craig A. Miller, Michele M. Szepanski et al., "Feasibility of Wolf Reintroduction to Olympic Peninsula," *Northwest Science* 78 (2004): 1; Olympic Natural Resources Center, *Should Wolves Be Reintroduced into Olympic National Park and Surrounding Lands?* (Forks, WA: 1998); Paul G. Sneed, "The Feasibility of Gray Wolf Reintroduction to the Grand Canyon Ecoregion," *Endangered Species UPDATE* 18(4) (2001): 153; see also John Elder, ed., *The Return of the Wolf: Reflections on the Future of Wolves in the Northeast* (Hanover, NH: University Press of New England, 2000); Nie, *Beyond Wolves*, 10–18.
43. The Muir quotation can be found in Donald Worster, *A Passion for Nature: The Life of John Muir* (New York: Oxford University Press, 2008), 424. Muir also memorably referred to Hetch Hetchy Valley as "no holier temple." Worster, *A Passion for Nature*, 425. Secretary of the Interior Donald Hodel's proposal to remove O'Shaughnessy Dam and restore Hetch Hetchy Valley is discussed in State of California and the Resources Agency, *Hetch Hetchy*

Restoration Study (Sacramento: State of California, 2006); Norimitsu Onishi, "Putting Bay Area's Water Source to a Vote," *New York Times*, Sept. 9, 2012. The Everglades restoration effort is examined in chapter 9.

44. On the Glen Canyon Dam Adaptive Management program, see www.usbr.gov/uc/rm/amp/index.html. On the Colorado River restoration efforts, see Robert W. Adler, *Restoring Colorado River Ecosystems: A Troubled Sense of Immensity* (Washington, DC: Island Press, 2007); William R. Lowry, *Repairing Paradise: Restoration of Nature in America's National Parks* (Washington, DC: Brookings Institution Press, 2009), 157–204; April Reese, "Colorado River Adaptive Management Needs Overhaul, Critics Say," *Land Letter*, May 7, 2009; April Reese, "USFWS Analysis Offers Key Findings from High-Flow Tests," *Land Letter*, Feb. 10, 2010.
45. See National Park Service, *Elwha River Ecosystem Restoration: Final Environmental Impact Statement* (1995); National Park Service, *Elwha River Ecosystem Restoration Implementation: Final Environmental Impact Statement* (1996); National Park Service, *Elwha River Ecosystem Restoration Implementation: Final Supplement to the Final Environmental Impact Statement* (2005); George R. Pess, Michael L. McHenry, Timothy J. Beechie, and Jeremy Davies, "Biological Impacts of the Elwha River Dams and Potential Salmonid Responses to Dam Removal," *Northwest Science* 82 (2008): 72; Philip M. Bender, "Restoring the Elwha, White Salmon, and Rogue Rivers: A Comparison of Dam Removal Proposals in the Pacific Northwest," *Journal of Land, Resources and Environmental Law* 17 (1997): 189; see also Scott Streater, "Dam Removals Will Restore Century-Old Salmon Runs in Olympic National Park," *Land Letter*, Mar. 31, 2011; William Yardley, "Removing Barriers to Salmon Migration," *New York Times*, July 29, 2011; National Park Service, "Elwha River Restoration," www.nps.gov/olym/naturescience/elwha-ecosystem-restoration.htm.
46. On the subject of global climate change, see Intergovernmental Panel on Climate Change, *Fourth Assessment Report, Climate Change 2007: Synthesis Report* (Geneva: 2007); U.S. Global Change Research Program, *Global Climate Change Impacts in the United States* (New York: Cambridge University Press, 2009); Thomas E. Gradel and Paul J. Cruzen, *Atmosphere, Climate, and Change* (New York: Scientific American Library, 1995); National Research Council, *Abrupt Climate Change: Inevitable Surprises* (Washington, DC: National Academies Press, 2002).
47. For analysis of how global climate change is affecting the national parks, see Stephen Saunders, Tom Easley, and Suzanne Farver, *National Parks in Peril: The Threats of Climate Disruption* (Louisville, CO: Rocky Mountain Climate Change Organization / Natural Resources Defense Council, 2009); Stephen Saunders and Tom Easley, *Losing Ground: Western National Parks Endangered by Climate Disruption* (Louisville, CO: Rocky Mountain Climate Change Organization / Natural Resources Defense Council, 2006); Catherine E. Burns, Kevin M. Johnston, and Oswald J. Schmitz, "Global Climate Change

and Mammalian Species Diversity in U.S. National Parks," *Proceedings of the National Academy of Sciences* 100 (2003): 11474; Impacts of Climate Change to National Parks, Hearing before the Subcommittee on National Parks of the U.S. Senate Committee on Energy and Natural Resources, 111th Cong., 1st Sess., Sen. Hearing 111-239, Oct. 28, 2009.

48. Indeed, the Park Service is already engaged in basic science research about the effects of climate change on park resources. See National Park Service, "Special Issue: Climate Change Adaptation and Communication," *Park Science* 28(1)(2011): 1–75; National Park Service, "Special Issue: Climate Change Science in the National Parks," *Park Science* 28(2)(2011): 1-95.
49. The Park Service has begun to outline an initial climate change strategy for managing its resources. See National Park Service, *Climate Change Response Strategy* (2010); see also National Park Service Advisory Board Science Committee, *Revisiting Leopold: Resource Stewardship in the National Parks* (2012), www.nps.gov/calltoaction/PDF/Leopold Report_2012.pdf. As part of the Department of the Interior, the Park Service is also governed by Interior's approach to climate change. See Secretary of the Interior, Order No. 3289, *Addressing the Impacts of Climate Change on America's Water, Land, and Other Natural and Cultural Resources* (September 14, 2009); Department of the Interior, *Interior's Plan for a Coordinated, Science-Based Response to Climate Change Impacts on Our Land, Water, and Wildlife Resources* (nd).
50. Criticism of Yellowstone's science program can be found in Frederic H. Wagner, *Yellowstone's Destabilized Ecosystem: Elk Effects, Science, and Policy Conflict* (New York: Oxford University Press, 2006), 307–16; Stephen Budiansky, *Nature's Keepers: The New Science of Nature Management* (New York: Free Press, 1995); Bruce Goldstein, "The Struggle over Ecosystem Management at Yellowstone," *BioScience* 42(3) (1992): 183–87; Alston Chase, *Playing God in Yellowstone: The Destruction of America's First National Park* (Boston: Atlantic Monthly Press, 1986), 232–61; see also Karl Hess Jr., *Rocky Times in Rocky Mountain National Park: An Unnatural History* (Niwot: University of Colorado Press, 1993).
51. Yellowstone's controversial 1998 agreement with Diversa, a biotechnology company, which opened the park's hot pools to commercial bioprospecting in return for royalty payments, arguably represents an instance when economic concerns triumphed over the Park Service's long tradition of public research geared toward basic (noncommercial) science and resource management concerns. The Diversa agreement was unsuccessfully challenged in litigation, which is found at Edmonds Institute v. Babbitt, 42 F. Supp. 2d 1 (D.D.C. 1999); 93 F. Supp. 2d 63 (D.D.C. 2000). For a general discussion and analysis of the Yellowstone-Diversa bioprospecting arrangement, see Holly Doremus, "Nature, Knowledge, and Profit: The Yellowstone Bioprospecting Controversy and the Core Purposes of America's National Parks," *Ecology Law Quarterly* 26 (1999): 401.

Chapter 8

1. On the military's role in early national park management, see H. Duane Hampton, *How the U.S. Cavalry Saved Our National Parks* (Bloomington: University of Indiana Press, 1971).
2. For a detailed account of this incident, see Richard Bartlett, *Yellowstone: A Wilderness Besieged* (Tucson: University of Arizona Press, 1985), 319–21; see also Aubrey Haines, *The Yellowstone Story: A History of Our First National Park* (Boulder: Colorado Associated University Press, 1977), 2:62–64.
3. Alfred Runte, *Yosemite: The Embattled Wilderness* (Lincoln: University of Nebraska Press, 1990), 86.
4. For an account of the Yellowstone bison restoration effort, see Haines, *The Yellowstone Story*, 67–77, 311–13; Hampton, *U.S. Cavalry Saved Our National Parks*, 165–67; Paul Schullery, " 'Buffalo Jones' and the Bison Herd in Yellowstone," *Montana: The Magazine of Western History* 26 (1976): 40.
5. See R. Gerald Wright, *Wildlife Research and Management in the National Parks* (Urbana: University of Illinois Press, 1992), 3.
6. 16 U.S.C. § 352. The park's enabling legislation is found at 16 U.S.C. §§ 347–355a; Congress subsequently changed the name to Denali National Park.
7. "Secretary Lane's Letter on National Park Service Management" (May 13, 1918), in *America's National Park System: The Critical Documents*, ed. Lary M. Dilsaver (Lanham, MD: Rowman and Littlefield, 1994), 48, 49.
8. For accounts of these early national park wildlife management policies, see Richard West Sellars, *Preserving Nature in the National Parks: A History* (New Haven: Yale University Press, 1997), 69–90; James A. Pritchard, *Preserving Yellowstone's Natural Conditions: Science and the Perception of Nature* (Lincoln: University of Nebraska Press, 1999), 23–74; Alice Wondrak Biel, *Do (Not) Feed the Bears: The Fitful History of Wildlife and Tourists in Yellowstone* (Lawrence: University Press of Kansas, 2006), 7–27.
9. For accounts of the Kaibab deer population explosion controversy, see Christine Young, *In the Absence of Predators: Conservation and Controversy on the Kaibab Plateau* (Lincoln: University of Nebraska Press, 2002); Thomas R. Dunlap, "That Kaibab Myth," *Journal of Forest History* 32(2) (1988): 60; James C. Foster, "The Deer of Kaibab: Federal-State Conflict in Arizona," *Arizona and the West* 12(3) (1970): 255. The court cases sustaining the authority of the public land agencies to shoot (cull) deer and other animals on federal lands for resource management purposes are Hunt v. United States, 278 U.S. 96 (1928) (U.S. Forest Service); New Mexico State Game Commission v. Udall, 410 F.2d 1197 (10th Cir. 1969) (National Park Service).
10. For a discussion of Grinnell's role in national park wildlife policy, see Runte, *Yosemite*, 133–35; Sellars, *Preserving Nature*, 47; Wright, *Wildlife Research*, 13–14.
11. The quote is from George M. Wright, Joseph S. Dixon, and Ben H. Thomp-

son, "Fauna of the National Parks of the United States: A Preliminary Survey of Faunal Relations in National Parks (Fauna Series No. 1, May 1932)," in Dilsaver, *America's National Park System*, 106, (hereinafter "Fauna Survey I"). For an account of Wright's life, views, and influence on national park policy, see Jerry Emory and Pamela Wright Lloyd, "George Melendez Wright, 1904–1936: A Voice on the Wing," *George Wright Forum* 17 (2000): 14; see also National Park Service, *George Melendez Wright, Scientist and Visionary*, http://www.nps.gov/yose/naturescience/upload/george-wright-fact-sheet.pdf.

12. See "Fauna Survey I," 108-110; George M. Wright and Ben H. Thompson, *Fauna of the National Parks of the United States: Wildlife Management in the National Parks* (Fauna Series No. 2, July 1934) (Washington, DC: Government Printing Office, 1935) (hereinafter "Fauna Survey II").
13. Frederic H. Wagner, Ronald Foresta, Richard Bruce Gill, Dale Richard McCullough, Michael R. Pelton, William F. Porter, and Hal Salwasser, *Wildlife Policies in the U.S. National Parks* (Washington, DC: Island Press, 1995), 20.
14. See Sellars, *Preserving Nature*, 93–101; see also Dilsaver, *America's National Park System*, 87–98, 122–31, 137–50, setting forth several critical Park Service policy documents from the 1930s.
15. The Park Service's wartime years are recounted in Janet A. McDonnell, "World War II: Defending Park Values and Resources," *Public Historian* 29(4) (2007): 15–33; John C. Miles, *Guardians of the Parks: A History of the National Parks and Conservation Association* (Washington, DC: Taylor and Francis, 1995), 131–53; see also Sellars, *Preserving Nature*, 150–55.
16. See Sellars, *Preserving Nature*, 195–201, 214–17; see also chapter 7 for additional discussion of the Leopold and National Academy reports.
17. A.S. Leopold, S.A. Cain, C.M. Cottam, I.N. Gabrielson, and T.L. Kimball, "Wildlife Management in the National Parks" (Mar. 4, 1963), in Dilsaver, *America's National Park System*, 237.
18. "Memorandum from Secretary of the Interior Stewart Udall to the National Park Service Director, on the Report of the Advisory Board on Wildlife Management" (May 2, 1963), in Dilsaver, *America's National Park System*, 251.
19. National Research Council, "A Report by the Advisory Committee to the National Park Service on Research," in Dilsaver, *America's National Park System*, 253.
20. "Memorandum from Secretary of the Interior Stuart Udall," 251.
21. National Park Service, "Administrative Policies for Natural Areas" (1968), in Dilsaver, *America's National Park System*, 354.
22. How these laws affect national park resource management policy is discussed in Robert B. Keiter, "Preserving Nature in the National Parks: Law, Policy, and Science in a Dynamic Environment," *Denver University Law Review* 74 (1997): 649, 680–82. The pertinent Endangered Species Act experimental population provision addressing controversial species reintroductions is found at 16 U.S.C. § 1539(j). Yellowstone wolf reintroduction is

described and analyzed in chapter 6.

23. On hunting in the national parks, see Wright, *Wildlife Research*, 46–53; on hunting to manage Grand Teton National Park's elk, see Wright, *Wildlife Research*, 168–73. The statute allowing regulated sport hunting in Grand Teton is Pub. L. 81–787, 64 Stat. 849, 851–52 (1950), codified at 16 U.S.C. § 406.
24. Leopold et al., "Wildlife Management," 243, 246–47.
25. The judicial quotation is from Michigan United Conservation Clubs v. Lujan, 949 F.2d 202, 207 (6th Cir. 1991); see also National Rifle Association v. Potter, 628 F. Supp. 903 (D.D.C. 1986). In addition, the courts have consistently sustained the Park Service's culling practices. *New Mexico State Game Commission*, 410 F.2d 1197; Davis v. Latschar, 83 F. Supp. 2d 1 (D.D.C. 1998), affirmed, 202 F.3d 359 (D.C. Cir. 2000); WildEarth Guardians v. National Park Service, 804 F. Supp. 2d 1150 (D. Colo. 2011). See generally Wright, *Wildlife Research*, 46–54.
26. On the Alaska subsistence provisions and national preserves, see Wright, *Wildlife Research*, 163–68.
27. See Sportsmen's Heritage Act, HR 4089, 112th Cong. (2012) (passed House of Representatives, Apr. 17, 2012); see also HR 1179, 110th Cong. (2008); S. 684, 110th Cong. (2008); and S. 917, 110th Cong. (2008) (all proposing to allow lethal means for elk population control in Rocky Mountain National Park); as well as HR 5137, 110th Cong. (2008) (proposing to open parts of New River Gorge National River to hunting).
28. John Hope Franklin, Robert S. Chandler, Margaret L. Brown, Sylvia A. Earle, Javier M. Gonzales, Charles R. Jordan, Shirley M. Malcolm et al., *Rethinking the National Parks for the 21st Century: A Report of the National Park System Advisory Board* (Washington, DC: Department of the Interior, 2001).
29. These quotations are found in National Park Service, *Management Policies* 4.1, 4.4.1, 4.1.5, 4.4.2.2 (2006).
30. Peter White and Keith Langdon, "The ATBI in the Smokies: An Overview," *George Wright Forum* 23(3) (2006): 18; see generally "The All Taxa Biodiversity Index (ATBI) for Great Smoky Mountains National Park," *George Wright Forum* 23(3) (2006): 17–60.
31. National Park Service Office of Science and Technology, "State of the Parks: A Report to the Congress," in Dilsaver, *America's National Park System*, 405, 406.
32. "National Parks for the 21st Century: The Vail Agenda," in Dilsaver, *America's National Park System*, 434.
33. Franklin et al., *Rethinking National Parks* (noting that national parks cannot survive as islands of biodiversity and recommending linking parks with other natural areas through wildlife corridors and greenways).
34. National Parks Second Century Commission, *Advancing the National Park Idea* (Washington, DC: National Parks Conservation Association, 2009), 26.
35. See National Parks Second Century Commission, *Advancing the National*

Park Idea, Science and Natural Resources Committee Report (Washington, DC: National Parks Conservation Association, 2009), 3–4, http://www.nps.gov/yose/naturescience/upload/Second.Century.Science.Report.pdf; Pritchard, *Preserving Yellowstone's Natural Conditions*, 251–306; Wagner et al., *Wildlife Policies*, 69–76; see also Franklin et al., *Rethinking the National Parks*.

36. On the evolution of Yellowstone's natural resource management policies, see Pritchard, *Preserving Yellowstone's Natural Conditions*.
37. See Alston Chase, *Playing God in Yellowstone: The Destruction of America's First National Park* (Boston: Atlantic Monthly Press, 1986). See also Wagner et al., *Wildlife Policies*; Karl Hess Jr., *Rocky Times in Rocky Mountain National Park: An Unnatural History* (Boulder: University of Colorado Press, 1993); Stephen Budiansky, *Nature's Keepers: The New Science of Nature Management* (New York: Free Press, 1995), 131–58.
38. Wright, *Wildlife Research*, 75.
39. Wright and Thompson, "Fauna Survey II," 85.
40. Leopold et al., "Wildlife Management in the National Parks," 237.
41. Wright, *Wildlife Research*, 78–79; Pritchard, *Preserving Yellowstone's Natural Conditions*, 201–50; Paul Schullery, *Searching for Yellowstone* (New York: Houghton Mifflin, 1997), 219–31; see also Mark S. Boyce, "Natural Regulation or the Control of Nature?" in *The Greater Yellowstone Ecosystem: Redefining America's Wilderness Heritage*, eds. Robert B. Keiter and Mark S. Boyce (New Haven: Yale University Press, 1991), 183–208.
42. For elk population data, see Frederic H. Wagner, *Yellowstone's Destabilized Ecosystem: Elk Effects, Science, and Policy Conflict* (New York: Oxford University Press, 2006), 15–28.
43. See Wagner et al., *Wildlife Policies*, 48–58; Charles E. Kay, "Yellowstone's Northern Elk Herd: A Critical Evaluation of the 'Natural Regulation' Paradigm" (PhD diss., Utah State University, 1990).
44. David R. Klein et al., *Ecological Dynamics on Yellowstone's Northern Range* (Washington, DC: National Academies Press, 2002), 4, 9. In addition, the NRC concluded on page 133 that "no major ecosystem component is likely to be eliminated from the northern range in the near or intermediate term." For another summary of northern range conditions and the natural regulation policy, see National Park Service, *Yellowstone's Northern Range: Complexity and Change in a Wildland Ecosystem* (1997).
45. For a critique of the Park Service's science capacity, see Wagner et al., *Wildlife Policies*, 307–16.
46. Wright, *Wildlife Research*, 80.
47. These potential changes in the northern range are detailed in Wagner, *Yellowstone's Destabilized Ecosystem*, 209–301. Indeed, Wagner suggests on page 10 of his book that the Yellowstone elk controversy may be moving into yet another (or fifth) phase.
48. National Park Service, *Final Environmental Impact Statement Elk and Vegeta-*

tion Management Decision for Rocky Mountain National Park Record of Decision (2008); WildEarth Guardians v. National Park Service, 804 F. Supp. 2d 1150 (D. Colo. 2011). For a critique of Rocky Mountain National Park's elk management policies, see Hess, *Rocky Times*, 15–49.

49. See National Park Service, *Theodore Roosevelt National Park Draft Elk Management Plan/Environmental Impact Statement Update: Preferred and Environmentally Preferable Alternatives* (2009); National Park Service, *Wind Cave National Park Elk Management Plan/EIS Record of Decision* (2009).
50. H.C. Frost, G.L. Storm, M.J. Batcheller, and M.J. Lavallo, "White-Tailed Deer Management at Gettysburg National Military Park and Eisenhower National Historic Site," *Wildlife Society Bulletin* 25(2) (1997): 462; D.C. Fulton, K. Skeri, E.M. Shank, and D.W. Lime, "Beliefs and Attitudes toward Lethal Management of Deer in Cuyahoga Valley National Park," *Wildlife Society Bulletin* 32(4) (2004): 1166–76; see also Joseph Berger, "Cull Expands as the Deer Chomp Away," *New York Times*, Oct. 18, 2009. Court decisions sustaining Park Service culling programs in eastern parks include Davis v. Latschar, 83 F. Supp. 2d 1 (D.D.C. 1998), affirmed, 202 F.3d 359 (D.C. Cir. 2000) (Gettysburg); Friends of Animals v. Caldwell, 2010 WL 4259753 (E.D. Pa. 2010), affirmed, 434 Fed. Appx. 72 (3rd Cir. 2011) (Valley Forge). For background on the Park Service's eastern deer management problems, see Wright, *Wildlife Research*, 83–88.
51. National Park Service, *Management Policies* 4.4.2 (2006).
52. For a brief history of early Yellowstone bison management, see text accompanying note 4 in this chapter and references cited therein; see also www.nps.gov/yell/naturescience/bison.htm.
53. The bison-brucellosis controversy is recounted and analyzed in Christina M. Cromley, "Bison Management in Greater Yellowstone," in *Finding Common Ground: Governance and Natural Resources in the American West*, eds. Ronald D. Brunner, Christine H. Colburn, Christina M. Cromley, Roberta A. Klein, and Elizabeth A. Olson (New Haven: Yale University Press, 2002), 116–58; Robert B. Keiter, "Greater Yellowstone's Bison: Unraveling of an Early American Wildlife Conservation Achievement," *Journal of Wildlife Management* 61(1) (1997): 1–11; Robert B. Keiter and Peter Froelicher, "Bison, Brucellosis, and Law in the Greater Yellowstone Ecosystem," *Land and Water Law Review* 28 (1993): 1.
54. See U.S. Department of the Interior and U.S. Department of Agriculture, *Record of Decision for the Final Environmental Impact Statement and Bison Management Plan for the State of Montana and Yellowstone National Park* (2000); Todd Wilkinson, "In Montana, Bison Plan Paused," *Christian Science Monitor*, June 4, 2008; Matthew Brown, "Bison Slaughter Nears Record," *Deseret Morning News*, Mar. 10, 2008. See also U.S. General Accounting Office, *Preliminary Observations on the Implementation of the Interagency Bison Management Plan* (Washington, DC: 2007).

55. Matt Schweber, "As Bison Return to Prairie, Some Rejoice, Others Worry," *New York Times*, Apr. 26, 2012; Matthew Brown, "Bison Could Roam Year-Round Outside of Yellowstone," *Denver Post*, July 23, 2012; Laura Petersen, "Ranchers, Enviros Square Off in Court over Montana Bison Relocation," *Greenwire*, Apr. 11, 2012.
56. See generally Thomas McNamee, *The Grizzly Bear* (New York: Alfred A. Knopf, 1984); Frank C. Craighead Jr., *Track of the Grizzly* (San Francisco: Sierra Club Books, 1979); John Craighead, Jay S. Sumner, and John A. Mitchell, *The Grizzly Bears of Yellowstone: Their Ecology in the Yellowstone Ecosystem* (Washington, DC: Island Press, 1995), 15–92.
57. On the biology of the grizzly bear, see Craighead, Sumner, and Mitchell, *The Grizzly Bears*; Department of the Interior, Fish and Wildlife Service, *Final Rule Removing the Yellowstone Distinct Population Segment of Grizzly Bears from the Federal List of Endangered and Threatened Wildlife*, 72 Fed. Reg. 14866, 14866-78 (Mar. 29, 2007) (hereinafter *2007 Yellowstone Grizzly Bear Delisting Rule*).
58. See Brian L. Kuehl, "Conservation Observations under the Endangered Species Act: A Case Study of the Yellowstone Grizzly Bear," *University of Colorado Law Review* 64 (1993): 607; Robert B. Keiter, "Observations on the Future Debate over 'Delisting' the Grizzly Bear in the Great Yellowstone Ecosystem," *Environmental Professional* 13 (1991): 248; see also U.S. Fish and Wildlife Service, *Grizzly Bear Recovery Plan* (Washington, DC: Author, 1982).
59. See Congressional Research Service, Library of Congress, *Greater Yellowstone Ecosystem: An Analysis of Data Submitted by Federal and State Agencies* (99th Cong., Comm. Print No. 6, Dec. 1986); U.S. Fish and Wildlife Service, *Interagency Grizzly Bear Guidelines* (1986); U.S. Fish and Wildlife Service, *Grizzly Bear Recovery Plan* (1993); see also Mark L. Shaffer, *Keeping the Grizzly Bear in the American West: A Strategy for Real Recovery* (Washington, DC: The Wilderness Society, 1992). See chapter 9 for further discussion of the failed federal Greater Yellowstone Ecosystem management effort.
60. U.S. Fish and Wildlife Service, *Grizzly Bear Recovery Plan* (1993); see Fund for Animals v. Babbitt, 903 F. Supp. 96 (D.D.C. 1995) (enjoining the Yellowstone grizzly bear delisting decision); Philip Kline, "Grizzly Bear Blues: A Case Study of the Endangered Species Act's Delisting Process and Recovery Plan Requirements," *Environmental Law* 31 (2001): 371; Todd Wilkinson, "Grizzly Wars," *High Country News* 30(21) (1998): 1; Louisa Willcox, "The Last Grizzlies of the American West: The Long Hard Road to Recovery," *Endangered Species Update* 14 (1997): 11–16.
61. *2007 Yellowstone Grizzly Bear Delisting Rule*; Interagency Conservation Strategy Team, *Final Conservation Strategy for the Grizzly Bear in the Greater Yellowstone Area* 26 (2007), www.fws.gov/mountain-prairie/species/mammals/grizzly/Final_Conservation_Strategy.pdf (hereinafter *Interagency Conservation Strategy*).
62. *Interagency Conservation Strategy*, 37; Greater Yellowstone Coalition v.

Servheen, 665 F.3d 1015 (9th Cir. 2011); see also Louisa Willcox, *Bear with Us: An Alternative Path to Grizzly Bear Recovery in the Lower 48 States* (Washington, DC: Natural Resources Defense Council, 2004); Craig M. Pease and David J. Mattson, "Demography of the Yellowstone Grizzly Bears," *Ecology* 80(3) (1999): 957.

63. But see *Interagency Conservation Strategy*, 15 (stating that "implementation of the management strategies requires continued cooperation between federal and state agencies").
64. The Park Service's exotics policy can be found at National Park Service, *Management Policies* 4.4.4 (2006). For an overview and analysis of these policies and implementation issues, see Wright, *Wildlife Research*, 92–110; Sellars, *Preserving Nature*, 258–61.
65. On the Yellowstone lake trout problem and suppression program, see Robert E. Gresswell, *Scientific Review Panel Evaluation of the National Park Service Lake Trout Suppression Program in Yellowstone Lake: Final Report* (Bozeman, MT: U.S. Geological Survey Northern Rocky Mountain Science Center, 2009); James R. Ruzycki, David A. Beauchamp, and Daniel L. Yule, "Effects of Introduced Lake Trout on Native Cutthroat Trout in Yellowstone Lake," *Ecological Applications* 13 (2003): 23; John D. Varley and Paul Schullery, *Yellowstone Fishes: Ecology, History, and Angling in the Park* (Mechanicsburg, PA: Stackpole Books, 1998); see also National Park Service, *Yellowstone National Park, Native Fish Conservation Plan Environmental Assessment and Finding of No Significant Impact* (2011).
66. On the Olympic National Park mountain goat controversy, see Douglas B. Houston, Edward G. Schreiner, and Bruce B. Moorehead, *Mountain Goats in Olympic National Park: Biology and Management of an Introduced Species* (1994); Wright, *Wildlife Research*, 101–5; R. Lee Lyman, *White Goats White Lies: The Abuse of Science in Olympic National Park* (Salt Lake City: University of Utah Press, 1998). See also Department of the Interior, National Park Service, *Notice of Suspension for Draft Environmental Impact Statement for Mountain Goat Management Within Olympic National Park, Washington*, 64 Fed. Reg. 47196 (Aug. 20, 1999).
67. See Wagner, *Yellowstone's Destabilized Ecosystem*, 317–34.

Chapter 9

1. On the Yellowstone expansion proposal and the early forest reserves, see Aubrey Haines, *The Yellowstone Story* (Boulder: Colorado Associated University Press, 1977), 1:263–69, 2:94–97; on early Yosemite wildlife protection concerns, see Alfred Runte, *Yosemite: The Embattled Wilderness* (Lincoln: University of Nebraska Press, 1990), 71–72. On the Sequoia expansion proposal, see H. Duane Hampton, *How the U.S. Cavalry Saved Our National Parks* (Bloomington: Indiana University Press, 1971), 157.
2. The Yellowstone expansion controversy is related in Haines, *The Yellowstone*

Story, 2:319–36.

3. On the Mount Rainier, Grand Canyon, and Glacier expansion proposals, see John C. Ise, *Our National Park Policy: A Critical History* (Baltimore: Johns Hopkins University Press, 1961), 123, 178, 237, 285, 336; on the Sequoia expansion, see Ise, *Our National Park Policy*, 396–404; Robert Shankland, *Steve Mather of the National Parks* (New York: Alfred A. Knopf, 1970), 173–75, 177–79. See generally Ise, *Our National Park Policy*, 274–76. The Glacier-Blackfeet relationship is explored in more detail in chapter 6.
4. George M. Wright and Ben H. Thompson, *Fauna of the National Parks of the United States: Wildlife Management in the National Parks* (Fauna Series No. 2, July 1934) (Washington, DC: Government Printing Office, 1935) (hereinafter *Fauna Survey II*). The first quotation is found in George M. Wright, Joseph S. Dixon, and Ben H. Thompson, "Fauna of the National Parks of the United States: A Preliminary Survey of Faunal Relations in National Parks (Fauna Series No. 1, May 1932)" (hereinafter "Fauna Survey I"), in *The Critical Documents: America's National Park System*, ed. Lary M. Dilsaver (Lanham, MD: Rowman and Littlefield, 1994), 104.
5. See Wright and Thompson, *Fauna Survey II*, 96–103, 109, 130 (the "transformers" quotation appears on page 109).
6. See generally Alfred Runte, *National Parks: The American Experience*, 2d ed. (Lincoln: University of Nebraska Press, 1987), 124–54.
7. The first quotation is from A.S. Leopold, S.A. Cain, C.M. Cottam, I.N. Gabrielson, and T.L. Kimball, "Wildlife Management in the National Parks" (Mar. 4, 1963), in Dilsaver, *America's National Park System*, 237, 238 (quoting from First World Conference on National Parks, Management of National Parks and Equivalent Areas). The Robbins report is found in National Research Council, "A Report by the Advisory Committee to the National Park Service on Research," in Dilsaver, *America's National Park System*, 253–54. The 1968 policy document is found in National Park Service, "Administrative Policies for Natural Areas," in Dilsaver, *America's National Park System*, 354.
8. An Act to Amend the Act of October 2, 1968, an Act to Establish a Redwood National Park in the State of California, Pub. L. 95-250, 92 Stat. 163 (1978), codified at 16 U.S.C. § 79a et seq. On the Redwood controversy, see Dale A. Hudson, "Sierra Club v. Department of the Interior: The Fight to Preserve Redwood National Park," *Ecology Law Quarterly* 7 (1979): 781; Tom Turner, *Wild by Law: The Sierra Club Legal Defense Fund and the Places It Has Saved* (San Francisco: Sierra Club Books, 1990), 65.
9. The Redwood amendment is found at 16 U.S.C. § 1a-1. The Redwood National Park litigation includes Sierra Club v. Dept. of the Interior, 376 F. Supp. 90 (N.D. Cal. 1974), 376 F. Supp. 284 (N.D. Cal. 1974), 398 F. Supp. 284 (N.D. Cal. 1975). Court decisions interpreting the Redwood amendment include Sierra Club v. Andrus, 487 F. Supp. 443 (D.D.C. 1980); National Rifle Association v. Potter, 628 F. Supp. 903 (D.D.C. 1985); Sierra Club v. Mainella,

459 F. Supp. 2d 76 (D. Colo. 2006); Bluewater Network v. Salazar, 721 F. Supp. 2d 7 (D.D.C. 2010).

10. See Conservation Foundation, *National Parks for the Future* (Washington, DC, 1972), 12, 21, 46; National Parks Conservation Association, *NPCA Adjacent Lands Survey: No Park Is An Island* (Washington, DC: 1979), as quoted in John C. Miles, *Guardians of the Parks: A History of the National Parks and Conservation Association* (Washington, DC: Taylor and Francis, 1995), 287. See also William E. Brown, *Islands of Hope: Parks and Recreation in Environmental Crisis* (Washington, DC: National Recreation and Park Association, 1971), 37–73.
11. National Park Service, "State of the Parks, 1980: A Report to Congress," partially reprinted in Dilsaver, *America's National Park System*, 405. For a summary and analysis of the various park protection legislative proposals, see Robert B. Keiter, "On Protecting the National Parks from the External Threats Dilemma," *Land and Water Law Review* 20 (1985): 355, 396–408.
12. The Newmark studies are in William D. Newmark, "Legal and Biotic Boundaries of Western North American National Parks: A Problem of Congruence," *Biological Conservation* 33 (1985): 197; William D. Newmark, "Extinction of Mammal Populations in Western North American National Parks," *Conservation Biology* 9(3) (1995): 512. On island biogeography, see Robert H. McArthur and Edward O. Wilson, *The Theory of Island Biogeography* (Princeton, NJ: Princeton University Press, 1967); David Quammen, *The Song of the Dodo: Island Biogeography in an Age of Extinctions* (New York: Scribners, 1996). On conservation biology in theory and practice, see Reed F. Noss and Alan Y. Cooperrider, *Saving Nature's Legacy: Protecting and Restoring Biodiversity* (Washington, DC: Island Press, 1994); Michael E. Soule and John Terborgh, eds., *Continental Conservation: Scientific Foundations of Regional Reserve Networks* (Washington, DC: Island Press, 1999).
13. The Greater Yellowstone Ecosystem concept is discussed in Paul Schullery, *Searching for Yellowstone: Ecology and Wonder in the Last Wilderness* (New York: Houghton Mifflin, 1997), 197–207; James A. Pritchard, *Preserving Yellowstone's Natural Conditions: Science and Perception in Nature* (Lincoln: University of Nebraska Press, 1999), 251–306; Douglas B. Houston, "Ecosystems of National Parks," *Science* 172 (1971): 648, 651; Rick Reese, *Greater Yellowstone: The National Park and Adjacent Wildlands* (Helena: Montana Magazine, 1991), 55–99. The role of carnivores and other large predators as "keystone species" for maintaining ecosystem health is explained in John Terborgh, James A. Estes, Paul Paquet, Katherine Ralls, Diane Boyd-Heger, Brian J. Miller, and Reed F. Noss, "The Role of Top Carnivores in Regulating Terrestrial Ecosystems," in Soule and Terborgh, *Continental Conservation*, 39–64.
14. The noted external threats problems are documented and examined in U.S. General Accounting Office, *Parks and Recreation: Limited Progress Made in Documenting and Mitigating Threats to the Parks* (Washington, DC: 1987); National Park Service, *Natural Resources Assessment and Action Program* (Wash-

ington, DC: 1988); Joseph L. Sax and Robert B. Keiter, "Glacier National Park and Its Neighbors: A Study in Federal Interagency Relations," *Ecology Law Quarterly* 14 (1987): 207; John C. Freemuth, *Islands under Siege: National Parks and the Politics of External Threats* (Lawrence: University Press of Kansas, 1991); David J. Simon, ed., *Our Common Lands: Defending the National Parks* (Washington, DC: Island Press, 1988).

15. See Conservation Foundation, *National Parks for the Next Generation: Visions, Realities, Prospects* (Washington, DC: 1985), xxxix, 141–55; National Park Service, *Management Policies: Management of the National Park System* 2:9 (1988); Congressional Research Service, *Greater Yellowstone Ecosystem: An Analysis of Data Submitted by Federal and State Agencies*, 99th Cong., 2d Sess. (Comm. Print No. 6, Dec. 1986); Robert B. Keiter, "Taking Account of the Ecosystem on the Public Domain: Law and Ecology in the Greater Yellowstone Region," *University of Colorado Law Review* 60 (1989): 923; Robert B. Keiter and Mark S. Boyce, eds., *The Greater Yellowstone Ecosystem: Redefining America's Wilderness Heritage* (New Haven: Yale University Press, 1991).
16. On the vision process, see Bruce Goldstein, "Can Ecosystem Management Turn an Administrative Patchwork into a Greater Yellowstone Ecosystem?" *Northwest Environmental Journal* 8 (1992): 285; Pamela Lichtman and Tim W. Clark, "Rethinking the 'Vision' Exercise in the Greater Yellowstone Ecosystem," *Society and Natural Resources* 7 (1994): 459; John Freemuth and R. McGreggor Cawley, "Science, Expertise and the Public: The Politics of Ecosystem Management in the Greater Yellowstone Ecosystem," *Landscape and Urban Planning* 40 (1998): 211.
17. For the "Vail Agenda," see "National Parks for the 21st Century: The Vail Agenda," partially reprinted in Dilsaver, *America's National Park System*, 434, 440, 442. On the Northwest Forest Plan, see Kathie Durbin, *Tree Huggers: Victory, Defeat, and Renewal in the Northwest Ancient Forest Campaign* (Seattle: The Mountaineers, 1996); Robert B. Keiter, *Keeping Faith with Nature: Ecosystems, Democracy, and America's Public Lands* (New Haven: Yale University Press, 2003), 81–113; Forest Ecosystem Management Assessment Team, *Forest Ecosystem Management: An Ecological, Economic, and Social Assessment Report of the Forest Ecosystem Management Assessment Team* (Washington, DC: U.S. Forest Service, 1993). On the Clinton 'administration's ecosystem management policies, see Keiter, *Keeping Faith with Nature*, 113–27; Interagency Ecosystem Management Task Force, *The Ecosystem Approach: Healthy Ecosystems and Sustainable Economies* (Washington, DC: 1995); Nels C. Johnson, Andrew J. Malk, Robert C. Szaro, and William T. Sexton, eds., *Ecological Stewardship: A Common Reference for Ecosystem Management*, 3 vols. (Oxford: Elsevier Science, 1999).
18. See National Park Service, *Management Policies* 1.5, 3.4 (2000); National Parks Omnibus Management Act, 16 U.S.C. §§ 5923, 5934; William C. Halvorson and Gary E. Davis, eds., *Science and Ecosystem Management in the National*

Parks (Tucson: University of Arizona Press, 1996). On the Clinton administration and ecological restoration, see Keiter, *Keeping Faith with Nature*, 127–70. It is noteworthy that the Park Service's management policies were revised again in 2006; these revisions adopt the same basic view of the external threats problem, but use the rubric of "cooperative conservation beyond park boundaries." National Park Service, *Management Policies* 1.6, 1.7, 3.4 (2006).

19. For a critical analysis of the G. W. Bush administration's public land policies, see Robert B. Keiter, "Breaking Faith with Nature: The Bush Administration and Public Land Policy," *Journal of Land, Resources and Environmental Law* 27 (2007): 195; John D. Leshy, "Natural Resources Policy in the Bush (II) Administration: An Outsider's Somewhat Jaundiced Assessment," *Duke Environmental Law and Policy Forum* 14 (2004): 347. See chapter 4 for an analysis of the 2006 *Management Policies* revision controversy.
20. The potential effect of global warming on the national parks and wildlife is discussed in Stephen Saunders, Tom Easley, and Suzanne Farver, *National Parks in Peril: The Threats of Climate Disruption* (Louisville, CO: Rocky Mountain Climate Organization / Natural Resources Defense Council, 2009); National Parks Conservation Association, *Climate Change and National Park Wildlife: A Survival Guide for a Warming World* (Washington, DC: 2009); Robert L. Peters, *Beyond Cutting Emissions: Protecting Wildlife and Ecosystems in a Warming World* (Washington, DC: Defenders of Wildlife, 2008); E. Jean Brennan, *Reducing the Impact of Global Warming on Wildlife: The Science, Management and Policy Challenges Ahead* (Washington, DC: Defenders of Wildlife, 2008); Stephen Saunders and Tom Easley, *Losing Ground: Western National Parks Endangered by Climate Disruption* (Louisville, CO: Rocky Mountain Climate Organization / Natural Resources Defense Council, 2006); see also Stephen Saunders, Charles Montgomery, and Tom Easley, *Hotter and Drier: The West's Changed Climate* (Louisville, CO: Rocky Mountain Climate Organization / Natural Resources Defense Council, 2008).
21. On mitigation and adaptation as strategies to address global climate change, see Intergovernmental Panel on Climate Change, *Climate Change 2007: Impacts, Adaptation and Vulnerability: Contribution of Working Group II to the Fourth Assessment Report of the Intergovernmental Panel on Climate Change (IPCC)* (Cambridge, UK: Cambridge University Press, 2010); Michael Gerrard and Katherine Kuh, eds., *The Law of Adaptation to Climate Change: U.S. and International Aspects* (Washington, DC: American Bar Association, 2012); Robert W. Adler, "Balancing Compassion and Risk in Climate Adaptation: U.S. Water, Drought, and Agricultural Law," *Florida Law Review* 64 (2012): 201.
22. National Park Service, "State of the Parks," 405. For a discussion of the Glacier threats, see Sax and Keiter, "Glacier National Park: A Study."
23. See generally Sax and Keiter, "Glacier National Park: A Study."

24. See Joseph L. Sax and Robert B. Keiter, "The Realities of Regional Resource Management: Glacier National Park and Its Neighbors Revisited," *Ecology Law Quarterly* 33 (2006): 233; Joseph L. Sax and Robert B. Keiter, "Glacier National Park and Its Neighbors: A Twenty-Year Assessment of Regional Resource Management," *George Wright Forum* 24(1) (2007): 23.
25. See Sax and Keiter, "Realities," 300–302. For a description of the ecological significance of the greater Glacier area, also known as the Crown of the Continent, see John L. Weaver, *The Transboundary Flathead: A Critical Landscape for Carnivores in the Rocky Mountains* (New York: Wildlife Conservation Society, 2001).
26. For more information on the Crown Managers Group, see www.crownmanagers.org; Roundtable on the Crown of the Continent, *The Initiatives*, http://www.crownroundtable.org/initiatives.html.
27. See Sax and Keiter, "Realities," 302–4. On the Crown of the Continent Ecosystem idea and management challenges, see Tony Prato and Dan Fagre, eds., *Sustaining Rocky Mountain Landscapes: Science, Policy, and Management for the Crown of the Continent Ecosystem* (Washington, DC: Resources for the Future, 2007).
28. Press Release, America's Great Outdoors: Salazar Tours Crown of Continent, Highlights Economic Benefits of Conservation (July 16, 2011), www.doi.gov/news/pressreleases/AMERICAS-GREAT-OUTDOORS-Salazar-Tours-Crown-of-Continent-Highlights-Economic-Benefits-of-Conservation.cfm. See generally Department of the Interior, Department of Agriculture, Environmental Protection Agency, and Council on Environmental Quality, *America's Great Outdoors: A Promise to Future Generations* (Washington, DC: 2011).
29. A more extended analysis of key factors that help promote regional or ecosystem management can be found in Sax and Keiter, "Realities," 300–309. See also Matthew McKinney, Lynn Scarlett, and Daniel Kemmis, *Large Landscape Conservation: A Strategic Framework for Policy and Action* (Cambridge: Lincoln Institute of Land Policy, 2010); Tony Prato and Daniel B. Fagre, "Sustainable Management of the Crown of the Continent," *George Wright Forum* 21(1) (2010): 77.
30. See Sax and Keiter, "Realities," 295–96.
31. Memorandum of Understanding and Cooperation on Environmental Protection, Climate Action and Energy between the Province of British Columbia and the State of Montana (Feb. 18, 2010), http://www.gov.bc.ca/igrs/attachments/en/MTEnvCoop.pdf.
32. Rob Chaney, "British Columbia Okays Law Protecting Flathead Basin," *Missoulian*, Nov. 16, 2011.
33. North Fork Watershed Protection Act of 2011, S. 233, 112th Cong. (2011); Mike Dennison, "Land Board OKs Protection of Flathead," *Helena Independent Record*, Mar. 19, 2010.

34. See Tristan Scott, "Oil, Gas Dilemma for Blackfeet Tribe: Revenue versus Environment," *Missoulian*, Aug. 7, 2011; Tristan Scott, "Oil, Blackfeet Women Join Together to Oppose Oil, Gas 'Fracking,' " *Missoulian*, June 20, 2012; "Blackfeet Tribe Weighs Risks, Rewards of Fracking on Montana Reservation," *Land Letter*, Dec. 1, 2011.
35. See Jen Gerson, "Logging in Bear Habitat Worries Environmentalists; Companies Say Effect Is Minimal," *Calgary Herald*, Jan. 4, 2012.
36. Sax and Keiter, "Realities," 305–9. On ecosystem management in the national parks, see Robert B. Keiter, "Ecosystem Management: Exploring the Legal-Political Framework," in *National Parks and Protected Areas: Their Role in Environmental Protection*, ed. R. Gerald Wright (Cambridge: Blackwell Science, 1996), 63–88; Halvorson and Davis, *Science and Ecosystem Management*.
37. Mary Doyle, "Implementing Everglades Restoration," *Journal of Land Use and Environmental Law* 17 (2001): 59, 65.
38. This account of the Everglades restoration effort is derived from several sources, including Michael Grunwald, *The Swamp: The Everglades, Florida, and the Politics of Paradise* (New York: Simon and Schuster, 2007); William R. Lowry, *Repairing Paradise: The Restoration of Nature in America's National Parks* (Washington, DC: Brookings Institution, 2009), 107–55; Judith A. Layzer, *Natural Experiments: Ecosystem-Based Management and the Environment* (Boston: MIT Press, 2008), 103–36; Kelly F. Taylor, "A Trickle of Cash for the River of Grass: Federal Funding of Comprehensive Everglades Restoration, A Critique and a Proposal," *University of Miami Law Review* 64 (2010): 1407; Alfred R. Light, "Tales of the Tamiami Trail: Implementing Adaptive Management in Everglades Restoration," *Journal of Land Use and Environmental Law* 22 (2006): 59.
39. 16 U.S.C. § 410c.
40. National Park Service, *Everglades and Dry Tortugas National Parks: Superintendent's Annual Report* (2010), 32.
41. Lowry, *Repairing Paradise*, 112.
42. An overview of the Everglades ecosystem can be found in Steven M. Davie and John C. Ogden, *Everglades: The Ecosystem and Its Restoration* (Delray Beach, FL: St. Lucie Press, 1994).
43. See United States v. Southern Florida Water Management District, 28 F.3d 1563 (11th Cir. 1992).
44. For a good summary of the efforts to establish minimum flow levels, see Lowry, *Repairing Paradise*, 116.
45. See National Park Service, Everglades National Park, South Florida Natural Resources Center brochure, www.nps.gov/ever/naturescience/sfnrcaboutus.htm.
46. The quotation is from Michael Grunwald, as recounted in Lowry, *Repairing Paradise*, 137; see also Lowry, *Repairing Paradise*, 133, for others making the same point.

47. Water Resources Development Act of 2000, Comprehensive Everglades Ecosystem Restoration Plan, Pub. L. 106-541, § 601(b), 114 Stat. 2571 (2000).
48. The CERP projects are outlined in Lowry, *Repairing Paradise*, 137–41; Layzer, *Natural Experiments*, 118–23.
49. See Light, "Tales of the Tamiami Trail," 73–75.
50. See National Park Service, Everglades National Park, South Florida Natural Resources Center brochure, www.nps.gov/ever/naturescience/sfnrcaboutus.htm.
51. Peter J. Balint, Ronald E. Stewart, Anand Desai, and Lawrence C. Walters, *Wicked Environmental Problems: Managing Uncertainty and Conflict* (Washington, DC: Island Press, 2011), 41–42.
52. Layzer, *Natural Experiments*, 116–17.

Chapter 10

1. On the history of the national park system, see Alfred Runte, *National Parks: The American Experience*, 2d ed. (Lincoln: University of Nebraska Press, 1997); Barry Mackintosh, *The National Parks: Shaping the System*, 3d ed. (Washington, DC: National Park Service, 2005), 65; John Ise, *Our National Park Policy: A Critical History* (Baltimore: Johns Hopkins University Press, 1961). For a brief history of Glacier National Park and Rocky Mountain National Park, see Ise, *Our National Park Policy*, 171–82, 212–18.
2. The Antiquities Act is found at 16 U.S.C. §§ 431–33.
3. See David Harmon, Francis P. McManamon, and Dwight T. Pitcaithley, *The Antiquities Act: A Century of American Archaeology, Historic Preservation, and Nature Conservation* (Tucson: University of Arizona Press, 2006); Hal Rothman, *America's National Monuments: The Politics of Preservation* (Lawrence: University Press of Kansas, 1989).
4. 16 U.S.C. § 2. The Organic Act is codified at 16 U.S.C. §§ 1–18.
5. "Secretary Lane's Letter on National Park Management" (May 13, 1918), in *America's National Park System: The Critical Documents*, Lary M. Dilsaver, ed., (Lanham, MD: Rowman and Littlefield, 1994), 48. The letter also noted that the size of new areas proposed for national park status did not matter, and it instructed the Park Service to "study existing national parks with the idea of improving them by the addition of adjacent areas which will complete their scenic purposes or facilitate administration." The letter specifically mentioned the Teton Mountains as a potential addition to Yellowstone and the Sierra summits and slopes as a possible addition to Sequoia.
6. See John C. Miles, *Guardians of the Parks: A History of the National Parks and Conservation Association* (Washington, DC: Taylor and Francis, 1995), 93–95; Thomas R. Vale, *The American Wilderness: Reflections on Nature Protection in the United States* (Charlottesville: University of Virginia Press, 2005), 188; Ise, *Our National Park Policy*, 437–39.
7. 16 U.S.C. § 1a-5.

8. National Parks Omnibus Management Act, Pub. L. 105-391, title III, § 303, 112 Stat. 3497, 3501 (1998), codified at 16 U.S.C. §§ 1a-5(b)–(c). In addition, the legislation directed the Park Service to consider whether the area under consideration might be adequately protected by another federal or state agency and to conduct any new park study in compliance with the National Environmental Policy Act. See generally Carol Hardy Vincent, *National Park System: Establishing New Units* (Washington, DC: Congressional Research Service, 1999).
9. On the early growth of the park system, see Runte, *National Parks*, 112–18; Mackintosh, *National Parks*, 20–27; Ise, *Our National Park Policy*, 218–70; Dayton Duncan and Ken Burns, *The National Parks: America's Best Idea* (New York: Alfred A. Knopf, 2009), 136–95, 217–28, 244–48.
10. On the 1933 presidential order transferring national monuments and military parks to the National Park Service, see Runte, *National Parks*, 219–20; Ise, *Our National Park Policy*, 352–53; Donald C. Swain, *Wilderness Defender: Horace M. Albright and Conservation* (Chicago: University of Chicago Press, 1970), 226–30; see also Harlan D. Unrau and G. Frank Williss, *Administrative History: Expansion of the National Park Service in the 1930s* (Washington, DC: National Park Service, 1983); Duncan and Burns, *National Parks*, 285–98. On the Grand Teton controversy, see Robert W. Righter, *Crucible for Conservation: The Creation of Grand Teton National Park* (Boulder: Colorado Associated University Press, 1982); Runte, *National Parks*, 142–44; Duncan and Burns, *National Parks*, 311–17.
11. See Mackintosh, *National Parks*, 46–64; Ise, *Our National Park Policy*, 369–404, 415–27; Runte, *National Parks*, 128–37, 140–42; Richard West Sellars, *Preserving Nature in the National Parks: A History* (New Haven: Yale University Press, 1997), 137–39; Duncan and Burns, *National Parks*, 274–80.
12. See Stewart L. Udall, *The Quiet Crisis* (New York: Holt Rinehart, 1963); George B. Hartzog Jr., *Battling for the National Parks* (Mt. Kisco, NY: Moyer Bell, 1988); see also www.nps.gov/history/history/online_books/director/hartzog.pdf.
13. See Mackintosh, *National Parks*, 64–83; Duncan and Burns, *National Parks*, 337–50.
14. See Samuel Trask Dana and Sally K. Fairfax, *Forest and Range Policy: Its Development in the United States*, 2d ed. (New York: McGraw Hill, 1980), 207–38. The Wilderness Act and its effect on national park policy are discussed in chapter 2.
15. National Park Service, *Part Two of the National Park System Plan: Natural History* (Washington, DC: 1972), 14–17; see Ronald A. Foresta, *America's National Parks and Their Keepers* (Washington, DC: Resources for the Future, 1984), 111–18; Craig L. Shafer, "History of Selection and System Planning for U.S. Natural Area National Parks and Monuments: Beauty and Biology," *Biodiversity and Conservation* 8 (1999): 189, 194.

16. On the 1970s park system expansion legislation, see Mackintosh, *National Parks*, 84–90; Foresta, *America's National Parks*, 80–84; Runte, *National Parks*, 229–35. On "park barrel politics," see Foresta, *America's National Parks*, 76–80; Mackintosh, *National Parks*, 85; Runte, *National Parks*, 233–35.
17. The ANILCA legislation is found at 16 U.S.C. §§ 3111–26. Under ANILCA, however, the new Alaskan national preserves were off-limits to energy development and mining, unlike earlier national preserve designations in Texas and Florida. Runte, *National Parks*, 256. See generally Daniel Nelson, *Northern Landscapes: The Struggle for Wilderness Alaska* (Washington, DC: Resources for the Future, 2004); Deborah Williams, "ANILCA: A Different Legal Framework for Managing the Extraordinary National Park Units of the Last Frontier," *Denver University Law Review* 74 (1997): 859; Roderick Nash, *Wilderness and the American Mind*, 3d ed. (New Haven: Yale University Press, 1982), 272–315; Duncan and Burns, *National Parks*, 353–71.
18. See Mackintosh, *National Parks*, 84–103. On the California desert legislation, see Frank Wheat, *California Desert Miracle: The Fight for Desert Parks and Wilderness* (San Diego: Sunbelt, 1999).
19. On the Park Service's commitment to the "national significance" criteria for new park designations, see National Park Service, *Management Policies* 1.3-1.3.4 (2006). The Ridenour quotations can be found in James M. Ridenour, *The National Parks Compromised: Pork Barrel Politics and America's Treasures* (Merrillville, IN: ICS Books, 1994), 16–18. On the park decommissioning controversy, see Howard Witt, "National Parks Face Survival of the Fittest," *Chicago Tribune*, Sept. 4, 1995; James Gerstenzang, "House Rejects Effort of Shrink Park System," *Los Angeles Times*, Sept. 20, 1995.
20. For more about the people and events involved in creating these national parks, see Ise, *Our National Park Policy*, 212–18 (Rocky Mountain), 128–35 (Crater Lake), 171–82 (Glacier), 248–62 (Great Smoky Mountains), 371–78 (Everglades); Duncan and Burns, *National Parks*, 32–33 (Rocky Mountain), 109 (Crater Lake), 116–20 (Glacier), 217–23 (Great Smoky Mountains), 274–80 (Everglades).
21. On Mather's expenditure of his own funds for the national parks, see Ise, *Our National Park Policy*, 57–60, 80. On the Rockefeller family's philanthropic contributions to the national park system, see Nancy Wynne Newhall, *A Contribution to the Heritage of Every American: The Conservation Activities of John D. Rockefeller, Jr.* (New York: Alfred A. Knopf, 1957); Robin W. Winks, *Laurence S. Rockefeller: Catalyst for Conservation* (Washington, DC: Island Press, 1997); on the Mellon family's national park philanthropy, see Hartzog, *Battling for the National Parks*, 197–201. See generally Tom Butler, *Wildlands Philanthropy: The Great American Tradition* (San Rafael, CA: Earth Aware, 2008).
22. On the state's role in establishing Great Smoky Mountains, Shenandoah, and Everglades National Parks, see Ise, *Our National Park Policy*, 248–67, 376–78,

508–11. On Cape Cod, see 16 U.S.C. § 459b et seq.; Runte, *National Parks*, 224–26; Hartzog, *Battling for the National Parks*, 256–57.

23. On the Forest Service's opposition to new national parks and development of its primitive area policy, see Ise, *Our National Park Policy*, 212, 643; Dana and Fairfax, *Forest and Range Policy*, 131–34, 155–58; Foresta, *America's National Parks*, 30–32, 118–21. On the BLM and wilderness, see 43 U.S.C. § 1782; John D. Leshy, "Wilderness and Its Discontents—Wilderness Review Comes to the Public Lands," *Arizona State Law Journal* 2 (1981): 361. Moreover, passage of the National Landscape Conservation System legislation in the Omnibus Public Lands Management Act of 2009 adds even more weight to the BLM's argument against relinquishing its lands for new national parks. Pub. L. No. 111-11, 123 Stat. 991, 1094–96 (2009), codified at 16 U.S.C. §§ 7201–3.

24. On the growth of these protected land systems, see Vale, *American Wilderness*; Robert B. Keiter, "Saving Special Places: Trends and Challenges for Protecting Public Lands," in *The Evolution of Natural Resources Law and Policy*, eds. Lawrence J. McDonnell and Sarah F. Bates (Chicago: American Bar Association, 2010), 253–57. For more on ecosystem management, see chapter 9.

25. On creation of the eastern national parks, see Ise, *Our National Park Policy*, 258–62, 376–78; on the Cape Cod National Seashore legislation, see 16 U.S.C. § 459b-4; Hartzog, *Battling for the National Parks*, 257–58. On the National Heritage Area concept, see www.nps.gov/history/heritageareas. Although not a landowner in National Heritage Areas, the Park Service ordinarily has the authority to approve the area management plan. It may or may not participate in drafting the plan. On the land trust movement and conservation easements, see Sally K. Fairfax, *Conservation Trusts* (Lawrence: University Press of Kansas, 2001).

26. On the potential effect of climate change on national parks, see Steven Saunders and Tom Easley, *Losing Ground: Western National Parks Endangered by Climate Disruption* (Washington, DC: Natural Resources Defense Council, 2006); National Parks Conservation Association, *Unnatural Disaster: Global Warming and Our National Parks* (Washington, DC: 2007), www.npca.org/assets/pdf/unnatural_disaster_2.pdf. On nature deficit disorder, see Richard Louv, *Last Child in the Woods: Saving Our Children from Nature-Deficit Disorder* (Chapel Hill, NC: Algonquin Books, 2006); Richard Louv, "Leave No Child Inside," *Orion Magazine*, Mar./Apr. 2007.

27. See U.S. General Accounting Office, *Limited Progress Made in Documenting and Mitigating Threats to the Parks* (Washington, DC: 1987); National Parks Conservation Association, *Parks in Peril: The Race against Time Continues* (Washington, DC: 1992); U.S. General Accounting Office, *Activities Outside Park Borders Have Caused Damage to Resources and Will Likely Cause More* (Washington, DC: 1994); see also "National Parks for the 21st Century: The Vail Agenda," partially reprinted in Dilsaver, *America's National Park System*, 434; National Park System Advisory Board, *Rethinking the National Parks for*

the 21st Century (Washington, DC: 2001), 5–6; National Parks Conservation Association, *The State of Our Parks: A Resources Index* (Washington, DC: 2008).

28. On ANILCA and ecosystems, see 16 U.S.C. § 3101(b); on the North Cascades legislation, see 16 U.S.C. § 90; David Louter, *Windshield Wilderness: Cars, Roads, and Nature in Washington's National Parks* (Seattle: University of Washington Press, 2006), 134–64; on the Redwood controversy, see 16 U.S.C. § 79c; Sierra Club v. Dept. of the Interior, 398 F. Supp. 284 (N.D. Cal. 1974); on the Big Cypress designation, see Runte, *National Parks*, 194–95; on the California desert legislation, see 16 U.S.C. § 410aaa et seq.; Frank Wheat, *California Desert Miracle*.
29. On the Maine Woods proposal, see Tux Turkel, "Looking for Land; Roxanne Quimby Is Ready to Speed Up Her Purchases in Northern Maine to Help Create a National Park," *Maine Sunday Telegram*, Dec. 7, 2003; Diana Bowley, "Roxanne Quimby Discusses Her Plans for Her Wildlands," *Bangor Daily News*, May 5, 2011; Douglas Rooks, "Maine's Next National Park? Roxanne Quimby Is Changing Minds about Her Vision for 70,000 Acres in Northern Maine," *Lewiston Sun Journal*, June 5, 2011. On the Valles Caldera proposal, see Valles Caldera National Preserve Management Act, S. 564, 112th Cong. (2011); Kelly Bastone, "A Golden Opportunity," *National Parks Magazine* (Fall 2010). On the Canyonlands proposal, see Robert B. Keiter, "Completing Canyonlands," *National Parks Magazine* (Mar./Apr. 2000), 27–31; Tori Ballif et al., *Canyonlands Completion: Negotiating the Borders* (Salt Lake City: University of Utah Honors College, 2009), www.canyonlandscompletion.com. See also Steve Scauzillo, "National Park Service Opens Door to More Parks and Recreation in San Gabriel Mountains, Valley, River," *Whittier Daily News*, Oct. 17, 2011; Steve Scauzillo, "Congress to Decide between Park Service or Forest Service," *Whittier Daily News*, Nov. 27, 2011; Paul J. Nyden, "National Park Service to Consider New Park in W.Va.," *West Virginia Gazette*, Nov. 28, 2011; Randi Minetor, "What's the New National Park? Part II: Mount St. Helens National Monument," *National Parks Examiner*, Feb. 19, 2012. See generally Craig L. Shafer, "The Unspoken Option to Help Safeguard America's National Parks: An Examination of Expanding U.S. National Park Boundaries by Annexing Adjacent Federal Lands," *Columbia Journal of Environmental Law* 35 (2010): 57.
30. On the economic benefits that accrue from national parks, see Dennis J. Stynes, *Economic Benefits to Local Communities from National Park Visitation and Payroll* (Washington, DC: National Park Service, 2011); National Parks Conservation Association, *The U.S. National Park System: An Economic Asset at Risk* (Washington, DC: 2006); see also chapter 5; note 21 references.
31. On the "restoration reserve" proposal, see Conservation Foundation, *National Parks for the Future* (Washington, DC: Conservation Foundation, 1972), 12, 21, 46; Hartzog, *Battling for the National Parks*, 260–61. On creation of the eastern national parks, see Ise, *Our National Park Policy*, 248–70; Daniel S.

Pierce, *The Great Smokies: From Natural Habitat to National Park* (Knoxville: University of Tennessee Press, 2000); Carlos C. Campbell, *Birth of a National Park in the Great Smoky Mountains*, 2d ed. (Knoxville: University of Tennessee Press, 1960); Dennis Elwood Simmons, *The Creation of Shenandoah National Park and the Skyline Drive 1924–1936* (PhD diss., University of Virginia, 1978).

32. On the Redwood expansion, see 16 U.S.C. § 79c; Runte, *National Parks*, 147–54; on the Weeks Act and eastern forest restoration, see Dana and Fairfax, *Forest and Range Policy*, 111–14; William E. Shands and Robert G. Healy, *The Lands Nobody Wanted* (Washington, DC: Conservation Foundation, 1977); on national wildlife refuge restoration efforts, see Gregory Mensik and Fred L. Paveglio, "Biological Integrity, Diversity, and Environmental Health Policy and the Attainment of Refuge Purposes: A Sacramento National Wildlife Refuge Case Study," *Natural Resource Journal* 44 (2004): 1161; Richard L. Schroeder, Jeanne I. Holler, and John P. Taylor, "Managing National Wildlife Refuges for Historic or Non-Historic Conditions: Determining the Role of the Refuge in the Ecosystem," *Natural Resources Journal* 44 (2004): 1185, 1195–1208.
33. See National Park Service, *Management Policies* 1.3.1 (2006).
34. See Chevron USA, Inc. v. Natural Resources Defense Council, 467 U.S. 837, 842–45 (1984); Motor Vehicle Manufacturers Association v. State Farm Mutual Automobile Insurance Company, 463 U.S. 29, 42 (1983).
35. For a brief description of the noted proposals, see Robert B. Keiter, *Keeping Faith with Nature: Ecosystems, Democracy, and America's Public Lands* (New Haven: Yale University Press, 2003), 190–92; see also Tara Weinman, "The Northern Rockies Ecosystem Protection Act: In Support of Enactment," *Public Interest Journal* 5 (1995): 287; Mike Bader, "The Need for an Ecosystem Approach for Endangered Species Protection," *Public Land Law Review* 13 (1992): 137; Harvey Locke, "Preserving the Wild Heart of North America," *Borealis* 5(1) (1994): 20; Douglas H. Chadwick, *The Yellowstone to Yukon Initiative* (Washington, DC: National Geographic, 2000); "The Wildlands Project Mission Statement," *Wild Earth* (Special Issue, 1992): 3.
36. Mountain States Legal Foundation v. Bush, 306 F.3d 1132 (D.C. Cir., 2002); Tulare County v. Bush, 306 F.3d 1138 (D.C. Cir. 2002); Utah Association of Counties v. Bush, 316 F. Supp. 2d 1172 (D. Utah 2004), appeal dismissed by 455 F.3d 1094 (10th Cir. 2006).
37. On the Giant Sequoia National Monument, see Proclamation No. 7295, 65 Fed. Reg. 24,095 (Apr. 15, 2000); California ex rel. Lockyer v. U.S. Forest Service, 465 F. Supp. 2d 942 (N.D. Cal. 2006) (enjoining a Forest Service fuels treatment proposal for the Giant Sequoia National Monument); on the Grand Canyon–Parashant National Monument, see Proclamation No. 7265, 65 Fed. Reg. 2825 (Jan. 11, 2000); on the Vermillion Cliffs National Monument, see Proclamation No. 7374, 65 Fed. Reg. 69,227 (Nov. 9, 2000). See generally Sanjay Ranchod, "The Clinton National Monuments: Protecting Ecosystems with the Antiquities Act," *Harvard Environmental Law Review* 25

(2001): 535.

38. The Antiquities Act does not designate any particular agency to oversee national monuments, but Congress has otherwise expressed its view in the National Parks Organic Act that the Park Service will ordinarily oversee national monuments as part of the national park system. 16 U.S.C. § 1. And, following President Franklin Roosevelt's 1934 reorganization order, the Park Service has, until recently, been the sole agency responsible for national monuments. But see Mark Squillace, "The Monumental Legacy of the Antiquities Act of 1906," *Georgia Law Review* 37 (2003): 473, 550–68 (suggesting that the president has very limited authority to modify existing national monument proclamations, but not addressing the administering agency question).
39. On the role of wildlife corridors, see Donald McKenzie, Ze'ev Gadalof, David L. Peterson, and Philip Mote, "Climatic Change, Wildfire, and Conservation," *Conservation Biology* 18 (2004): 890; National Parks Conservation Association, *Climate Change and National Park Wildlife: A Survival Guide for a Warming World* (Washington, DC: 2009); Robert L. Peters, *Beyond Cutting Emissions: Protecting Wildlife and Ecosystems in a Warming World* (Washington, DC: Defenders of Wildlife, 2008). On the Upper Green River wildlife controversy, see Janice L. Thompson, Tim S. Schaub, Nada Wolff Culver, and Peter C. Aengst, *Wildlife at a Crossroads: Energy Development in Western Wyoming, Effects of Roads on Habitat in the Upper Green Valley* (Washington, DC: The Wilderness Society, 2005). On the Bridger-Teton forest plan amendment, see U.S. Department of Agriculture, *Bridger-Teton National Forest, Decision Notice and Finding of No Significant Impact, Pronghorn Migration Corridor Forest Plan Amendment* (2008); David N. Cherney, "Securing the Free Movement of Wildlife: Lessons from the American West's Longest Land Mammal Migration," *Environmental Law* 41 (2011): 599.
40. On federal-state wildlife corridor initiatives, see Western Governors' Association, *Protecting Wildlife Corridors and Crucial Wildlife Habitat in the West, Policy Resolution* (Feb. 27, 2007), www.blm.gov/pgdata/etc/medialib/blm/wy/information/NEPA/pfodocs/anticline/revdr-comments/eg.Par.89268.File.dat/02Bio-attach14.pdf; Western Governors' Association, *Western Wildlife Habitat Council Established* (2008), www.westgov.org/wga/publicat/wildlife08.pdf; U.S. Department of the Interior, U.S. Department of Agriculture, U.S. Department of Energy, and Western Governors' Association, *Memorandum of Understanding Regarding Coordination among Federal Agencies and States in Identification and Uniform Mapping of Wildlife Corridors and Crucial Habitat* (2009). The National Trails Act is found at 16 U.S.C. §§ 1241–49. For a legal overview of wildlife corridors, see Robert L. Fischman and Jeffrey B. Hyman, "The Legal Challenge of Protecting Animal Migration as a Phenomena of Abundance," *Virginia Environmental Law Journal* 28 (2010): 173; papers from the Symposium on Animal Migration Conservation published at *Environmental Law* 41 (2010): 277–679.

41. National Park Service, *Management Policies* 1.6, 4.1 (2006).
42. On ecosystem management, see Interagency Ecosystem Management Task Force, *The Ecosystem Approach: Healthy Ecosystems and Sustainable Economies* (Washington, DC: 1996); Wayne A. Morrissey, Jeffrey A. Zinn, and M. Lynne Corn, *Ecosystem Management: Federal Agency Activities* (Washington, DC: Congressional Research Service, 1994); Nels C. Johnson, Andrew J. Malk, Robert C. Szaro, and William T. Sexton, eds., *Ecological Stewardship: A Common Reference for Ecosystem Management* (Oxford: Elsevier Science, 1999); Keiter, *Keeping Faith with Nature*. On the Utah controversy, see Katie Howell, "Oil and Gas: BLM, Park Service Squabble over Lease Sale near Utah Parks," *Land Letter*, Nov. 13, 2008; U.S. Department of the Interior, *Final BLM Review of 77 Oil and Gas Lease Parcels Offered in BLM-Utah's December 2008 Lease Sale* (Washington, DC: 2009).
43. These coordination ideas are developed further in Keiter, *Keeping Faith with Nature*, 309; Robert B. Keiter, "The National Park System: Visions for Tomorrow," *Natural Resource Journal* 50 (2010): 71, 103–4. On using cultural resource preservation strategies, see National Parks Second Century Commission, *Advancing the National Park Idea* (Washington, DC: National Parks Conservation Association, 2009), 6; Hartzog, *Battling for the National Parks*, 259–60. On extending federal ecosystem management principles or strategies onto nearby private lands, see Keiter, *Keeping Faith with Nature*, 208–18; Robert B. Keiter, "Ecosystems and the Law: Toward an Integrated Approach," *Ecological Applications* 8 (1998): 332, 336–38.
44. U.S. Census Bureau News, *An Older and More Diverse Nation by Midcentury*, Aug. 14, 2008, www.census.gov / newsroom / releases / archives / population / cb08-123.html.
45. For the legislation creating existing urban parks, see 16 U.S.C. § 460kk (Santa Monica Mountains); 16 U.S.C. § 460bb (Golden Gate); 16 U.S.C. § 460cc (New York's Gateway National Recreation Area). On national heritage areas, although the Park Service is not a landowner, it ordinarily has the authority to approve the area management plan and may participate in drafting the plan. See www.nps.gov / history / heritageareas.
46. On minority use of national parks, see National Parks Second Century Commission, *Advancing the National Park Idea*, 3; Frederic I. Solop, Kristi K. Hagen, and David Ostergren, *The National Park Service Comprehensive Survey of the American Public: Ethnic and Racial Diversity of National Park System Visitors and Non-Visitors Technical Report* (Flagstaff: National Park Service and Northern Arizona University, 2003); see also Myron F. Floyd, "Managing National Parks in a Multicultural Society: Searching for Common Ground," *George Wright Forum* 18 (2001): 41; Rebecca Stanfield, Robert E. Manning, Megha Budruk, and Myron Floyd, "Racial Discrimination in Parks and Outdoor Recreation: An Empirical Study," *Proceedings of the 2005 Northeastern Recreation Research Symposium* (Bolton Landing, NY: U.S. Department of

Agriculture, Northeastern Research Station, General Technical Report NE-341, 2005), 247; Jack Goldsmith, "Designing for Diversity," *National Parks* 68 (1994): 20; Audrey Peterman and Frank Peterman, *Legacy on the Land: A Black Couple Discovers Our National Inheritance and Tells Why Every American Should Care* (Atlanta: Earthwise Productions, 2009). On the Yosemite buffalo soldiers, see www.nps.gov/yose/historyculture/buffalo-soldiers.htm; Shelton Johnson, *Gloryland: A Novel* (San Francisco: Sierra Club Books, 2009).

47. See chapter 6 for additional discussion about Native Americans and the national parks.
48. See Miles, *Guardians of the Parks*, 93–95.
49. 16 U.S.C. § 1a-5.
50. See National Park Service, *Management Policies* 1.3.1 (2006). According to the Park Service, "suitability" focuses on whether the resource type is already adequately represented in the national park system or elsewhere, whereas "feasibility" addresses whether the area is of sufficient size and configuration to ensure adequate resource protection and whether it is capable of efficient administration. National Park Service, *Management Policies* 1.3.2, 1.3.3 (2006).
51. See Craig R. Groves et al., "Planning for Biodiversity Conservation: Putting Conservation Science into Practice," *BioScience* 52 (2002): 499; Jordan S. Rosenfeld, "Functional Redundancy in Ecology and Conservation," *OIKOS* 98 (2002): 156; Shahid Naeem, "Species Redundancy and Ecosystem Reliability," *Conservation Biology* 12 (1998): 39; National Parks Second Century Commission, *Advancing the National Park Idea*, 4–6; see also Craig R. Groves, *Drafting a Conservation Blueprint: A Practitioners Guide to Planning for Biodiversity* (Washington, DC: Island Press 2003).

Chapter 11

1. 16 U.S.C. § 1.
2. See "Secretary Lane's Letter on National Park Management" (May 13, 1918) in *America's National Park System: The Critical Documents*, ed., Lary M. Dilsaver, (Lanham, MD: Rowman and Littlefield, 1994), 48–52.
3. See chapter 2 for additional discussion and analysis of the wilderness concept in the national parks.
4. See chapters 3, 4, and 5 for additional discussion and analysis of the national parks as tourist destinations, recreational playgrounds, and commercial commodities.
5. See generally chapters 6 through 9 for further elaboration on these dimensions of the national park idea.
6. For more on the Organic Act and its enduring significance, see Robert B. Keiter, "Revisiting the Organic Act: Can It Meet the Next Century's Conservation Challenges?" *George Wright Forum* 28(3) (2011): 240–53.
7. "Secretary Lane's Letter," 48.

8. See Robin Winks, "The National Park Service Act of 1916: 'A Contradictory Mandate'?" *Denver University Law Review* 74 (1997): 575–624.
9. See National Rifle Association v. Potter, 628 F. Supp. 903, 909 (D.D.C. 1986); Fund for Animals v. Norton, 294 F. Supp. 2d 92, 105 (D.D.C. 2003); Bicycle Trails Council of Marin v. Babbitt, 82 F.3d 1445, 1453 (9th Cir. 1996); Mausolf v. Babbitt, 125 F.3d 661 (8th Cir. 1997); see also Greater Yellowstone Coalition v. Kempthorne, 577 F. Supp. 2d 183 (D.D.C. 2008); Southern Utah Wilderness Alliance v. National Park Service, 387 F. Supp. 2d 1178 (D. Utah 2005).
10. National Park Service, *Management Policies* 1.4 et seq. (2006).
11. The Grand Canyon wilderness designation controversy is discussed and analyzed in chapter 2.
12. 16 U.S.C. §§ 1531–42; see Mausolf v. Babbitt, 125 F.3d 661 (8th Cir. 1997).
13. 16 U.S.C. § 1539(j). The Yellowstone wolf reintroduction is examined in chapter 7.
14. 16 U.S.C. §§ 1271–87. The Yosemite Valley and Merced River controversies are examined in chapter 3.
15. To some extent, these new directions are beginning to emerge in National Park Service policy documents, particularly the director's *Call to Action* document and the recently released scientists' report updating the 1963 Leopold report. See National Park Service Advisory Board Science Committee, *Revisiting Leopold: Resource Stewardship in the National Parks*, www.nps.gov/calltoaction/PDF/LeopoldReport_2012.pdf.
16. On the idea that the National Park Service needs to enter a new post–Leopold report era, see William C. Tweed, *Uncertain Path: A Search for the Future of National Parks* (Berkeley: University of California Press, 2010), 188–89. As part of the National Park Service's centennial preparations, the director issued a "Call to Action" that included preparing a new Leopold report, which has now been issued by the agency. National Park Service, *A Call to Action: Preparing for a Second Century of Stewardship and Engagement* (2011); National Park Service Advisory Board Science Committee, *Revisiting Leopold*.
17. See chapter 5 for more on the role of railroads and local communities in the national park establishment process.
18. For more on the enclave approach to nature conservation, see Joseph L. Sax, "Nature and Habitat Conservation and Protection in the United States," *Ecology Law Quarterly* 20 (1993): 47–56; see also chapter 9.
19. See chapter 10 for more on expansion of the national park system.

INDEX

ABOUT THE AUTHOR

Robert B. Keiter is the Wallace Stegner Professor of Law, University Distinguished Professor, and founding director of the Wallace Stegner Center for Land, Resources, and the Environment at the University of Utah S.J. Quinney College of Law. He holds a law degree with honors from Northwestern University and a bachelor's degree with honors from Washington University. He has taught at the University of Wyoming, Boston College, and Southwestern University, and he served as a Senior Fulbright Scholar at Tribhuvan University in Kathmandu, Nepal. His books include *Keeping Faith with Nature: Ecosystems, Democracy, and America's Public Lands* (2003); *Reclaiming the Native Home of Hope: Community, Ecology, and the West* (1998); and *The Greater Yellowstone Ecosystem: Redefining America's Wilderness Heritage* (1991). Keiter has served on the boards of the National Parks Conservation Association, the Greater Yellowstone Coalition, the Sonoran Institute, and the Rocky Mountain Mineral Law Foundation. In 2008, the National Parks Conservation Association honored him with its National Parks Achievement Award. He lives outside Salt Lake City in Emigration Canyon with his wife, Linda, and their two Labrador retrievers. In his spare time, he can be found wandering the West's national parks and remote mountain ranges.

Island Press | Board of Directors